PLANT SPECIATION

OTHER BOOKS BY THE SAME AUTHOR

Natural History of the Phlox Family (*1959*)

The Origin of Adaptations (*1963*)

The Architecture of the Germplasm (*1964*)

Flower Pollination in the Phlox Family. With Karen A. Grant (*1965*)

Hummingbirds and Their Flowers. With Karen A. Grant (*1968*)

Plant Speciation

VERNE GRANT

Columbia University Press

NEW YORK AND LONDON

1971

This book is dedicated
to
George Gaylord Simpson

Preface

This book deals with speciation phenomena in higher plants. Speciation is treated as a process consisting of many facets and affected by many factors. We begin with a discussion of the nature of plant species (in Part I), then go on to consider the primary divergence of species (Part II), and their refusion in hybridization (Part III). This sets the stage for a discussion of secondary or hybrid speciation and the genetic systems involved therein (Parts III and IV). Finally (in Part V) we discuss the evolution of hybrid complexes.

In other words, the book deals at some length with the divergence and refusion of plant species, and considers more briefly the bearing of these processes on plant macroevolution.

Plant Speciation is complementary to my earlier book, *The Origin of Adaptations* (1963). The latter work dealt with evolutionary mechanisms in diploid sexual organisms, animals as well as plants. But, since many higher plants are neither diploid nor sexual, some characteristically botanical phenomena of evolution were necessarily left out of consideration in that earlier work. In *Plant Speciation* we undertake the complementary task of describing evolutionary processes and patterns across the board in higher plants as they occur in the various diverse types of genetic systems.

The research reading and original research for the present book have spanned a period of twenty-five years when I worked successively at five different institutions. The formative years, as far as this book is concerned, were those spent at the University of California, Berkeley, and the Carnegie Institution of Washington, Stanford, California. What was started there was worked out during the next seventeen years at the Rancho Santa Ana Botanic Garden, Claremont, California.

The composition of the book itself, including not only the actual writing but also much of the theoretical work behind the writing, was begun in the Institute of Life Science, Texas A & M University, and finished at the Boyce Thompson Arboretum, University of Arizona.

Plant Speciation owes a major debt to three earlier treatises, namely, Darlington's *Recent Advances in Cytology* (1932, 1937), Gustafsson's *Apomixis in Higher Plants* (1946–1947), and Stebbins' *Variation and Evolution in Plants* (1950). In addition the book is indebted to a long list of other workers and other works, and I hope that the literature citations in the text reflect these intellectual debts properly.

Karen A. Grant critically read the entire manuscript and discussed innumerable knotty problems with me. Her comments and encouragement throughout the period of composition have been invaluable or, rather, indispensable.

V. G.

Boyce Thompson Arboretum
University of Arizona

Contents

PLANT SPECIATION

PART I

Nature of Species

CHAPTER 1

Plant Reproduction

Introduction · *Departures from Random Cross-fertilization* · *Vegetative Propagation* · *Agamospermy* · *Autogamy* · *The Panmictic Unit* · *Dispersal* · *Vicinism* · *Racial Differentiation*

Introduction

One of the fundamental characteristics of living material is the power of reproduction. This power is vested in organized units of very diverse size and complexity, ranging from DNA macromolecules through chromosomes and cells to individual organisms and breeding populations. In this book we are concerned with one of the more inclusive units of reproduction in the world of life, namely, species. The concept of species as reproductive units is an age-old one, subsumed in the statement that like begets like.

In most animals and in many plants, reproduction is inseparably connected with sex. But higher organisms have evolved from unicellular forms in which the sexual process and the reproductive process are distinct and even antagonistic. Reproduction in a protozoan is achieved by fission: one cell divides into two. Sexuality in the same organisms is expressed by conjugation: two cells fuse into one. The whole life cycle consists of alternating phases of cell division, leading to an increase in population size, and sexual conjugation, resulting in a decrease in numbers of individuals.

The basic distinctness of sex and reproduction persists in covert form in higher animals and plants. Sexuality in higher organisms means that two individuals are required to carry out a process of reproduction which, in the absence of sexuality, one individual alone could accomplish by some method of budding.

The distinctness of the two processes can be seen from the physiological as well as from the populational standpoint in higher plants. Many or most perennial plants possess methods of vegetative propagation along with methods of sexual reproduction. Both methods entail a drain on the food supply of the parental plants. The potentially conflicting demands of sexual reproduction and of vegetative propagation for a limited supply of parental food are reconciled in many plants, as Salisbury has pointed out (1942, Ch. 17), by an inverse correlation between the opposing methods of reproduction. Abundant seed production is often combined with slight vegetative propagation and, conversely, copious vegetative reproduction is often associated with reduced seed output (Salisbury, 1942).

Thus, in some varieties of the strawberry, removal of the runners brings about a marked increase in the number of fruits, whereas other ever-fruiting varieties produce few or no runners. The periwinkle (*Vinca major*) usually spreads widely by long runners but produces few fruits; if, however, the runners are cut off and kept from forming anew, the plants go on to set seeds (Salisbury, 1942).

The importance of reproduction per se is obvious enough. It is the indispensable condition for the perpetuation of the species through a succession of generations and hence over any prolonged period of time. And it is the only means of multiplication, increase in numbers, and colonization of new territories.

The function served by sex which accounts for its establishment in the life cycle, even though it alters the reproductive efficiency of the species, is less obvious and was in fact a much debated question in biology before the rise of classical genetics. It is now recognized that the sexual process, involving an alternation of cross-fertilization and meiosis, is a mechanism for bringing about gene recombination. Gene recombination is in turn the chief source of hereditary variations in a sexually reproducing species. And hereditary variations are the necessary raw materials for response of the species to changing environmental conditions.

It follows that vegetative and other nonsexual means of reproduction are advantageous in a stable environment to which the species is already well adapted, since the formation of a fraction of poorly adapted progeny is largely avoided by these methods, whereas the production of seedling progeny by the sexual process is advantageous in relation to unstable or changing environments (Salisbury, 1942, Ch. 17). Both types of environmental challenges exist and both types of reproductive response are called for. The overall breeding system of a species which is successful over a long period of time is likely to represent a combination of sexual and nonsexual processes.

Sexual and nonsexual methods of reproduction are in fact combined in various ways in plants. Each method serves its function in the life of the species. But each method also has its characteristic effect on the nature and structure of the species.

Departures from Random Cross-Fertilization

It is useful for purposes of analysis to postulate an ideal set of sexual reproductive conditions in a hypothetical species, and then to consider the various deviations from this ideal condition which are found in real species.

Let us assume that our hypothetical species consists of a randomly mating or panmictic population. In other words, not only does the species reproduce exclusively by cross-fertilization, but also the cross-fertilization takes place under conditions such that the union of gametes in pairs is governed by chance alone. Any female gamete produced by the population is assumed to have an equal chance of being fertilized by any male gamete. Consequently, under panmixia the individuals constituting the descendant generation in the population represent the products of different pairs of gametes drawn at random from the gamete pool produced by the parental generation.

Departures from panmixia take four common forms in real plant species. (1) Cross-fertilization takes place preponderantly between neighboring individuals (vicinism), and leads to mating between relatives, or inbreeding. (2) Reproduction is by self-fertilization (autogamy), the closest kind of inbreeding. (3) The new plants arise from buds produced by the mother plant (vegetative propagation). (4) Or the new generation of plants arises from seeds which develop on the mother plant without fertilization (agamospermy).

The first two reproductive methods listed above (vicinism and autogamy) involve the sequence of fertilization and meiosis, and are thus methods of sexual reproduction. In the last two methods (vegetative propagation and agamospermy), fertilization and meiosis are circumvented, and the reproduction is therefore asexual, or apomictic as it is called in higher plants.

Another and perhaps even more meaningful distinction can be drawn between the first reproductive method listed and all the others. Only in the first case does cross-fertilization occur. In the remaining three cases, fertilization is either bypassed entirely or takes place between gametes produced by a single parental plant. Accordingly, reproduction is biparental in the first case, and uniparental in the other three.

All the individuals derived by uniparental reproduction from a single parental individual are referred to as members of a clone. Uniparental reproduction in plants thus leads to the formation of a clone or array of clones. A population composed of one or more clones obviously differs in an important respect from a true breeding population, the individuals of which are tied together by mating bonds.

Random mating in a large population is an idealized condition useful as a standard of reference. Probably no real, wide-ranging plant species comes very close to this condition. Plant species with biparental reproduction exhibit some degree of vicinism in the known cases. Extreme departures from random mating in the direction of strict uniparental reproduction are found, however, in some real plant species. Vegetative propagation, agamospermy, and perhaps autogamy may replace biparental reproduction completely in some plant groups.

It is far more usual, however, to find some combination of uniparental and biparental reproduction in the same plant. Such plants are then intermediate between strictly biparental organisms and strictly uniparental ones. Indeed, as Gustafsson (1946–1947) has pointed out, a series of transitions connects the truly sexual and the various uniparental methods of reproduction in plants.

Vegetative Propagation

Vegetative reproduction takes place in many ways: by surface stolons and runners, by underground rhizomes and tubers, by offset buds on corms and bulbs, by adventitious buds on cut stems or fallen leaves, and by vegetative propagules arising within a flower or inflorescence (vivipary).

Vegetative propagation is very widespread in perennial angiosperms, occurring in all major groups. Species of perennials which lack the ability to reproduce vegetatively are exceptional. Gustafsson (1946–1947, p. 272) quotes figures showing that about 80% of all angiosperm species in certain Scandinavian floras have some means of vegetative propagation, and about 50% of the perennial angiosperms in the same floras have these methods developed to such an extent that they can spread rapidly by vegetative reproduction.

The redwood tree (*Sequoia sempervirens*) has the ability to sprout from the root crown, an ability, incidentally, which is exceptional among conifers (Jepson, 1923). When an old tree dies, the crown sprouts grow up into saplings and finally into adult trees arranged in a circle around the original parental stump. In time the members of the second-growth tree ring may give way to third-growth rings of their own. Inasmuch as a redwood tree may live to an age of about 1300 years (Jepson, 1923, p. 15), the life span of the clones has to be reckoned in millenia.

Jepson (1923, p. 160) estimates that about 80% of the mature trees in the redwood forest originated from crown sprouts and about 20% from seeds. In our experience the products of asexual and of sexual reproduction have a highly nonrandom spatial distribution in and around the grove. Reproduction in the dense central parts of some groves is virtually exclusively by crown sprouting, whereas seedlings succeed in establishing themselves only in open, peripheral areas.

The aspen, *Populus tremuloides*, produces much viable seed, in some years at least, and also produces vigorous suckers from the root crown. In the mountains of Utah the germination of aspen seeds is negligible under the present climatic conditions of low early-summer rainfall, and in this area the aspen apparently reproduces entirely by vegetative means. Since the climatic regime of summer drought commenced about 8000 years ago in Utah, it is possible that some clones of aspen are as much as 8000 years old. In the course of centuries or millenia of vegetative propagation an ecological race of the aspen with an early leafing habit has succeeded in colonizing the higher elevations of the Utah mountains which were formerly covered with ice and snow (Cottam, 1954).

Festuca rubra is a perennial herb which spreads vegetatively by rhizomes and also reproduces sexually, being wind-pollinated and self-incompatible. Harberd (1961) analyzed the composition of populations of this grass in Scotland on the basis of samples taken within large quadrats. Plants possessing different phenotypic characters and exhibiting cross-compatible relationships were considered to represent different genotypes. Cross-incompatible plants with the same phenotypic characters, on the other hand, were held to be clonal divisions of the same genotype.

From the phenotypic characters and breeding behavior of his samples, Harberd (1961) concluded that a quadrat 100 yards square contained relatively few genotypes, but many plants belonging to the same clone. One particular genotype was found to be spread over an area more than 240 yards in diameter. As Harberd notes, a clone must be hundreds or perhaps a thousand years old to cover an area of this size. Some other clones were small. The populations sampled in *Festuca rubra* thus consist mainly of a few genotypes which have spread clonally over large areas, and in addition of some genotypes which have formed only small clones. Cross-pollination by wind in a self-incompatible grass with this clonal structure must usually lead to incompatible unions (Harberd, 1961).

A related perennial species, *Festuca ovina*, exhibits a different pattern. *Festuca ovina* spreads much more slowly than *F. rubra*. In *F. ovina*, Harberd (1962) found numerous genotypes within a quadrat 10 yards square. Here, as compared with *F. rubra*, the balance between asexual and sexual reproduction appears to be shifted more toward the latter mode.

Multiplication by vegetative means alone is illustrated by the well-known case of *Elodea canadensis* (Hydrocharitaceae) in Europe. In eastern North America where it is native, this dioecious pondweed reproduces both sexually by seeds and vegetatively by the breaking off of shoots and the formation of winter buds. Female plants were introduced from North America into Britain about 1840. In the absence of male flowers in the alien territory the Canadian pondweed could only reproduce itself vegetatively. Nevertheless, between 1840 and 1880 it managed to spread through the inland waters of Europe (see Gustafsson, 1946–1947, p. 46).

Potentilla anserina in Eurasia and North America contains sexual tetraploids $(2n = 4x = 28)$ and some asexual hexaploids. Rousi (1965) found that the hexaploid populations are completely seed-sterile but reproduce vigorously by runners.

Agamospermy

In agamospermy an individual plant produces viable seeds containing embryos which have arisen without fertilization. The embryological details are complex and vary from case to case (see Chapter 18). The new embryo in the seed may develop from an unreduced egg in the embryo sac (parthenogenesis), from some cell or nucleus other than the egg in the embryo sac (apogamety), or from some somatic cell in the ovule (adventitious embryony). Agamospermous seed formation may take place without pollination, or may require pollination (pseudogamy). In the latter case the pollen stimulates the growth of the endosperm which is necessary for the normal development of the seed and its embryo (Gustafsson, 1946–1947).

The various embryological pathways are alike in bypassing both meiosis and fertilization in the cell lines leading to the new embryo. The result, apart from certain exceptional processes, is the formation of seeds containing embryos which are genetically identical with the maternal parent.

Agamospermy is widespread in higher plants. This condition is found in many members of the Gramineae, Compositae, and Rosaceae, and in a host of smaller families. Familiar examples occur in Hieracium, Taraxacum, Crepis, Citrus, and Poa. Agamospermous plants usually have a perennial growth habit. Their sexual relatives, where known, are invariably cross-fertilizing by means of self-incompatibility, dioecism, or some other outcrossing breeding system, suggesting that agamospermous plants have been derived from strongly outcrossing ancestors (Gustafsson, 1946–1947).

In some plant groups agamospermy replaces sexual reproduction completely (obligate agamospermy). In other plants some seeds form by agamospermous processes and some by sexual processes (facultative agamospermy). As with vegetative propagation, no sharp line can be drawn between sexual and asexual reproduction, but instead the two modes are bridged by transitional conditions (Gustafsson, 1946–1947).

The effect of agamospermy on the variation pattern in a facultatively agamospermous population is to break up the variability into a series of groups consisting of identical individuals which differ from one another by minor characteristics. The population is composed of swarms of more or less discrete agamospermous microspecies.

Autogamy

Among hermaphroditic angiosperms we find a spectrum of breeding systems ranging from obligate outcrossing at one extreme to virtually complete autogamy at the other. Hermaphroditic flowers may be completely self-incompatible, setting no seeds after self-pollination, as in *Gilia capitata capitata* (Polemoniaceae); or the self-incompatibility may be incomplete, as in *G. capitata tomentosa*, where many self-pollinated flowers produce just a few seeds (Grant, 1950a). Conversely, self-compatibility may be incomplete and partial. *Cheiranthus cheiri* (Cruciferae) is fully self-compatible in artificial selfings but, when exposed to equal amounts of self- and cross-pollination, most (92%) of the seeds set are products of cross-pollination (Bateman, 1956).

Completely self-compatible angiosperms then vary in details of the floral mechanism which promote outcrossing, selfing, or mixtures of both. The rate of outcrossing in natural populations of *Clarkia unguiculata* (Onagraceae), which has strongly protandrous flowers with exserted sex organs, is estimated from progeny tests to be 96% (Vasek, 1965). The floral mechanism of *C. exilis*, by comparison, permits considerable spontaneous self-pollination, and two populations of this species have 43% and 45% outcrossing (Vasek, 1964, 1967).

In many self-compatible plants the stamens and stigma of the same flower stand close together and mature simultaneously, so that self-pollination normally occurs automatically. It is generally agreed by students of autogamous plants that the autogamy is usually not so complete as to exclude some outcrossing. Self-pollination may predominate, but some cross-pollination is brought about occasionally by insects or wind, and the breeding system is therefore more properly referred to as predominant autogamy (Allard and Jain, 1962; Kannenberg and Allard, 1967).

One can observe rare insect visitations to the flowers of several predominantly autogamous species of Polemoniaceae in western North America (V. and K. A. Grant, 1965). Progeny tests indicate that the rate of outcrossing ranges from 0 to 16% in the predominantly autogamous *Galeopsis tetrahit* (Labiatae), and from 1 to 12% in the likewise mainly autogamous *Avena fatua* (Gramineae) (Müntzing, 1930a; Iman and Allard, 1965).

We could expect to find some instances in which autogamy is complete or nearly so. *Polemonium micranthum* (Polemoniaceae) may be such a case. The tiny flowers undergo self-pollination in the bud and later open for a few hours after pollination has taken place. There is no apparent opportunity for insect pollination to occur. Our attempts to cross-pollinate the plants artificially were unsuccessful; the only progeny obtained from these artificial crosses were products of selfing. The evidence available suggests that some populations of *P. micranthum* are virtually completely inbreeding (V. and K. A. Grant, 1965, pp. 21–23).

Autogamy or, more particularly, predominant autogamy, is a common and widespread condition in angiosperms, especially among annual herbs (Gustafsson, 1946–1947, p. 294; Stebbins, 1950, Ch. 5). In the family Polemoniaceae this is the most frequent single method of pollination, being known from breeding tests in some 45 species and predicted from field observations in 30 additional species. Here the autogamous habit is found exclusively, as far as known, among the annual members of the family (V. and K. A. Grant, 1965).

The breeding system influences the variation pattern of the population. Given two or more homozygous individuals differing with respect to two or more genes, outcrossing generates a much greater amount of individual variability by recombination, and maintains a heterozygous condition for at least one gene in most of these recombination types. Continued selfing has the opposite effect. Given an original population composed of individuals heterozygous for two or more genes, the proportion of heterozygous individuals decreases at a regular rate in each generation of selfing, and the original array of heterozygous and homozygous recombination types becomes sorted out into a smaller number of homozygous types. The strictly autogamous population is expected to consist mainly of a series of true-breeding pure lines.

The expected results of selfing may or may not be realized in actual populations of autogamous plants, depending on other factors. In the first place, autogamy is usually not obligate, as we have already seen, and occasional outcrosses between different biotypes in a predominantly autogamous population will regenerate new variability at periodic intervals (Harlan, 1945; Stebbins, 1957a; Grant, 1958). Furthermore, continued selfing will not lead to a decline in the proportion of heterozygotes at the expected rate if the heterozygous types have a selective advantage over the homozygotes (Hayman and Mather, 1953; Jain and Allard, 1965).

The *Oenothera biennis* group provides a classical example of permanent heterozygosity for chromosome translocations and for genes in plants having a predominantly autogamous breeding system (see reviews by Cleland, 1962; and Grant, 1964b, pp. 185 ff.). *Paeonia californica* is another such case (J. L. Walters, 1942; Grant, 1964b, pp. 191 ff.). Recent studies by Allard and co-workers provide evidence of persistent genic heterozygosity due to heterozygous advantage in several predominantly autogamous plants. Such evidence has been found in *Secale cereale* (Jain and Allard, 1960; Allard and Jain, 1962; Jain and Jain, 1962), *Phaseolus lunatus* (Allard and Workman, 1963; Harding, Allard, and Smeltzer, 1966), and *Avena fatua* (Iman and Allard, 1965; Jain and Marshall, 1967).

The actual composition of populations in many autogamous plant species is thus a good deal more complex than would be expected on the basis of extrapolations from the pure-line concept, as has been emphasized by Allard and others (Allard and Jain, 1962; Kannenberg and Allard, 1967). This is not to say that the simple models are unrealistic in every case, however. The populations of some autogamous species approach a simple composition, consisting of one or a few homozygous biotypes; those of other species contain a greater store of variability and a higher frequency of outcrossing, approaching in these respects the populations of regularly outcrossing species.

The Panmictic Unit

In a large continuous population of a biparental organism the panmictic unit is the group in which gametes unite at random or, in short, the randomly interbreeding group (S. Wright, 1943a). The panmictic unit is coextensive with the whole population only if the parents of any given individual are drawn at random from the whole population. This coextensiveness of the panmictic unit with the whole population depends on the range of dispersal of the organism being great relative to the areal extent of the whole population (S. Wright, 1943a).

If the individuals belonging to a large continuous population normally migrate over a wide range, the panmictic unit can approach the size of the whole population. But, if the individuals in a large continuous population are sedentary, and always have a short range of dispersal, interbreeding will take place exclusively between neighboring individuals. The panmictic unit is then a small local group within the whole population. And, if the normal range of dispersal has wide latitude or variance, with much short-range and some long-range dispersal, the panmictic units will be of intermediate size (S. Wright, 1943a).

The size of the panmictic unit affects the distribution of genetic variations within the whole population. Leaving natural selection out of consideration, the variations tend to become spread out evenly by interbreeding on a wide scale. Conversely, interbreeding within a series of small subpopulations promotes local differentiation.

Wright (1943a) showed on the basis of mathematical considerations that, if the panmictic unit consists of about 10 breeding individuals, much local racial differentiation can develop within a large continuous population, apart from the action of natural selection. If the panmictic unit consists of 100 breeding individuals, racial differentiation is expected to develop on a regional but not on a local scale. And, if the random breeding group contains 1000 or more individuals, there is only slight racial differentiation over great distances, insofar as the distribution of variations is determined by the breeding structure of the species. A small amount of long-range dispersal will retard regional differentiation within a large continuous population (S. Wright, 1943a).

Estimates of the size of the panmictic unit are available for a few biparental animals and plants. In the Rusty lizard (*Sceloporus olivaceus*) the random breeding group is estimated to consist of 225 to 270 individuals, and is probably closer to the lower figure (Blair, 1960; Kerster, 1964). An early estimate of the size of the panmictic unit in *Drosophila pseudoobscura* is 500 to 1000 breeding individuals; more recent studies point to a lower number (Dobzhansky and Wright, 1943; B. Wallace, 1966). The snail *Cepaea nemoralis* has effective breeding groups as small as 236 individuals in some populations and as large as 8400 individuals in others (Lamotte, 1951; Falconer, 1961). The outcrossing annual plant, *Linanthus parryae*, was at first thought to have panmictic units of only 14 to 27 individuals, but more recent evidence indicates that the earlier estimate should be revised upward (Epling and Dobzhansky, 1942; S. Wright, 1943b; Epling, Lewis, and Ball, 1960). In *Phlox pilosa*, finally, the random breeding group is estimated to contain 75 to 282 individuals (Levin and Kerster, 1968).

The situation in some snail populations seems to approach panmixia. The conditions in the other cases reviewed above represent a considerable departure from panmixia. The population structure in the Rusty lizard, the Phlox, some snail populations, and probably some Drosophila populations is one which would promote regional differentiation, but not local differentiation. Local differentiation on the basis of the pattern of interbreeding alone is a possibility, though admittedly a debated one, in *Linanthus parryae*.

Dispersal

Sedentary plants are capable of dispersal at two stages in their life cycle. Pollen is carried by wind, insects, birds, or other agents, and fruits or seeds are likewise transported by various physical or biotic agents. The normal radius of dispersal of the pollen and seeds, of course, varies greatly from species to species.

The botanist who has observed seed dissemination in nature soon comes to recognize the following general pattern. Most of the seeds produced by a maternal plant, which escape being consumed by animals, lodge fairly close to the parental individual, but some seeds are dispersed over longer distances.

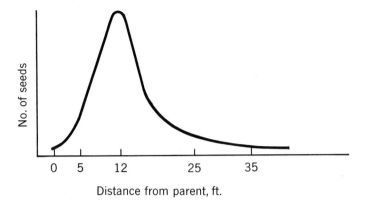

Figure 1 Number of seeds carried by wind to various distances from parent plant in *Verbascum thapsus*. (From *Weeds and Aliens*, by Salisbury, copyright 1961, Collins, London, by permission.)

General observations of this sort have been supported by quantitative evidence in some cases. Salisbury (1961, pp. 99 ff.) counted the number of seeds of *Verbascum thapsus* which were carried by wind to various distances from the parent plant. The main bulk of the wind-borne seeds fell to the ground about 12 feet from the seed parent. The numbers of seeds fell off rapidly beyond the 12-foot radius. The dispersal pattern is shown graphically in Figure 1. A similar pattern of dispersion is found for the wind-borne fruits of *Senecio jacobea* (Salisbury, 1961).

Several studies indicate that the pollen grains of outcrossing plants are likewise mostly carried to neighboring flowers or female cones. Colwell (1951) measured the amounts of pine pollen dispersed by wind to various distances from a source point. The greatest bulk of the pollen fell downwind at a distance of 10 to 30 feet from the source. Thereafter the amount of pollen diminished rapidly. At a distance of 150 feet from the source very little pollen was found (Colwell, 1951).

Pollen dispersal by bees has been studied by sprinkling blue dye in newly opened flowers of cotton (*Gossypium hirsutum* and *G. arboreum*) and following its subsequent distribution in the cotton field (Stephens and Finkner, 1953). Dye particles were found up to 80 feet away from the source after bee visitations. However, the concentration of dye diminished in a uniform manner away from the source point, suggesting a corresponding concentration of bee activity within a given neighbor-hood of flowers (Stephens and Finkner, 1953).

Vicinism

The dispersal pattern of seeds described above, if continued over successive generations, will result in related plants coming to stand as close neighbors. The similar pattern of pollen dispersion will then have the effect that cross-pollination is preponderantly between neighboring related individuals. The net result is a considerable degree of inbreeding in an outcrossing plant population.

The effects can be seen where two varieties of an outcrossing species of crop plant are grown in adjacent plots in an experimental field under similar or practically identical environmental conditions and then, after open pollination, show little or no intervarietal contamination. Moderate-sized plots of wind-pollinated maize (*Zea mays*) separated by a distance of 50 feet showed only 1% contamination (Bateman, 1946). Intervarietal contamination was reduced to a similar low level by the same isolation distance between plots of insect-pollinated turnips (*Brassica rapa*) and again between plots of insect-pollinated radish (*Raphanus sativus*) (Bateman, 1946). In the outcrossing radish, con-tamination was only 2% between varieties growing 15 feet apart (Crane and Mather, 1943).

Similar instances of vicinism have been observed in natural popula-tions. Here, however, environmental differences are usually noted between the subpopulations, and therefore differential or disruptive selection cannot be ruled out as a possible complicating factor. We return to this aspect of the problem in the next section.

Linanthus parryae (Polemoniaceae) is a self-incompatible insect-pollinated annual of the Mojave Desert and adjacent mountains. Populations in the western Mojave Desert are polymorphic for flower color. Blue- and white-flowered forms are intermixed in some local areas and segregated in others. There are some pure blue colonies growing close or adjacent to all white colonies which have persisted unchanged in this respect during many years of observation (Epling and Dobzhansky, 1942; Epling, Lewis, and Ball, 1960). Both the blue- and the white-flowered forms grow on the same sandy plain, and no differences are apparent between their respective environments.

Racial Differentiation

A large continuous outcrossing population could theoretically become differentiated into a series of local races, on the basis of the pattern of interbreeding alone and without the action of selection, if the range of dispersal is short and the panmictic units are accordingly small (S. Wright, 1943a). This theoretical conclusion is supported by some observational evidence as reviewed in the preceding sections.

In nature, however, subpopulations living in different parts of the species area are normally exposed to different environmental conditions and hence to differential or disruptive selection. Natural selection acts in combination with the breeding structure of the species.

Under conditions of random mating on a wide scale in a large population, the effects of selection for adaptation to any given local environment tend to be swamped out by the continual influx of genes from other localities. Broad regional races but not distinct local races can be maintained by selective pressures of moderate intensity. Conversely, local races can be maintained under wide outcrossing only by very strong disruptive selection.

But, if a population of the same total size is subdivided by its breeding pattern into small panmictic units, selection can operate on the genetic variations in each local area without interference from gene flow from other areas. This situation is favorable for the formation and maintenance, by selection, of a series of local races adapted to their respective environments. It is also a favorable situation for the evolutionary advance of the species as a whole, since any panmictic unit possessing characteristics which have general adaptive value, not merely local significance, may expand at the expense of other less favorably endowed panmictic units (S. Wright, 1943a, 1949).

Many cases are known in outcrossing plants where different local races occur in adjacent sites, separated by distances of several hundred feet, and still preserve their distinct racial characteristics from generation to generation. The ecological conditions in the neighboring colonies are observably different in these cases. Therefore it is probable that the localized racial differentiation is due to the action of disruptive selection in separate, small breeding groups.

The foregoing pattern of local racial differentiation has been observed in the following outcrossing species:

Gilia capitata	(Grant, 1950a, 1952a)
Clarkia xantiana	(H. Lewis, 1953)
Galium pumilum	(Ehrendorfer, 1953)
Pinus monticola	(Squillace and Bingham, 1958)
Potentilla glandulosa	(Clausen and Hiesey, 1958)
Agrostis tenuis	(Bradshaw, 1959)
Helianthus annuus X bolanderi	(Stebbins and Daly, 1961)
Diplacus longiflorus	(Beeks, 1962)
Escholtzia californica	(Cook, 1962)
Collinsia heterophylla	(Weil and Allard, 1964)
Pinus albicaulis	(J. Clausen, 1965).

The example of *Agrostis tenuis* reveals some significant features. This outcrossing and wind-pollinated grass has a more or less continuous distribution in the rolling hills of Wales. Its environment varies geographically with respect to conditions of soil, altitude, climate, and grazing. Population samples show hereditary or ecotypic differences correlated with the environmental differences. The populations evidently respond to local environmental selection in spite of continuous distribution and cross-pollination by wind. This local differentiation could be due in some cases to weak or moderate disruptive selection combined with relatively short-range pollen dispersal (Bradshaw, 1959).

Some other cases in *Agrostis tenuis* require a different combination of forces for their explanation. In lead-mining districts of Wales this grass grows on soils containing much lead and also on adjacent soils without toxic amounts of lead. There are lead-tolerant and lead-intolerant ecotypes on the two classes of soils. In some places the two races occur in very close proximity and well within the known range of wind pollination of one another. Here the racial differences must be maintained by disruptive selection strong enough to overcome the effects of interracial crossing (Jowett, 1964; Jain and Bradshaw, 1966).

CHAPTER 2

The Biological Species

Introduction · An Example · The Biological Species as a Corollary of Sexual Reproduction in a Biotic Community · Sibling Species · Biological Species vs. Taxonomic Species

Introduction

The biological species is a fundamental unit of organization in biparental organisms. It is the reproductively isolated system of breeding populations. In other words, the biological species is the sum total of interbreeding individuals and, hence, the most inclusive unit of normal biparental reproduction. The biological species concept is reached in several ways.

It was reached originally by the early naturalists from direct observation of nature. It was apparent to these observers, as it has been to their modern followers in population biology, that living organisms do not form a continuum, but fall into a series of more or less discrete clusters of interbreeding individuals. Thus the members of the cat family are seen to fall into distinct groups such as lions, tigers, bobcats, and domestic cats. Each group breeds true to type or, in the language of the early naturalists, begets its own like, and hybrids between the groups are either nonexistent or so rare as to call for special attention. This discontinuous pattern of relationship and of variation is found in many animal families and among some higher plants. The discrete breeding groups were the species.

As regards species in the plant kingdom we can do no better than quote the views of Cesalpino in *De plantis libri* (1583), of John Ray in *Historia plantarum* (1686), and of Linnaeus in *Critica botanica* (1737) and *Philosophia botanica* (1751).

Cesalpino stated, with reference to species, that like always produces like (cf. Sachs, 1906). John Ray went on to state that plants which spring from the same seed and produce their kind again through seed belong to the same species (Sachs, 1906). "No more certain criterion of a species exists," he said, "than that it breeds true from seed within its own limits" (Darlington, 1937b).

And Linnaeus followed with statements contrasting species with varieties. "The Author of Nature, when He created species, imposed on his Creations an eternal law of reproduction and multiplication within the limits of their proper kinds. He did indeed in many instances allow them the power of sporting in their outward appearance, but never that of passing from one species to another." (*Critica botanica*; cf. Ramsbottom, 1938.) But varieties are a different matter. "There are as many varieties as there are different plants produced from the seed of the same species." (*Philosophia botanica*; Ramsbottom, 1938.)

It would take us too far afield to trace the subsequent history of the species concept in any detail here (but see Mayr, 1957, and Grant, 1963, pp. 336 ff.). Suffice it to say that the species concept of the early naturalists has been retained in principle but developed and clarified by modern population biologists. That concept in its modern form is known as the biological species concept and, by extension, the inclusive breeding groups themselves are designated biological species.

If biological species are distinct reproductive groups, barriers to hybridization should exist between them, and the finding of such barriers would constitute a second argument for the biological species concept. The postulated barriers, or reproductive isolating mechanisms, have indeed been found to exist. The sterility of the mule is legendary, and the early plant hybridizers sometimes referred to their sterile hybrids as plant mules or mule plants. Buffon in 1749, followed by Kant in 1775 and by many later students, was to use sterility of progeny as the main objective criterion for distinguishing between separate species. However, the situation is not quite this simple.

Interbreeding between biological species is prevented by many kinds of isolating mechanisms. Some pairs of species form sterile hybrids, to be sure, but other combinations can be made to produce fertile hybrids artificially, yet do not normally hybridize naturally owing to various external barriers such as aversions to mating or mechanical difficulties in crossing.

An Example

Gilia capitata and *G. tricolor* are related species belonging to the same section of the genus Gilia (Polemoniaceae). They are herbaceous plants with an annual life cycle, diploid chromosome condition ($2n = 18$), and predominantly cross-fertilizing breeding system.

Gilia tricolor occurs in the valleys and foothills of central and northern California. *Gilia capitata* occurs in the same areas and also ranges farther north, farther south, and farther west to the coast line (Figure 2). In many localities, individual plants of the two species grow side by side or close together. The two species, in other words, are sympatric.

The morphological characteristics of representative specimens of the two species are shown in Figures 3 and 4. The plants are large and tall in *Gilia capitata* and relatively small in *G. tricolor*. The inflorescence is a dense head in *G. capitata* and a loose cyme in *G. tricolor*. The sepals of the calyx have narrow midribs in *G. capitata* and broad, green bands in *G. tricolor*. The corolla shape is different in the two species. And in *G. capitata* the corolla is of one color, blue-violet, while in *G. tricolor* it is three-colored, with purple spots and an orange tube contrasting with a blue-violet background.

This is not to say that either species is uniform. In fact, much geographical variation exists within *Gilia capitata*. The characteristics of the flowers, capsules, and seeds vary racially as shown in Figure 5

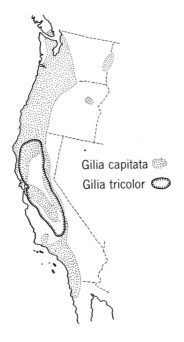

Gilia capitata

Gilia tricolor

Figure 2 Distribution areas of *Gilia capitata* and *G. tricolor* on the Pacific slope of North America.

along a transect from coastal sand dunes to interior mountains in the San Francisco Bay area. There is complete intergradation between these and other racial forms in *G. capitata* (Figure 6). Geographical variation occurs in *G. tricolor* also but is less developed here. But the variations are circumscribed within each species. No intermediate forms between *Gilia capitata* and *G. tricolor* are known. No hybrids between these two species have ever been found.

Experimental hybridizations have been carried out within each species and between them. Seven strains belonging to five geographical races of *Gilia capitata* were intercrossed. The races cross with one another in all combinations and, in most combinations, cross with ease. The interracial F_1 hybrids are highly fertile or semifertile with 50% to 90% good pollen and abundant seeds. These give rise to F_2 generations which are generally viable but include a small proportion of subvital plants (Grant, 1950a, 1952a). A northern and a southern race of *G. tricolor* also proved to be interfertile (Grant, 1952b).

Repeated attempts to cross *Gilia capitata* with *G. tricolor*, on the other hand, have always failed. The interspecific cross-pollinations lead

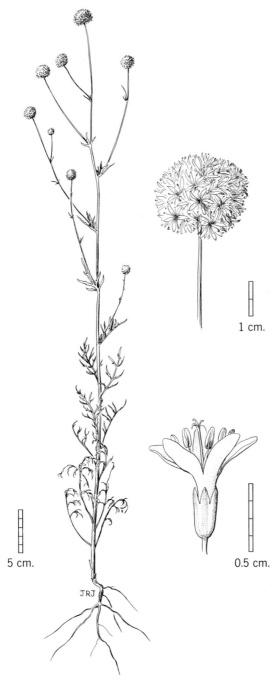

5 cm.

1 cm.

0.5 cm.

Figure 3 *Gilia capitata* (Polemoniaceae). The plant shown belongs to subspecies *G. c. capitata.*

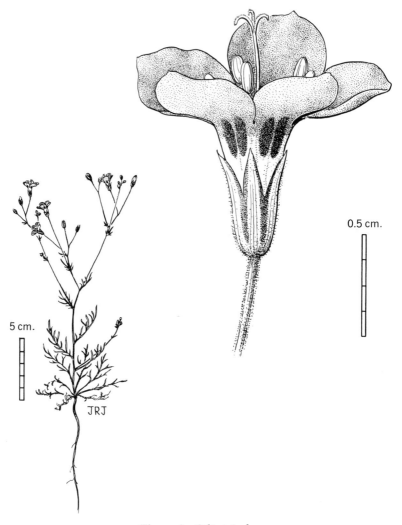

0.5 cm.

5 cm.

JRJ

Figure 4 *Gilia tricolor.*

to the formation of abortive seeds, but no sound seeds and no F_1 hybrids are produced. In addition it is known that *G. tricolor* will not cross with any other related species of Gilia. *Gilia capitata* crosses artificially with difficulty with some other related species, but then the hybrids are highly sterile (Grant, 1952b, 1954b).

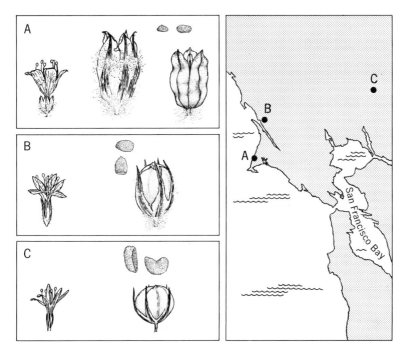

Figure 5   Flowers, capsules, and seeds of three races of *Gilia capitata* on a transect from the coast to the interior mountains north of San Francisco Bay, California. (A) *Gilia capitata chamissonis.* (B) *G. c. tomentosa.* (C) *G. c. capitata.*

Thus *Gilia tricolor* and *G. capitata* are separate biological species. Sexual reproduction takes place between members of *G. tricolor* and leads to intergrading variations. Interbreeding takes place in a similar way and has similar effects in *G. capitata*. But *G. tricolor* and *G. capitata* cannot and do not interbreed with one another owing to the operation of strong, reproductive isolating mechanisms.

The Biological Species as a Corollary of Sexual Reproduction in a Biotic Community

The biological species concept can be reached by a third line of reasoning. It can be shown that the organization of breeding populations

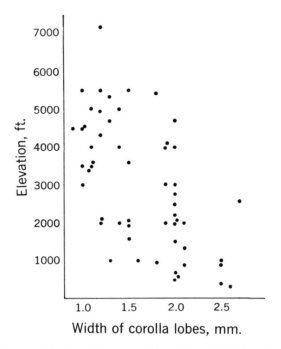

Figure 6 Geographical variation in width of corolla lobes in a series of populations of *Gilia capitata* ranging from the San Joaquin Valley to middle elevations in the Sierra Nevada, California. (Grant, 1950a.)

into separate biological species is a necessary accompaniment of the sexual process in a heterogeneous environment. The premise has been laid down by Dobzhansky (1937a) and Cain (1954). As Cain has put it (1954, p. 130):

> One has only to imagine the consequences of unrestrained hybridization between all living animals to appreciate the extreme importance of the process of speciation. If all specific barriers were suddenly and miraculously removed, the result would be an appalling welter of hybrids with every possible combination of characters. No single individual would be properly adapted to any one mode of life, and many of the characters of each might be adapted for well-nigh incompatible purposes.

And Dobzhansky (1937a) has stated that the "maintenance of life is possible only if the gene patterns whose coherence is tested by natural selection are prevented from disintegration due to unlimited hybridization." Let us now develop an argument based on this premise.

Any inhabitable local area on the earth's surface normally contains diverse ecological niches which are occupied by different sympatric populations of organisms. Each population is adapted for its particular niche in the biotic community by a combination of physiological and morphological characteristics. Each character combination is based in turn on a combination of genes.

Now sexual reproduction is a mechanism for bringing about gene recombination. This is not an absolute good. Gene recombination has advantageous effects in some situations and disadvantageous effects in others. The adaptive properties of different possible gene combinations vary widely according to the particular genes involved and their interactions.

The individuals belonging to a given breeding population may vary in several hereditary characters, yet they can intercross to produce viable and fertile progeny. The population is polymorphic for an array of coadapted genes. In terms of a simple example, the gene A is present in two allelic forms, A_1 and A_2, and a separate gene B is likewise present in two forms, B_1 and B_2. The nine diploid genotypes which can be produced by the sexual process from these two polymorphic genes $(A_1A_1 \, B_1B_1, A_1A_1 \, B_1B_2, \ldots, A_2A_2 \, B_2B_2)$ exhibit individual differences; nevertheless they are all normal representatives of their population. They are all adapted for living and reproducing more or less successfully in their population's ecological zone. The good adaptive characteristics of the several recombination types are not a fortunate coincidence, moreover, but result from natural selection having acted through past generations in this breeding population to preserve the alleles with good combining ability.

A second population living in a different ecological niche in the same community has, like the first population, an adaptive character combination which is subject to variation within circumscribed limits. But it is polymorphic for a different array of coadapted genes. Let us say that it contains the alleles A_8 and A_9 and B_8 and B_9, which combine to form the nine diploid genotypes, $A_8A_8 \, B_8B_8, A_8A_8 \, B_8B_9, \ldots, A_9A_9 \, B_9B_9$. These genotypes, despite individual differences, are all fitted for life in their population's niche.

The consequences of crossing between members of the two populations are very different as regards the adaptedness of the offspring. Consider, for example, the nine types of F_2 progeny resulting from the cross of A_1A_1 B_1B_1 $\times$ A_8A_8 B_8B_8. The two parental types reappearing in the F_2 generation are certainly viable, and the F_1 type may be viable. The remaining six genotypes, representing untested recombinations of the parental genes, are likely to be poorly adapted for either parental environment or for any other available environment. If the genes A and B are unlinked, these inviable or subvital recombination types can be expected to constitute 10/16 of the F_2 zygotes.

For other combinations of parents from the two populations, i.e., A_1A_1 B_2B_2 $\times$ A_8A_8 B_9B_9, etc., the Mendelian proportion of recombination types in F_2, defined here as non-P and non-F_1 types, is the same, namely, 62.5%. In general, therefore, the progeny resulting from hybridization between two populations differing with respect to a simple adaptive gene combination are preponderantly recombination types of reduced fitness.

The proportion of subvital recombination types rises rapidly as the number of gene differences between the parental populations increases. Suppose that two populations differ allelically in five unlinked genes, A to E, which determine different adaptive modes in each population. Then the cross of $A_1A_1 \cdots E_1E_1$ $\times$ $A_8A_8 \cdots E_8E_8$ produces F_2 zygotes, only 3.3% of which are genotypically like the parents or F_1s. The remaining 96.7% of the F_2 zygotes are recombination types which are expected to be more or less subvital. If the parental populations differ with respect to an adaptive gene combination based on ten independent genes, 99.9% of their F_2 zygotes are subvital recombinations.

In other terms, 966,797 of every million F_2 zygotes derived from hybridization between populations differing in an adaptive gene combination composed of five independent genes will possess an ill-adapted genetic constitution. And 999,022 per million F_2 zygotes derived from hybridization between populations differing in an adaptive character combination determined by ten independent genes will be likely to fail in any available environment.

Actual populations will, of course, deviate from these simplified models in various ways. Some factors like linkage will act to lessen the burden of hybridization. Other factors, such as the great complexity of the genetical differences between sympatric species, increase that burden. The latter factors probably outweigh the former. Therefore our numerical examples probably do not give an exaggerated estimate of the potential burden of hybridization between physiologically differentiated, sympatric populations.

The enormous loss of reproductive potential resulting from interbreeding between differentiated sympatric populations stands in marked contrast to the generally beneficial results of interbreeding within populations. Those individuals which cross with other members of the same population will usually leave vastly greater numbers of descendants than the individuals which hybridize with foreign populations. Under these conditions any hereditary aversions or blocks to hybridization will be favored by natural selection and will spread through each sympatric population.

We thus arrive at the conclusion that sexual reproduction, if not confined within the limits of separate populations in a biotic community, would lead to the breaking up of the adaptive gene combinations within each population. The continued existence of the sympatric populations is threatened in proportion to the freedom of interbreeding between them. Conversely, the stable biotic communities are those composed of reproductively isolated breeding populations or, in other words, of biological species.

Sibling Species

The essential characteristic of biological species, as we have seen, lies in their breeding relationships. The ability of individuals to exchange genes successfully, that is, to cross freely and produce fertile and viable progeny, characterizes them as members of the same biological species, whereas the inability to exchange genes freely and successfully is the mark of separate biological species. All else, including morphological difference, is superstructure.

Related biological species usually do differ in external morphological characters, as well as in underlying physiological traits, for many morphological features are components of the adaptive character combinations of the respective species. Traditional taxonomic methods of classification and identification rest on this common association of external morphological characters with physiogenetic traits.

But the correlation does not always hold. We find cases of good biological species which are virtually indistinguishable morphologically. Such cryptic species are termed sibling species.

A classical example is that of *Drosophila pseudoobscura* and *D. persimilis*. These morphologically similar flies were originally considered to be members of a single species, *D. pseudoobscura*, until certain "intraspecific" crosses revealed the existence of hybrid sterility barriers. The intersterile entities were then found to form sympatric populations over a vast area in western North America without hybridizing, and were accordingly recognized as separate species (Dobzhansky and Epling, 1944; Dobzhansky, 1951).

A search by several workers resulted eventually in the finding of minor morphological differences in the male genitalia and wings, as well as some behavioral, physiological, and cytological differences (see review by Mayr, 1963, pp. 34–35). The character differences are very slight, and the flies are difficult to identify by ordinary taxonomic procedures.

However, the flies readily recognize one another. When males and females of *Drosophila pseudoobscura* and *D. persimilis* are intermixed in a population cage in the laboratory, they mate exclusively or predominantly in intraspecific combinations; and only rarely do species-foreign females and males copulate under ordinary laboratory conditions (review in Grant, 1963, pp. 362–63). Likewise in nature, where the opportunities for hybridization are widespread, interspecific copulations are extremely rare and species hybrids are unknown (Dobzhansky, see Grant, 1963, p. 390). Obviously Drosophilas can discriminate between *D. pseudoobscura* and *D. persimilis* whether taxonomists can or not.

Sibling species are found in other species complexes in Drosophila, and in many other groups of animals (review in Mayr, 1963, pp. 33 ff.). Sibling species also occur in many genera of higher plants (Grant, 1957). Here the phenomenon is often associated with polyploidy.

A tetraploid species with the genomic constitution $AABB$ will resemble morphologically, and may intergrade with, the ancestral diploid species AA for two reasons. First and foremost, the A genome is in the tetraploid species. Second, the tetraploid is likely to segregate some individuals like the A parental type, especially if it is a segmental allotetraploid. The tetraploid will resemble its other diploid parent, species *BB*, for the same reasons. The diploid species *AA* and *BB* will produce more or less sterile F_1 hybrids because of their chromosomal differentiation; and the hybrids between the tetraploid species *AABB* and either diploid, being triploid, will also be sterile. What appears to be a continuously intergrading population system on external morphology, and may be treated as a single species by the taxonomist, is thus a group of three intersterile sibling species (Grant, 1964a).

By way of illustration we consider a case in the *Gilia transmontana* group, which has been carefully analyzed by Day (1965). The plants are small annual herbs of the Mojave Desert and adjacent areas in western North America. They are predominantly autogamous but are cross-pollinated by insects to a small but biologically significant extent. Their general appearance is shown in Figures 7 and 8.

The *Gilia transmontana* group consists of not three but five interrelated diploid and tetraploid species. One of the tetraploid species, *G. transmontana*, is derived from the two diploids, *G. minor* and *G. clokeyi;* the other tetraploid, *G. malior*, stems from the diploids, *G. minor* and *G. aliquanta* (Day, 1965). It will be noted that the two tetraploid species have one diploid ancestor (*G. minor*) in common. All five species were regarded as variant forms of a single species in the earlier taxonomic treatments.

Artificial hybridizations show that the five species, although fertile themselves, are highly intersterile in all combinations. In nature they grow sympatrically in various combinations without interbreeding (V. and A. Grant, 1960; Grant, 1964a; Day, 1965).

Day (1965) has grown and compared a series of strains representing the known range of variation of each species under uniform environmental conditions in the greenhouse. She finds that the species do differ slightly but consistently in almost every plant part (see Figures 7, 8, and 9). Thus *Gilia aliquanta* is distinguished by its slightly larger flowers (Figure 8E); but the other four species have flowers of a similar small size (Figure 8, A–D). *Gilia minor* has smaller seeds than the other species (Figure 9). The leaves, the corolla colors, the capsules, and other features also show small interspecific differences.

But these external phenotypic differences are not such as to permit identification in every instance. Consider seed size again. The seeds range in weight from 10–20 mg in *Gilia minor*, 20–30 mg in *G. transmontana*, and 30–85 mg in *G. clokeyi*. These three species differ in

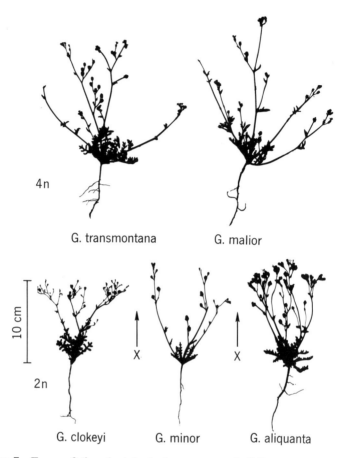

Figure 7 Form of the plant body in a group of sibling species in Gilia. (Day, 1965.)

average seed size (Figure 9, A–C) but form a continuum when the range of variation within the several species is considered (Day, 1965).

In general, *Gilia transmontana* bridges the morphological gap between its diploid ancestors, *G. minor* and *G. clokeyi*, so that clear-cut distinctions between the three species are hard to find. Similarly, *G. malior* exhibits apparent intergradation with *G. minor* and *G. aliquanta*. And, since the *G. minor* genome is present in both *G. transmontana* and *G. malior*, these two tetraploid species have many morphological features in common.

1 cm.

Figure 8 Flowers of the five sibling species of Gilia shown in Figure 7. (A) *Gilia clokeyi* (2x). (B) *G. transmontana* (4x). (C) *G. minor* (2x). (D) *G. malior* (4x). (E) *G. aliquanta* (2x). (Day, 1965.)

The problems of taxonomic identification in the *Gilia transmontana* group are, however, irrelevant to the question of the biological composition of this group. *Gilia transmontana, malior, minor, clokeyi,* and *aliquanta* form five separate breeding groups, and thus are good biological species whether we can always identify them by their external morphological characters or not.

Biological Species vs. Taxonomic Species

The early naturalists studied chiefly the higher animals and plants within a local fauna and flora. Hence they were dealing with sympatric populations of sexually reproducing organisms distinguished by prominent morphological differences. The species which they recognized could be defined equally well by breeding relationships or by morphological discontinuities. The inclusive unit of biparental reproduction and the basic unit of taxonomic classification were synonymous for them. In other words, no distinction was made, and no distinction was needed, between biological species and taxonomic species.

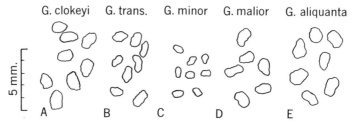

Figure 9 Seeds of the same five sibling species of Gilia. (Day, 1965.)

Similarly, in many groups of organisms which have been well studied by modern methods, there is good agreement between the criterion of breeding relationships and the criterion of morphological discontinuity. In such groups also, as they are currently understood, the biological species and the taxonomic species are synonymous.

It should be noted, however, that the criteria of breeding relationships and of morphological distinction do not necessarily always coincide. Inevitably, therefore, with the extension of biological research since the period of the early naturalists, numerous cases have come to light in which the two criteria do not coincide. Which criterion then should be given priority? On the answer to this question there has been no satisfactory general agreement among biologists. Instead, the question has been answered in terms of divergent practices carried out by divergent schools of workers.

One situation in which the breeding and the morphological criteria come into conflict is that involving sibling species. These entities are real biological species, but they are not readily identifiable by their morphological character differences and hence are not good taxonomic species.

This is not to say that related sibling species cannot be distinguished by their phenotypic features. In fact they can in every group of sibling species that has been analyzed thoroughly. But the amount of effort required to make identifications in such groups is much greater than the norm in routine taxonomic practice.

Let us reconsider the common pattern in higher plants of related diploid and tetraploid species introduced in the preceding section. External morphological characters alone are likely to provide unreliable criteria for distinguishing between three species with the genomic constitutions: *AA*, *BB*, and *AABB*. But a knowledge of the chromosome number when added to the morphological evidence is usually sufficient for purposes of identification. The two diploid species, *AA* and *BB*, are the morphological extremes within the group, and ordinarily they can be distinguished from one another by external characters. The morphologically intermediate species, *AABB*, can be distinguished reliably from either diploid by chromosome number alone.

Thus *AA*, *BB*, and *AABB* can be identified by a combination of external and internal phenotypic features. But the first step in identification is the determination of the chromosome number. This step, however, by requiring a microscopic examination of living plant material, goes well beyond the procedures which are standard for routine identification in the higher plants.

Another situation which has brought the morphological and the breeding criteria of species into conflict has resulted from the extension of a universal system of classification and nomenclature to all organisms. It is axiomatic according to the International Code of Botanical Nomenclature that every plant belongs to a species. Not every plant, however, is sexual, and not every plant group consists of individuals linked by mating bonds into breeding populations.

The association between morphological resemblances or differences and breeding relationships may break down in uniparental organisms as it does in sibling species groups. In the latter we found separate breeding populations with closely similar morphologies. In uniparental organisms prominent morphological differences may exist between classes of individuals without indicating their membership in different breeding groups.

In summary, the history of the species concept involves branching lines of thought. From a common starting point in the period of the early naturalists, the species concept has developed along two partially independent lines. One line has led to the modern biological species concept. The other main line has been the development of a concept of species which is useful in taxonomy. As a result, different meanings have become attached to the term species as it is used in population biology and in museum taxonomy. Many discussions of species fail to make the distinction between the two usages. It is commonplace to criticize the biological species concept from the viewpoint of the taxonomic species concept without recognizing the difference in viewpoints. Needless to say, such discussions perpetuate confusion of thinking.

Attempts have also been made to clarify the issue by restricting the technical term species to only one of its traditional usages. Gilmour and

Heslop-Harrison (1954) and Sonneborn (1957) would keep the term species for formal taxonomy and propose new terms (hologamodeme, syngen) to designate the inclusive breeding group in nature. I once made the opposite suggestion that we use the word species in its biological sense exclusively and employ a different term (binom) for the unit of formal classification (Grant, 1957). These proposals are logical but unhistorical. Undoubtedly they would clarify discussions of the species problem, if they were widely adopted. But they stand little chance of becoming widely adopted because, in each case, they ignore long-standing historical claims on the term species by one or another large school of workers.

A way out of the semantic impasse which achieves the desired end of clarifying discussions, yet avoids the pitfall of going against long-established usages, is to recognize, and distinguish between, *biological species* and *taxonomic species* (Cain, 1954, Ch. 7; Grant, 1963, p. 342). The distinction between the two types of species has proved its usefulness in the presentation and discussion of the problem in this section.

Consider again the case of sibling species—so often disputed by taxonomists—in the light of this distinction. The individual sibling species are good biological species and can be recognized as such by workers interested in their genetical or ecological behavior. At the same time, the whole sibling species group can be treated as a single taxonomic species for purposes of routine identification.

We are confronted here with a new form of the old dichotomy between natural and artificial systems of classification. The former reflects genetic relationships but may not be convenient to use; the latter is useful in practical taxonomic work but may not express biological relationships. The dichotomy between these approaches to classification has usually involved the higher categories. But we find it repeated again at the species level wherever the taxonomic species recognized in ordinary practice are not equivalent to the biological species (Grant, 1964a).

The distinction between biological and taxonomic species goes far toward resolving differences in approach between different schools of workers. But this distinction also leaves some important problems unresolved. A third species concept, that of the so-called evolutionary species, is necessary in order to deal with these remaining problems, as we shall see in the next chapter.

CHAPTER 3

The Evolutionary Species

Introduction

The common animals and plants with which the early naturalists dealt were mostly biparental organisms, and they were mostly grouped into distinct species. The species concept of the naturalists was the historical and logical forerunner of the modern biological species concept, as developed by a number of students in the period from about 1929 to 1957.

The biological species concept applies to biparental organisms exclusively. Uniparental organisms are not included in this concept and, furthermore, have been explicitly excluded by some authors (i.e., Dobzhansky, 1937b, Ch. 10; Grant, 1957). The argument is that, where there are no breeding populations, there can be no biological species, for the latter is the sum total of interbreeding groups. In uniparental organisms, in short, we have arrays of clones but not true species according to the biological species concept.

It is only fair to say that this argument has not met with universal approval. Uniparental organisms do exist in the world. And the biological species concept has not dealt with them in a constructive manner. To be sure, many of the counter arguments have confused taxonomic considerations with biological ones, thus missing the real issue.

But two authors, Simpson (1951, 1961) and Meglitsch (1954), have sought and found a biological basis for species groupings common to uniparental and biparental organisms. This common denominator is the genotypic similarity between related individuals: a similarity that is brought about by community of descent and stabilizing selection in a given environment, and which leads reciprocally to a certain integration of the similar individuals into populations in their environment (Simpson, 1951; 1961, Ch. 5; Meglitsch, 1954).

These forces and their interactions are common to uniparental and biparental organisms. They provide a basis for the concept of the evolutionary species, as Simpson (1961) terms it, which is more general than the biological species concept.

The Evolutionary Species

The evolutionary species as defined by Simpson (1961, Ch. 5) is a population system which possesses the following characteristics. (1) It is a lineage, an ancestral-descendant sequence of populations existing in space and in time. (2) The lineage evolves separately from other lineages or, in other words, from other species. (3) It has its own "unitary evolutionary role," that is, it fits into its own particular ecological niche in a biotic community. (4) And it has its evolutionary tendencies, being susceptible to change in evolutionary role during the course of its history (Simpson, 1961).

These characteristics are found in both uniparental and biparental organisms. The concept of the evolutionary species therefore embraces a greater diversity of breeding systems, and is consequently more general, than the concept of biological species in biparental groups, as already noted.

The unitary evolutionary role of a species—its occupation of an ecological niche of its own in nature for which it is especially adapted—is accomplished by a combination of processes.

The processes common to both uniparental and biparental organisms are: (1) inheritance of similar genes from common ancestors; (2) spread of certain genes throughout a population system or species by natural selection and migration; and (3) inhibition of spread of these genes to other species as a result of their failure to migrate to or reproduce in a different environment. The foregoing processes all act to collect related individuals, which carry similar genes, into populations living in the same environment, whether or not these individuals reproduce sexually (Meglitsch, 1954; Simpson, 1961).

In sexual organisms the unity of the species is promoted by two additional processes not found in strictly uniparental groups. One is interbreeding, which facilitates the spread of genes within a population system. The other process is reproductive isolation, which inhibits the spread of these genes to other species (Simpson, 1961).

Sexually reproducing organisms also have the potentiality, absent in strictly uniparental forms, of engaging in natural interspecific hybridization, which detracts from the integrity of the biological species involved. Viewed from the standpoint of the evolutionary species concept, however, the important question is not whether two species hybridize, but whether two hybridizing species do or do not lose their distinct ecological and evolutionary roles. If, despite some hybridization, they do not merge, then they remain separate species in the evolutionary perspective (Simpson, 1961).

The Species Problem in Uniparental Organisms

We have seen that there are population-building forces common to both uniparental and biparental organisms. This is the basis for extending the species concept to both classes of organisms. There are also integrative forces peculiar to the biparental forms. And this suggests that the biological species is or can be a better integrated unit than the species in asexual organisms.

In fact, discrete species groupings are usually recognizable in sexually reproducing groups of organisms, as exemplified by most genera and families of mammals and birds, but are conspicuously absent in many asexual groups, as exemplified by large segments of the genera Rubus and Hieracium. This general correlation between sexuality and a species organization has been a compelling argument for the restriction of the concept of true species to sexual organisms alone (Dobzhansky, 1937b, Ch. 10).

But there are some comparisons other than that between, let us say, the cat family and the genus Rubus which can and should be made. And there are factors other than sexuality which should be taken into consideration in these comparisons. Let us return to the case of Rubus and other plant groups with a similar variation pattern.

The situation in many apomictic plant groups is complicated by the fact that asexual methods of reproduction are combined with natural hybridization. Let two sexual species hybridize. Their F_1 hybrids and various later-generation segregates, if viable, can then perpetuate themselves and multiply by asexual means. A complex array of hybrid clones develops around the original sexual species. The variations in the plant group are not circumscribed within definite limits, corresponding to species, but encompass two or more parental species and all their hybrid clones.

The variation pattern in parts of Rubus, Crepis, Hieracium, Taraxacum, and other apomictic groups is a huge network consisting of several sexual species and their various asexual hybrid derivatives. Such groups are notoriously difficult taxonomically. The biological species concept breaks down in these asexual hybrid complexes, and so, for that matter, does any alternative taxonomic concept of species. Any system of classification into subgroups which can be devised is largely arbitrary.

We note that natural interspecific hybridization is an important factor contributing to the obliteration of discrete species groups in apomictic hybrid complexes. But natural hybridization also obscures species lines in many sexual plant groups, as we shall see later.

The breakdown of the biological species organization as a result of reversion to asexual reproduction in hybrid derivatives occurs widely in plants which reproduce asexually by vegetative or agamospermous means. Autogamous plant groups, on the other hand, are often organized into good, though cryptic, biological species, even where there is some hybridization. It is probable that the low rate of outcrossing which occurs in most predominantly autogamous plants is sufficient to link the individuals into breeding populations and these into biological species (Grant, 1957; Beaudry, 1960).

The effect of uniparental reproduction per se on species organization, uncomplicated by the effects of natural interspecific hybridization, could perhaps be elucidated by an analysis of the variation pattern in parasexual or predominantly asexual groups of bacteria. Are the arrays of clones in such bacteria grouped into species-like units separated from one another by genetic discontinuities? The variation pattern is difficult to interpret in terms of the presence or absence of discontinuities, but apparently genetic discontinuities have developed between certain strains of bacteria (Stanier, 1955; Ravin, 1961). The population biology of bacteria and other nonsexual microorganisms warrants much further study.

Degree of Integration

The unity and integrity of an evolutionary species depend on the likeness and relationship of its component individuals. Likeness stems in the first place from community of inheritance. This is a factor common to both uniparental and biparental organisms (Meglitsch, 1954; Simpson, 1961, Ch. 5). Biparental organisms possess an additional factor promoting group cohesion, namely, interbreeding.

We can say that the individuals belonging to the same population and to the same species are linked by bonds of parenthood in uniparental organisms, whereas in biparental organisms the conspecific individuals are linked together by bonds of parenthood and also by mating bonds. The linkages resulting from interbreeding provide a means of integration in the populations and species of biparental organisms which does not exist in uniparental groups.

We have also seen in Chapter 1 that the distinction between biparental and uniparental reproduction does not cover the situation entirely. Existing plant species form a spectrum of breeding systems ranging from wide outcrossing at one extreme through various intermediate conditions to strict uniparental reproduction at the other extreme. It follows that the degree of integration reached in these species should vary also and, in a correlated manner, from a maximum in widely outcrossing species to a relatively loose condition in predominantly uniparental groups.

In other words, if we follow Meglitsch (1954) and Simpson (1961) in recognizing a species organization in uniparental as well as biparental organisms, then as a corollary we also have to recognize various degrees of integration in species with different reproductive biologies. Well-integrated biological species represent one condition found in nature, while the breakdown of a biological species organization in asexual hybrid complexes is another. Between these extremes we find various intermediate degrees of integration represented in the species of higher plants, as exemplified by hybridizing sexual species and predominantly autogamous species.

The Species as a Unit in Macroevolution

Various molecular biologists, in attempting to deal with organic evolution, have implicitly reduced this phenomenon to a resultant between just two forces, mutation and selection (i.e., Beadle, 1963; Jukes, 1966; and, to a lesser degree, Anfinsen, 1959). By the same token, this school of workers largely ignores the existence of species as units and speciation as a process in evolution. Since a multiplicity of species has in fact arisen in the course of evolution, any treatment of the subject which fails to take their formation into account is suspect.

The combination of mutation and selection will account for biochemical evolution, for the most primitive stages of biological evolution, and for some very simple evolutionary changes in higher organisms. But the mutation-selection theory does not provide an adequate explanation of the evolution of the more complex organic structures and functions. The development of plants and animals probably could not have taken place on this planet in the time available as a result of mutation and selection alone.

The characteristics by which organisms are adapted to their environmental conditions are based, in all but exceptionally simple cases, not on single genes, but on combinations of genes, usually complex combinations. The sexual process arose relatively early in evolutionary history, among unicellular forms, as a mechanism for producing combinations and recombinations of different genes. At the same time, divergent specializations, based on different gene combinations, arose in response to different sets of environmental conditions within a common area. The advantageous effects of sex could be combined with the preservation of divergent specializations only by the erection of barriers to hybridization, as we have seen in the preceding chapter.

In short, we have to go well beyond mutation and selection in our search for an adequate explanation of organic evolution. The development of complex adaptations required sexual reproduction, and this in turn called for the formation of reproductively isolated species.

There has been no doubt in the minds of professional evolutionists concerning the central place of species in evolution, ever since Darwin entitled his book *On The Origin of Species*. It is widely agreed among modern evolutionary biologists that the diversity of species in the world is a system which permits the continued maintenance of different specialized ways of life. On the relation of species to phyletic evolution, however, there has been less unanimity, three main viewpoints having been put forward.

Huxley (1942, 387–89) regarded speciation as a process generating diversity but having no bearing on progressive evolutionary trends. Mayr (1963, p. 621) later pointed out that, while most species are indeed evolutionary dead ends, a small but significant minority of them hit upon a combination of favorable characteristics and ecological opportunities which enable them to become the progenitors of new dominant groups. There is no denying the validity of this view. In addition, I have suggested (Grant, 1963, pp. 566–68) that speciation may be inextricably involved in some cases of phyletic evolution, where a progression takes place as a series of successive speciational steps.

Species Biology

The natural or evolutionary species is one of the basic units of organization of living material. As such, and in company with other biological units, the species possesses certain general properties of its own, which can be discovered and elucidated by scientific study. By scientific study we mean that species should be regarded not only as an inexhaustible number of particular entities with particular features, as in purely descriptive taxonomy, but also and more importantly from the generalizing standpoint of theoretical biology.

Whether the evolutionary species has received its fair share of analytical scientific study is open to serious debate. As compared with molecular biology, cytology, morphology, general physiology, population genetics, and such-like fields dealing with other levels of biological organization, the investigation of species is still in its infancy. As a result the species is one of the most poorly understood of all basic units of biological organization. Species biology is not even recognized officially in most quarters as an accredited field of study (Mayr, 1963, p. 11).

The reasons for this situation are partly historical. It is suggested here that the underlying reasons for the lag in development of species biology may also be partly psychological or psycho-genetical.

The diverse major branches of biology call for very different methods of research and for very different modes of thinking. The mathematical population geneticist, the experimental biochemist, the descriptive morphologist, the museum taxonomist—to name a few—all have distinctive and divergent approaches. A descriptive cytologist and an experimental cytogeneticist will look at the same set of chromosomes from quite different points of view. Likewise, descriptive systematic biologists and analytical population biologists can look at the same species and see them in entirely different lights.

Undoubtedly the various characteristic thought patterns in biology are products of training. Probably they have a hereditary basis too. Men and women with inherent aptitudes for analytical work would not be expected to go into or remain in purely descriptive fields, and vice versa. In any case, the population of biologists is observably polymorphic for mental traits, whether these are products of nurture or nature or both.

It is very interesting that the best analytical minds in biology have been drawn in disproportionate numbers to the investigation of the microscopic units of organization. Conversely, the species has traditionally been the province chiefly of descriptive-minded taxonomists whose writings, from Linnaeus to the present day, show little use of analytical methods and little grasp of abstract concepts, such as are needed in order to put species biology on a scientific basis.

The situation is aggravated by the much greater complexity of the biological phenomena to be analyzed at the species level than at the various microscopic and submicroscopic levels of organization. As a result of this disparity between levels, a degree of analytical ability which is good for dealing with the smaller biological units may be quite inadequate to handle populations and species. More than one reductionist biologist who has won distinction, if not a Nobel prize, for his work on microscopic or molecular processes has later turned his attentions unsuccessfully to the processes of organic evolution, as the recent literature shows only too clearly.

Yet species biology has not been without its spokesmen since Darwin and Wallace in the last century. Among important modern works dealing with the biology of plant and animal species we can mention: *Apomixis in Higher Plants* (Gustafsson, 1946–1947); *Variation and Evolution in Plants* (Stebbins, 1950); *Stages in the Evolution of Plant Species* (J. Clausen, 1951); *Animal Species and Their Evolution* (Cain, 1954); *Principles of Animal Taxonomy* (Simpson, 1961); and *Animal Species and Evolution* (Mayr, 1963).

The ability to recognize and deal analytically with species phenomena thus seems to be rare in the polymorphic population of biologists. This may be why our understanding of speciation processes has lagged behind other fields of biology. But equally important is the fact that the rare combination of analytical and compositionist abilities is present at all. This relatively rare type of worker has been responsible for the development of population genetics and population ecology in recent decades and will be responsible for the future development of species biology.

CHAPTER 4

The Species Situation
in Plants

Introduction

In the preceding chapters we have seen that a wide range of conditions exists among higher plants as regards the mode of organization of related individuals into population systems. Well-integrated biological species exist in the plant kingdom, as in animals, but they do not represent the only form of population organization in plants. At the opposite extreme are some large and variable population systems, developed on the scale of a taxonomic section or subgenus, which are not subdivided into discrete species units. Between these extremes we find species which display various intermediate levels of integration.

The questions which arise next, and which concern us in this chapter, are: How frequent are the various modes of species organization in plants? What factors are responsible for their development?

In order to approach these questions we have to establish some standard of reference. It is useful to take the sets of discrete and well-integrated biological species found in many or most higher animals, as exemplified for instance in the cat family, as such a standard of comparison. Then we can consider the similarities to, and the departures from, this standard which occur in plants.

Departures from a well-developed biological species organization take two common forms in plants. In biparental plants there may be a level of reproductive isolation between two or more breeding groups which is intermediate between the level characteristic of biological species and that of races. Or biparental reproduction may be replaced by some means of uniparental reproduction. The departure along either line may be moderate or extreme. Corresponding to the two types of deviation from the standard condition, in their extreme development, we can recognize two main kinds of "nonspecies," namely, semispecies and microspecies.

We shall see that, in some floras at least, these types of nonspecies are not uncommon relative to biological species. Their fairly widespread occurrence is attributed partly to the nature of speciation processes and partly to the fundamental characteristics of plants.

Semispecies as a Product of Gradual Speciation

A common mode of speciation in cross-fertilizing plants and animals involves the gradual divergence of two or more population systems. At an early stage of divergence the population systems are related to one another as geographical or ecogeographical races. They occupy adjacent territories and interbreed and intergrade freely in their zones of contact. In later stages of divergence the population systems have

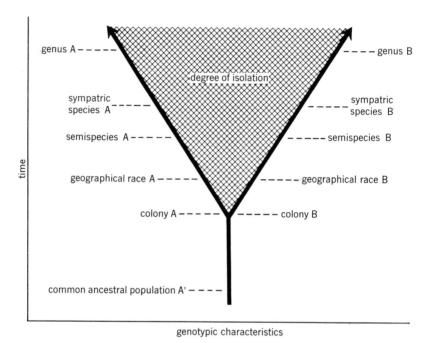

Figure 10 Stages of evolutionary divergence. (From *The Origin of Adaptations*, by Grant, copyright 1963, Columbia University Press, New York, by permission.)

the relationships of sympatric species, living in the same territory without interbreeding.

Between the early and late stages there are various intermediate stages of divergence and reproductive isolation, when the populations are neither good races nor good species but are connected by a reduced amount of interbreeding and gene flow. The population systems in such intermediate stages of divergence are fittingly referred to as semispecies (see Grant, 1963, p. 343, for review and references).

We shall discuss geographical speciation as a process later (Chapter 8). Here we are interested in one result of this process. It is an inevitable consequence of gradual and continuous evolutionary divergence from geographical race to sympatric species that some population systems will be found at any given time in the intermediate stage of semispecies (see Figure 10).

In many plant genera we find disjunct allopatric populations which live in similar ecological habitats in their respective areas, are differentiated morphologically to such an extent that taxonomists give them different species names, but prove to be interfertile when crossed experimentally. An example is provided by *Catalpa bignonioides* of the eastern United States and *C. ovata* of China (Stebbins, 1950, p. 200). *Platanus racemosa* and *P. wrightii* of stream-fed valleys and canyons in California and Arizona, respectively, are another example. The separate population systems in these and similar cases are more than races and less than biological species. They stand at the semispecies level of divergence.

An even more interesting type of intermediate situation is that of overlapping rings of races. A series of intergrading races occurs along a geographical transect. The terminal races of the series, which are the most differentiated, then coexist sympatrically without interbreeding to any considerable extent. One and the same population system thus exhibits both racial and species relationships in different parts of its geographical area.

The best-known cases are in birds and other animals (see Mayr, 1963, pp. 507 ff. for review). Among plants of the western American flora examples are found inter alia in the *Gilia capitata* group, *Ipomopsis aggregata* group, *Aquilegia chrysantha* group, *Diplacus longiflorus* group, and *Pinus ponderosa* group (Grant, 1950a, 1952a, 1952c, unpubl.; Beeks, 1962; Haller, 1962).

Diplacus puniceus (Scrophulariaceae) in its extreme form is a tall, red-flowered shrub of the relatively moist coastal zone of southern California (Figure 11A). *Diplacus longiflorus* is an orange-flowered bush in the chaparral belt of the arid interior foothills (Figure 11B). On a broad front in southern California from the coastline to the interior mountains, complete intergradation occurs between *D. puniceus* and *D. longiflorus*. In parts of the Santa Ana Mountains, however, where habitats suitable for *D. puniceus* and *D. longiflorus* occur close together, populations of these two entities coexist without intergrading (Beeks, 1962). Thus *D. puniceus* and *D. longiflorus* behave as races in many areas, but as sympatric species in some localities; and, when both types of relationships are taken into consideration, the two population systems are seen to fall into the intermediate category of semispecies.

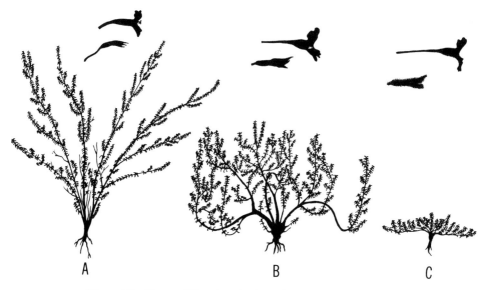

Figure 11 Form of the plant body, calyx, and corolla in three semispecies of *Diplacus* (Scrophulariaceae). (A) *D. puniceus.* (B) *D. longiflorus.* (C) *D. calycinus.* (Beeks, 1962.)

The fact that not all systems of breeding populations fall into two clear-cut alternative categories, races and species, but that some have intermediate and transitional properties, was emphasized by Darwin (1859), who concluded from this that species are units which can be defined only in an arbitrary way. Darwin's view has been followed by many later students down to the present time. We admit the facts but not the inference.

It is a characteristic feature of biological units in general that they multiply by gradual and time-consuming processes. The division of one chromosome, cell, or individual into two daughter chromosomes, cells, or individuals proceeds through intermediate stages of separation. Yet no one thinks that chromosomes, cells, and individuals, once formed, are arbitrarily delimited units of organization. By the same token, one biological species can pass through intermediate stages in the process of subdivision into two or more new species without compromising the status of the species as nonarbitrary units in their terminal and well-developed stages (Grant, 1957).

Semispecies as a Product of Natural Hybridization

The process of evolutionary divergence is not inexorable. Within wide limits it is subject to reversal. Semispecies, therefore, can develop not only as entities on the way to becoming species, but also as products of hybridization between previously well-isolated species.

Natural hybridization and gene flow can take place between biological species, even though they are highly intersterile or well-isolated in other ways, as long as the breeding barriers are less than 100% effective. The barriers to interbreeding between related species of plants belonging to the same genus or section are in fact often capable of being breached to some extent.

Interspecific hybridization is a potential threat to the integrity of the parental species as distinct breeding groups. Whether the potential threat becomes an actual fact or not is another question. Interspecific hybridization may lead to end results of various sorts (see Part III). Some of these results of hybridization do not affect the distinctness of the species involved, and hence do not concern us now.

We are interested here in the hybrid formation of new populations which bridge the breeding barrier, the morphological gap, and the ecological differentiation between the originally distinct parental species. Such bridging populations of hybrid origin fall into two general classes: hybrid swarms and introgressive populations. Their formation reverses the process of speciation and converts the formerly distinct species into semispecies if the hybridization occurs on an extensive scale.

A hybrid swarm is a complex mixture of parental forms, F_1 hybrids, backcross types, and segregation products. It exhibits, as expected, a very high degree of individual variability. Many examples have been reported, only a few of which can be mentioned here.

Hybrid swarms between *Aquilegia formosa* and *A. pubescens* occur at various points of contact between these two semispecies in the Sierra Nevada of California (J. Clausen, 1951; Grant, 1952c). Hybrid swarms of *Juniperus virginiana* and *J. scopulorum* have been described from Kansas and Nebraska (Fassett, 1944). *Geum rivale* and *G. urbanum* produce hybrid swarms in many parts of England and Europe (Marsden-Jones, 1930; Gajewski, 1957). Colonies of three and four species of Baptisia (Leguminosae) and their hybrids are found on the gulf coast of Texas (Alston and Turner, 1962, 1963a, 1963b).

When the hybrids backcross repeatedly to one parental species, an introgressive population arises which resembles the recurrent parent but varies in the direction of the opposite species (Anderson, 1949, 1953). If the introgression goes far enough, it may obliterate the morphological and ecological distinctions between the original species, which thereupon become reduced to semispecies.

The Irises of the Mississippi delta in southern Louisiana have been studied by Riley (1938) and others. In undisturbed natural habitats in this area, *Iris fulva* and *I. giganticaerulea* occur sympatrically without hybridizing. *Iris giganticaerulea* (formerly *I. hexagona giganticaerulea*) grows in the waterlogged soil of marshes, and *I. fulva* in the drier soil of neighboring banks and woods. Where the natural habitats have been disturbed by man, with the formation of pastures and ditches, hybridization between the two Irises has occurred and introgressive populations have grown up (Riley, 1938).

Riley selected four colonies for detailed analysis. Colony G is a pure population of *Iris giganticaerulea* growing in an undisturbed marsh. Colonies H_1 and H_2 are more or less hybridized populations growing in a pasture 500 feet from G. Colony F consists of pure *I. fulva* in natural oak woods a few miles from G. The uncontaminated populations (G and F) provide a standard of comparison for the hybridized populations (H_1 and H_2).

In each colony, Riley collected 23 plants. Seven floral characters were measured on each plant. The measurements were next converted into score values on an arbitrary scale in which the extreme *fulva*-like condition was set at 0 and the *giganticaerulea*-like condition at either 2, 3, or 4. The scores for all seven characters were then combined to form hybrid index values for each plant which ranged from 0 to 17. The frequency distribution of the index values was plotted as a histogram for each colony; the results are shown in Figure 12.

The histograms show that colony H_2 is essentially like *Iris giganticaerulea* in its morphological characteristics but contains a couple of hybrid plants. Colony H_1, however, is an introgressive population with a range of variation which largely spans the morphological gap between *I. giganticaerulea* and *I. fulva* (Riley, 1938).

Randolph and co-workers have recently reexamined this case

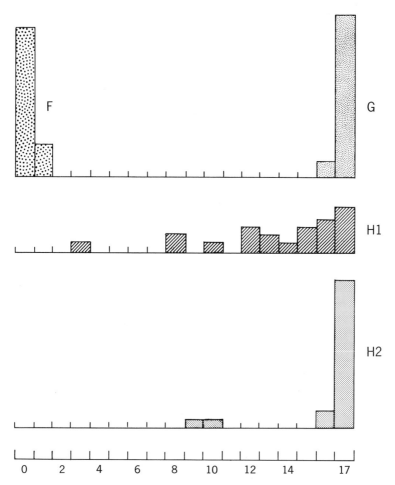

Figure 12 Frequency distribution of hybrid index values for 23 plants in each of four colonies of Iris in southern Louisiana. (F) *Iris fulva*. (G) *I. giganticaerulea*. (H1) A strongly introgressive population. (H2) A slightly introgressive population. (Riley, 1938.)

(Randolph, Nelson, and Plaisted, 1967). On the basis of a thorough analysis of variation in thirteen characters in numerous allopatric and sympatric populations of the two species, they conclude that the effects of introgression are localized in disturbed habitats within the zone of sympatry. Each Iris species in a large part of its distribution area remains uninfluenced, as far as can be detected, by genes from the other species (Randolph, Nelson, and Plaisted, 1967).

Nevertheless, introgression does bridge the morphological and ecological gap between the species locally. When the variation patterns of the two species are considered throughout their whole area of distribution, they are seen to be connected by intergrades in certain zones of contact and hybridization.

The Syngameon

The definition of the biological species as the most inclusive breeding group obviously does not hold up in cases of naturally hybridizing species and semispecies. Where limited gene exchange is taking place between otherwise isolated species or semispecies, the most inclusive unit of interbreeding is a community of species.

A situation of this sort was described by Lotsy (1925, 1931) in a segment of the genus Betula. On the European continent there are many forms of Betula, of various and disputed taxonomic rank, which hybridize on a wide scale. Lotsy (1925) concluded that "we have in Betula one very large pairing community, one syngameon."

DuRietz (1930, pp. 367 ff.) reviewed many similar cases in Nothofagus in New Zealand, Geum in Europe, Melandrium in Sweden, Salix in northern Europe, etc. All these groups represent "large and highly polymorphic 'hybrid'-syngameons in which the species have got more or less lost . . ." (op. cit., p. 384).

Several later workers have developed the concept of the syngameon (Cuénot, 1951; Grant, 1957; Beaudry, 1960). The syngameon can be defined in terms of modern concepts as "the sum total of species or semispecies linked by frequent or occasional hybridization in nature; [hence] a hybridizing group of species . . . ." (Grant, 1957, p. 67). Where interspecific hybridization occurs in nature, therefore, the most inclusive breeding group is not the species but the syngameon.

Syngameons have been described in various species groups in western North America. Pinus, Quercus, Aquilegia, Gilia, Iris, and Diplacus all furnish examples (Grant, 1952c, 1957; V. and A. Grant, 1960; Lenz, 1959a; Beeks, 1962). The Cerastium alpinum group is a syngameon composed of five or more basic species in the circumboreal region

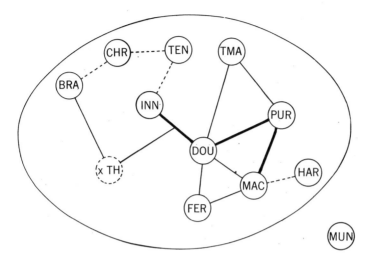

Figure 13   The structure of the syngameon in the Pacific coast Irises (Iris series Californicae). All the species and semispecies in this group except *I. munzii* are linked by natural hybridization, in the combinations shown by the connecting lines. The relative amount of hybridization is indicated by the breadth of the lines: bars = extensive hybridization, solid lines = moderate hybridization, broken lines = slight hybridization. (Lenz, 1959a.)

Abbreviations of Iris species and semispecies are as follows. BRA = *I. bracteata*. CHR = *I. chrysophylla*. DOU = *I. douglasiana*. FER = *I. fernaldii*. HAR = *I. hartwegii*. INN = *I. innominata*. MAC = *I. macrosiphon*. MUN = *I. munzii*. PUR = *I. purdyi*. TEN = *I. tenax*. TMA = *I. tenuissima*. x TH = *I. thompsonii*.

(Hultén, 1956). The most complex syngameon which has been analyzed in detail is perhaps the one in the Pacific coast Irises of North America (Iris series Californicae). The structure of this syngameon as worked out by Lenz (1959a) is shown in Figure 13. Another in Pacific coast oaks is shown in Figure 18 (in Chapter 7).

Arrays of Microspecies

The term microspecies (or jordanon) has been applied by botanists to any uniform population which is slightly different morphologically from related uniform populations. Thus considered, the term covers a wide range of population phenomena, being synonymous with pure lines, with some local races, with sibling species in certain cases, etc.; in fact it has been used in these and other senses. To the extent that the term has been widely and loosely used, it has of course lost preciseness of meaning, and has been replaced by more useful terms.

There is, however, a common element in the actual application of the term microspecies which warrants its retention. Botanists have tended to apply this term in actual usage to slightly differentiated populations in plant groups with uniparental reproduction.

Thus microspecies have been described and recognized in the *Erophila verna* group, which is autogamous, and in Rubus section Moriferus, which is agamospermous. *Erophila verna* and Rubus section Moriferus are certainly very different in systematic affinities, life form, and breeding system. It is significant, in view of the obvious differences between these groups, that they possess enough similarities in their variation pattern to have evoked the application of a common concept of microspecies. The common element in the variation pattern to which the term microspecies can be applied, moreover, is correlated with two other common properties, namely, natural hybridization and uni-parental reproduction.

These common properties have not been recognized or even been known by all the botanists, particularly the earlier ones, who have utilized the term microspecies. Their usage represents the largely intuitive recognition of a certain type of population unit in a certain type of variation pattern. As such it provides a precedent for formulating a more definite concept of microspecies.

Microspecies are defined here as populations in predominantly uniparental plant groups which are uniform themselves and are slightly differentiated morphologically from one another. They are often restricted in distribution to a relatively small geographical area, although this is not always or necessarily the case. They are usually of hybrid origin, either directly or indirectly, and often retain a permanent hybrid constitution.

Microspecies fall into four main classes according to the mode of reproduction: (1) clonal microspecies, reproducing by vegetative propagation, as in Opuntia; (2) agamospermous microspecies, reproducing by agamospermy, as in Rubus Moriferus; (3) heterogamic microspecies, reproducing by the *Oenothera biennis* or *Rosa canina* genetic systems; (4) autogamous microspecies, predominantly auto-gamous and chromosomally homozygous, as in Erophila. It will be noted that the uniparental reproduction is asexual in classes 1 and 2, and sexual in classes 3 and 4.

The plants are permanent hybrids in the case of agamospermous and heterogamic microspecies. They are usually highly heterozygous and may be interspecific hybrids in clonal microspecies. Autogamous microspecies, in contrast with the other three classes, do not retain a strongly hybrid constitution, and may be complete or nearly complete homozygotes; but, like the other classes of microspecies, they often originate as products of hybridization.

In most actual cases the autogamous microspecies is not exclusively autogamous but is partially outcrossing and otherwise possesses a normal sexual cycle. Such an autogamous microspecies is therefore a type of biological species. The concept of autogamous microspecies, which is reached by one line of reasoning here, is largely synonymous with the concept of sibling biological species, as presented in Chapter 2. The pattern of sibling species in autogamous plant groups is discussed further in Chapter 7.

Clonal and agamospermous microspecies may be, or may become, evolutionary species, but they do not belong in the general category of biological species. They create an even more difficult species problem than does the class of autogamous microspecies. Heterogamic micro-species which are also hybrid in constitution, though sexual, likewise pose a difficult species problem.

In these cases the hybrid individuals build up uniform populations by apomictic or heterogamic methods of reproduction. The buildup passes through a stage when the microspecies is narrowly restricted in distri-bution, and may, in the evolutionarily successful cases, reach a stage where the microspecies is geographically widespread. The process of population growth from hybrid individual through small clone and endemic microspecies to uniparental evolutionary species is continuous. At some undefinable point in the latter stages of this gradual process the asexual or heterogamic microspecies emerges as a full-fledged evolutionary species. And in these latter stages, though not at the beginning, it warrants recognition as a taxonomic species.

In any given plant group, natural hybridization between three or more original species in different combinations, followed by uniparental reproduction of the hybrids, leads to the formation of numerous new asexual or heterogamic microspecies. The first generation of microspecies can then hybridize again to yield second-order microspecies. The plant group generates new diversity within itself by the combined processes of hybridization and uniparental reproduction carried out extensively in time and space.

In such plant groups it is common to find large numbers of localized microspecies together with a few geographically widespread microspecies or species of hybrid constitution. The whole array of asexual or heterogamic microspecies, including both the widespread and the endemic members, is superimposed phylogenetically on the original biological species in the group. And the superstructure of derived hybrid microspecies obscures the morphological discontinuities between the ancestral biological species.

This pattern is seen in many agamic hybrid complexes, such as those in Rubus and Crepis, as will be described later in Chapter 21. Similar patterns occur also in heterogamic complexes (i.e., Oenothera) and clonal complexes (i.e., Opuntia) (see Chapters 22 and 23).

Magnitude of the Species Problem

In biparental plants, hybrid semispecies are common, at least in some floras. In the western American flora, the only flora which I can claim to know reasonably well, the impression which I have gained from studying various genera of outcrossing plants is that syngameons composed of hybrid semispecies are often more common than well-defined biological species. This would be the case, for example, in Quercus, Ceanothus, Diplacus, Iris, and the outcrossing members of Gilia in the American west. Other genera of outcrossing plants in the same area, like Pinus, Aquilegia, and Ipomopsis, include a strong minority of hybrid semispecies.

Agamic hybrid complexes are likewise quite common, especially but not exclusively in northern floras. Examples are found in such genera as Antennaria, Crepis, Hieracium, Taraxacum, Poa, Ranunculus, Rubus, and Potentilla. Gustafsson (1946–1947, pp. 306 ff.) lists 73 genera of plants from the world flora which contain some agamospermous microspecies. Some other genera, like Allium, Festuca, and

Saxifraga, contain hybrid forms which reproduce uniparentally by vivipary.

Deviations from the standard of good biological species are thus normal in higher plants, and extreme departures, as seen in the breakdown of species organization, are not uncommon. Hybridization, with or without uniparental reproduction, is involved in the cases which have been studied thoroughly. Herein lies a most important difference between plants and many groups of higher animals, in which good species seem to be the rule and syngameons or agamic complexes are exceptional (Grant, 1957; Mayr, 1963).

Causes and Effects of the Species Situation in Plants

On comparing well-studied groups of higher plants with many groups of birds, mammals, and such insect genera as Drosophila, we find that discrete biological species are, in general, less frequent and other deviant species conditions are more characteristic in the plants. It follows that the causal factors determining the various modes of species organization should also be expected to differ in the two kingdoms. Therefore a comparative approach to the species problem may shed some light on its causes in higher plants.

An obvious and basic difference between plants and animals lies in the degree of complexity of the individual organism. The development of the relatively simple plant body is commensurate with an open system of growth by which new parts are built up in series. The animal body is a vastly more complex and delicately balanced organization and, furthermore, is one which must develop as a whole and by a closed system of growth. It is logical to suppose that the gene systems controlling growth and development differ correspondingly in complexity and integration in the two types of organisms.

This single premise will account for the two main factors contributing to the partial or complete breakdown of species organization in many plant groups. It will account for the frequency of interspecific hybridization and of uniparental reproduction.

A species possessing a highly integrated and finely balanced genotype will suffer relatively worse effects from interspecific hybridization than will a species with a simpler and more loosely coordinated set of genes. The relatively simple physiological-morphological organization of the plant body, reflecting a correspondingly relatively simple genotype, probably gives plants a greater tolerance for interspecific gene exchange than is present in most animal groups (Grant, 1957). From the open system of growth of plants, then, stems the possibility of the various forms of vegetative and agamospermous reproduction by which the hybrids and hybrid progeny can increase in numbers.

As Gustafsson has put it (1946–1947, p. 302), "Phanerogamous plant species repeatedly fall a victim to their own characteristics: their loose system of growth, their vegetative dispersal. These properties continually work to rebuild the population, change it in the polyploid and apomictic direction."

The fundamental characteristics of plants have not only affected the nature of plant species, but they have also had a profound effect on plant macroevolution. As we shall see later, plant phylogeny has involved the repeated anastomosing of previously separate lines. If a phylogenetic tree is the extension of the normal pattern of animal speciation, plant speciation has often led to the formation of a phylogenetic web.

CHAPTER *5*

The Genetic Basis of Species Differences

Introduction · Multiple Gene Systems · Multifactorial Linkage · The M-V Linkage · Chromosomal Rearrangements · Block Inheritance · The Integrated Genotype

Introduction

Studies of inheritance in interspecific and interracial crosses have been carried out by many workers in many plant groups. These studies show that related species or races are differentiated in respect to various types of gene systems, cytoplasmic factors, and linkage systems. The evidence has been reviewed elsewhere (Grant, 1964b) and does not warrant detailed repetition here. It will suffice for our present purposes to give a brief summary with emphasis on principles.

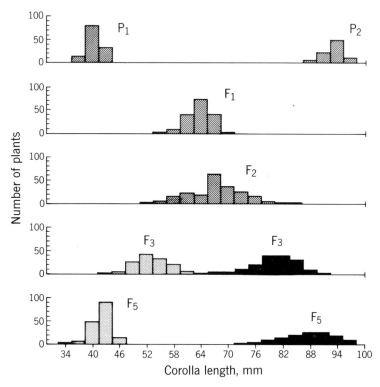

Figure 14 Frequency distribution of individuals with different lengths of corolla in an intervarietal cross in *Nicotiana longiflora*. (Drawn from data of East, 1916, data simplified.)

Multiple Gene Systems

The vast majority of character differences between races or interfertile species of plants are controlled by multiple gene systems of one sort or another. A classical example is the difference in corolla length between two varieties of *Nicotiana longiflora* as studied by East (1916).

The normal form of *Nicotiana longiflora* has a corolla tube nearly 10 cm long; there is in addition a short-tubed form. These differences are genetically determined and have a high heritability. East (1916) used inbred and true-breeding lines with long and short corollas as parents. The main results of the cross are shown in simplified form in Figure 14.

It will be seen that the F_1 generation was intermediate in phenotype between the parents and comparable to them in variability. The F_2 generation, however, although still intermediate as to mode, showed a wider range of continuous variations in flower length and a higher coefficient of variability than the parental and F_1 generations. These facts are consistent with the idea that the F_2 is segregating for several independent genes with additive effects.

East (1916) tested this hypothesis by growing advanced generations derived from selected F_2 individuals with short or long corolla tubes. He found, as expected on his hypothesis, a shift in the mode, that is, a response to selection in the F_3 families, along with a general decrease in variability from F_3 to F_5.

East was thinking in terms of multiple genes with cumulative effects on a quantitative character. His example in Nicotiana and some other early cases were in line with this simple concept. Other early and later examples, however, revealed more complex gene interactions.

Thus the components of a multiple gene system within a given race or species may not have additive effects alone on a quantitative character; but instead some components have additive, and others negative, effects on character development. When two such internally balanced, oppositional gene systems are combined by interracial or interspecific hybridization, the F_2 generations, and under some circumstances also the F_1 hybrids, show transgressive segregation, exceeding the range of either parental type in quantitative expression of the character (see Grant, 1964b, pp. 90 ff.).

Comparable epistatic interactions may occur between gene systems controlling some qualitative difference between races or species like presence or absence of pigmentation. For example, *Viola tricolor* has a dark spot in the flowers; the related *V. arvensis* normally lacks spots. J. Clausen (1926, 1951) showed that both species possess complementary genes for spot formation. But *V. arvensis* carries in addition one or two inhibitors of spot formation. Progeny of the interspecific cross sometimes display complex segregations for this character (J. Clausen, 1926).

An interesting type of epistatic interaction is that in which wholly novel and unpredictable characters arise in the progeny of a species hybrid. *Antirrhinum majus* and *A. glutinosum* (Scrophulariaceae) both have typical snapdragon flowers, and so do their F_1 hybrids and most of their F_2 progeny, but some F_2 individuals have flowers like those of a distantly related genus, Rhinanthus (Lotsy, 1916). Likewise, in Digitalis (Scrophulariaceae), an interspecific cross led to some progeny with a novel floral character, namely, spurs, not found in the parental species (Schwanitz, 1957).

Novel phenotypes have been observed in hybrid progenies in a number of plant groups besides Antirrhinum and Digitalis. They thus represent a general phenomenon and have been referred to as macrorecombinations (see Grant, 1963, p. 470, for brief review with references). Macrorecombinations are significant evolutionarily in showing that interspecific hybridization can engender not only intermediate forms, but also products which are unlike either parental species.

Multifactorial Linkage

Chromosome numbers in diploid angiosperms range from $n = 2$ to $n = 12$, 13, or 14. The modal haploid numbers are $n = 7$ to 9 in herbaceous Dicotyledons, $n = 8$ to 9 and 11 to 14 in woody Dicotyledons, and $n = 7$ in Monocotyledons (see Chapter 15).

If, therefore, the multiple factors determining two character differences between two races or species are distributed more or less randomly throughout the chromosome set, some of the genes for each character should be borne on the same chromosome(s). In short, the two characters are expected to show partial or multifactorial linkage. A third quantitative character is likely to be partially linked in the same manner to one or both of the preceding two characters. By extension, all multifactorial characters differentiating the two races or species are expected to be tied together, directly or indirectly, in an interlocking system of weak linkages. Such a linkage system does not prevent, but it does strongly restrict, recombination in interracial and interspecific hybridization (Clausen and Hiesey, 1958, 1960; Grant, 1956a; 1964b, pp. 137 ff.).

The expected multifactorial linkages have been found in several hybridization experiments. The best data are those obtained by Clausen and Hiesey (1958) for an interracial cross in *Potentilla glandulosa* $(2n = 14)$. They studied the correlations between 14 characters, of which 11 were known to be determined by two or more genes, in a large segregating F_2 population.

Sixty-seven of the 91 pairs of characters showed weak but statistically significant correlation. No character segregated independently of all other characters but, on the contrary, each character was correlated with many or most of the other 13 characters. The correlations between certain pairs of characters, such as petal length and sepal length, are probably due to pleiotropy; but many other combinations of developmentally unrelated characters, like pubescence and seed weight, are also correlated in inheritance, and these cases point to linkage (Clausen and Hiesey, 1958).

Multifactorial linkage thus gives rise to a correlation between characters which Clausen and Hiesey (1960) aptly describe as genetic coherence.

The M-V Linkage

Studies of species hybrids in several plant genera provide evidence for linkage between genes determining various morphological characters and genes for viability. This widespread and probably general condition is referred to as the *M-V* linkage (Grant, 1967). The term morphological genes, as used in this connection, refers to any genes controlling visible phenotypic traits including coloration of plant parts. By viability genes is meant any genes affecting growth and vigor.

Viability genes may have pleiotropic effects on morphological characters and vice versa, as in the case of albino seedlings. We wish to be able to rule out such cases of pleiotropy. Therefore we confine our attention to phenotypic characters that are expressed in the flowers or fruits, but not detectably in the vegetative phase of the plants, and which are correlated with differences in viability. The observed correlations in these instances suggest linkage between separate morphological and viability genes rather than pleiotropy.

We are dealing, then, with evidence for allelic differences between species or races of plants in linked M and V genes. The simplest possible situation is a linked pair of genes, M-V, without important gene interactions. One parental species would have the constitution $M_a V_a / M_a V_a$, and the other the constitution $M_b V_b / M_b V_b$. Assume that V_a and V_b have different environmental optima for growth. Then the progeny of the hybrid, when grown in an environment favoring one of the V alleles, will give an altered monofactorial segregation ratio for the morphological marker M. The various, more complex setups involving partial or multifactorial linkage between two or more V genes and two or more M genes will give similar results (Grant, 1967).

The interfertile species, *Mimulus cardinalis* and *M. lewisii* (Scrophulariaceae), occur at lower and higher elevations, respectively, in the California mountains. The seedlings of a coastal strain of *M. cardinalis* survive well in a lowland garden (at Stanford) but suffer much mortality in an alpine garden (Timberline). A high-mountain race of *M. lewisii*, conversely, has good seedling survival at Timberline but high seedling mortality at Stanford (Nobs and Hiesey, 1957).

The two species differ inter alia in flower color, which is pink in *Mimulus lewisii* and orange in *M. cardinalis*. This character is inherited as a single-gene difference with pink dominant over orange. Nobs and Hiesey (1958) found that the F_2 progeny of the interspecific cross, when grown in a generally favorable environment, had approximately a 3 to 1 ratio of pink- and orange-flowered plants.

But an F_2 progeny grown at Timberline gave a ratio of 8.7 pink to 1 orange, and a replicate F_2 grown at Stanford in summer gave 1.7 pink to 1 orange. Nobs and Hiesey next compared the ratios obtained in the same garden at Stanford under cool winter vs. hot summer conditions. The winter-grown F_2 population had a 4.4 to 1.0 ratio of pink and orange as compared with a 1.7 to 1.0 ratio in the summer-grown F_2 (Nobs and Hiesey, 1958).

These altered segregation ratios were associated in each case with a high seedling mortality. The mortality evidently involves a selective elimination of *cardinalis* types in an alpine or cool coastal environment and of *lewisii* types in a warm coastal environment. Furthermore, the genes controlling the general physiological adaptation of the seedlings to climatic conditions must be linked with the flower color gene in order to give the observed skewness of the phenotypic ratios (Nobs and Hiesey, 1958).

Similar evidence has been obtained from an interracial cross in *Potentilla glandulosa* and from interspecific crosses in Gossypium, Lycopersicon, and Phaseolus. This evidence suggests that the related races or species in question differ allelically with respect to linked M and V genes (Lamprecht, 1941, 1944; Stephens, 1949, 1950; Clausen and Hiesey, 1958; Rick, 1963; Grant, 1967).

Chromosomal Rearrangements

Segmental rearrangements are initiated by breaks in the chromosomes and completed by reunion of the broken ends in new ways. The number and distribution of the breaks and the mode of reunion determine the type of rearrangement, as shown in Figure 15.

One break near the end of one chromosome can give rise to a terminal deficiency. Two breaks in one chromosome can lead to an interstitial deletion, a duplication, a paracentric inversion, or a pericentric inversion. Three breaks in one chromosome may give a transposition. If the broken ends of two nonhomologous chromosomes rejoin, small reciprocal translocations or whole arm translocations result. The reunion of the broken ends of three or more nonhomologous chromosomes, finally, leads to successive translocations (Figure 15).

In many plant groups, cytological studies of related species and their F_1 hybrids show that the species in question differ with respect to inversions, translocations, and other structural rearrangements. Two classical examples will be described briefly for purposes of illustration.

Avery (1938) made a thorough cytogenetic analysis of a hybrid triangle in Nicotiana involving *N. alata*, *N. langsdorffii*, and *N. bonariensis*. These species have nine pairs of chromosomes which fall into different classes on size and morphology within a given complement and to some extent between species complements.

At meiosis in the F_1 hybrid of *Nicotiniana alata* X *langsdorffii* the A_1, B_1, and D chromosomes form a chain, indicating that the parental species differ by two successive translocations on these chromosomes. The pairing configuration in *N. alata* X *bonariensis* shows that these species differ by a reciprocal translocation in the A_1 and E_3 chromosomes and by two successive translocations in the A_2, B_2, and E_3 chromosomes. From these findings it is possible to predict that *N. langsdorffii* and *N. bonariensis* are differentiated with respect to a series of translocations involving all chromosomes of the complements except C and E_2. The pairing configurations observed in the hybrid of *N. langsdorffii* X *bonariensis* confirmed this expectation (Avery, 1938).

An equally ingenious cytogenetic analysis of *Lilium martagon*, *L. hansonii*, and their F_1 hybrid was carried out by Richardson (1936). These lily species both have twelve pairs of chromosomes. The pairing

1. Terminal deficiency

Std.   A B C D   E F G H I

Rear.   A B C D   E F G

2. Interstitial deletion

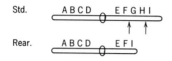

Std.   A B C D   E F G H I

Rear.   A B C D   E F I

3. Duplication

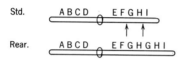

Std.   A B C D   E F G H I

Rear.   A B C D   E F G H G H I

4. Paracentric inversion

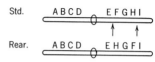

Std.   A B C D   E F G H I

Rear.   A B C D   E H G F I

5. Pericentric inversion

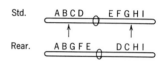

Std.   A B C D   E F G H I

Rear.   A B G F E   D C H I

6. Transposition

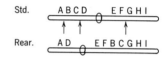

Std.   A B C D   E F G H I

Rear.   A D   E F B C G H I

7. Small reciprocal translocation

Std.   A B C D   E F G H I   M N O P   Q R S T

Rear.   A B C D   E F G H S T   M N O P   Q R I

8. Whole arm translocation

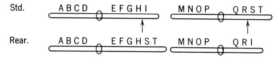

Std.   A B C D   E F G H I   M N O P   Q R S T

Rear.   A B C D   Q R S T   M N O P   E F G H I

9. Successive translocations

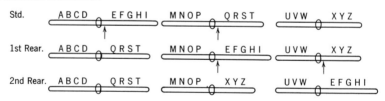

Std.   A B C D   E F G H I   M N O P   Q R S T   U V W   X Y Z

1st Rear.   A B C D   Q R S T   M N O P   E F G H I   U V W   X Y Z

2nd Rear.   A B C D   Q R S T   M N O P   X Y Z   U V W   E F G H I

Figure 15 Types of chromosomal rearrangements. The structurally altered chromosome (labeled Rear.) is compared in each case with one and the same standard arrangement (Std.). The arrows mark the breakage points on the standard chromosome which can give rise to a given structural rearrangement. The centromere is shown as a median or submedian oval. (From *The Architecture of the Germplasm*, by Grant, copyright 1964, John Wiley & Sons, New York, by permission.)

configurations in the hybrid indicate that the parental species differ by small paracentric inversions on six of the twelve chromosomes (Richardson, 1936).

Interspecific differences in segmental arrangement of the chromosomes have been found in many plant genera and are particularly common in annual herbs of open habitats. Familiar examples occur in Brassica, Clarkia, Crepis, Elymus, Galeopsis, Gossypium, Gilia, Layia, Madia, and Nicotiana.

Block Inheritance

Athough it has been possible to assay the genic contents of homologous chromosomes belonging to interfertile species or races in a number of plant groups, we know very little about the genetic factors borne on the rearranged segments which differentiate chromosomally intersterile species. Indeed, the latter problem has rarely been tackled, because the complicating effects of hybrid sterility on factorial analysis have made it seem almost unsolvable. Nevertheless, we now have some suggestive indications concerning the genic contents of rearranged segments in several plant groups (Grant, 1966a).

The evidence comes from hybridization experiments involving pairs of species in Godetia (now Clarkia, Onagraceae), Gossypium (Malvaceae), Triticum (Gramineae), Zea (Gramineae), and Gilia (Polemoniaceae). In each case the evidence is necessarily indirect and the analysis complex. I have reviewed the various stories in some detail elsewhere (Grant, 1966a), and will only summarize the main points here.

Hiorth (1940, 1942) and Håkansson (1947) showed that the species *Godetia amoena* and *G. whitneyi* (now treated as members of the genus Clarkia) have the same chromosome number ($2n = 14$) but differ by one or more translocations which give rise to a chromosome chain of varying length in the F_1 hybrid. The end arrangements of the homeologous chromosomes in the two species which concern us here can be designated as follows: in *G. whitneyi*, 1—2; in *G. amoena*, 1—6 (Håkansson, 1947).

These species also differ in the position of the petal spot, which is basal in *Godetia amoena* and central in *G. whitneyi*. This floral difference is controlled by a single gene F represented by the alleles F_b in *G. amoena* and F_x in *G. whitneyi*. The F gene is located in the 1—arm of the homeologous chromosomes. The F_b allele of *G. amoena* can cross over rarely from the *amoena* 1—6 chromosome to the *whitneyi* 1—2 chromosome in translocation heterozygotes. Hence the gene for petal spot is borne on the translocated segment (Håkansson, 1947).

Gossypium hirsutum and *G. barbadense* differ by small chromosomal rearrangements on the basis of several lines of cytogenetic evidence (Stephens, 1949, 1950; Gerstel, 1953; Gerstel and Sarvella, 1956; Rhyne, 1958). The two species also differ by sets of modifier genes for various morphological characters which are inherited in interspecific backcrosses as though they are grouped in linked blocks (Stephens, 1949, 1950). Stephens has therefore suggested that the gene blocks may be coextensive with the segmental rearrangements. Furthermore, the morphological differences, including some which are affected by the blocks of modifier genes, are correlated in inheritance with viability differences, as indicated by a consistent deficiency of the marker allele from the nonrecurrent parent in segregating interspecific backcross progenies (Stephens, 1949, 1950). The evidence from Gossypium thus suggests that linked morphological and viability genes may be borne on structurally rearranged segments.

The speltoid mutants in hexaploid wheat, *Triticum aestivum* (or *T. vulgare*), form a syndrome of inflorescence and shoot characters. These characters mutate from the normal to the speltoid condition in a block, there being no recombination between the individual characters, and furthermore the mutational change always occurs from normal to speltoid and never in the opposite direction (Frankel and Munday, 1962). The speltoid characters appear to be borne on a segment, designated Q, in the long arm of chromosome IX, which is occasionally lost as a result of a deletion (Sears, 1944, 1959; MacKey, 1954). Where the Q segment is present, the phenotype is that of normal hexaploid wheat and, when this segment is deleted, the speltoid characters develop.

The Q segment is also present in a form of tetraploid wheat known as *Triticum carthlicum* in the *T. dicoccum* group, but is absent in most other tetraploid species of wheat. This suggests that the Q segment was introduced from *T. carthlicum* into *T. aestivum* during the evolutionary history of wheat (Frankel and Munday, 1962). It follows that *T. carthlicum* and some other tetraploid species differ in respect to the presence or absence of this Q segment with its associated morphological genes.

The chromosomes of *Zea mays* and *Z. mexicana* usually pair normally in the F_1 hybrid, but nevertheless apparently differ by some small rearrangements (Mangelsdorf, 1958; Ting, 1967). Mangelsdorf (1958) produced backcross derivatives consisting mainly of homozygous maize chromosomes but containing substituted segments of *Z. mexicana* in various parts of the genome. These backcross derivatives were highly mutable with a breeding behavior like that of Bar-eye in Drosophila, suggesting that the mutations may be products of unequal crossing-over in the small nonhomologous regions. The mutations affect endosperm, stature, and other characters, usually deleteriously (Mangelsdorf, 1958).

Gilia malior and *G. modocensis* differ with respect to numerous segmental rearrangements which contribute in large measure to the high sterility of their F_1 hybrid. These species also differ in viability factors which lead to much hybrid breakdown in F_2 to F_6, and they differ in many morphological characters. Selection in the hybrid progeny for viability under environmental conditions favorable to the *modocensis* types, and for fertility, led to the establishment of three vigorous and fertile lines, two of which were segmentally homologous and interfertile with *G. modocensis*. This result is very difficult to explain unless some of the viability genes are located on some of the rearranged segments. The viable derived lines also resembled *G. modocensis* in most of their morphological characters, suggesting that some morphological genes are linked with viability genes on the same rearranged segments (Grant, 1966b).

We thus have evidence for the occurrence of morphological genes on rearranged segments in several different plant groups. In some of the same cases the morphological genes appear to be linked with viability genes. There does not seem to be any basic difference in the M-V linkage system as it is found between chromosomally homologous species and between structurally differentiated ones.

The Integrated Genotype

Each part or quantitative character of an organism is a product of developmental processes controlled by a system of genes, as we have already seen. It follows that the individual components of the gene system must have harmonious interactions if the organ or character is to develop normally. Furthermore, the successful functioning of the organism as a whole requires a coordination between its separate parts and characters. Consequently, the different gene systems belonging to a given genotype have been selected for their harmonious interactions.

The functional value and the selective value of harmonious gene interactions have been recognized by many students. Bridges' (1922) concept of genetic balance is its best-known formal expression. As Clausen, Keck, and Hiesey (1945, p. 62) put it, "Natural species consist of individuals whose genes are in internal balance so that a harmonious physiologic and morphologic development is assured generation after generation."

The extent to which genes are internally balanced within a species is best revealed by the hybrid breakdown resulting from interspecific crosses. The F_2 and later-generation progeny of species crosses usually contain arrays of dwarf, sterile, subvital, or otherwise aberrant individuals. *Gilia malior*, *G. modocensis*, and their F_1 hybrids are vigorous. But 42% of the seedlings in the F_3 to F_6 generations were weak, stunted, or semilethal, and much additional inviability is inferred to exist in the embryo stage from failure of seed germination (Grant, 1966c).

The disharmonious interactions between the genes of different species may also express themselves in the F_1 generation. A good example is the red-leaved dwarf hybrids found intermixed with vigorous plants in the F_1 of *Gossypium arboreum* X *hirsutum*. This semilethal condition results from the complementary action of two genes, Rl_a and Rl_b, carried by particular strains of the parental species (Gerstel, 1954).

CHAPTER 6

Reproductive Isolation

---

*Main Forms of Isolation · External Reproductive Isolation ·
Internal Reproductive Isolation · Conclusions*

Main Forms of Isolation

Hybridization and gene exchange between species are prevented or
reduced by isolating mechanisms of many different kinds. They can be
divided into three main classes: spatial, environmental, and reproductive.

Spatial or geographical isolation exists between any two allopatric
plant species whose respective geographical areas are separated by gaps
greater than the normal radius of dispersal of their pollen or seeds.
This would be the case, for example, in *Platanus occidentalis* and *P.
orientalis*, which occur in the eastern United States and the eastern
Mediterranean region, respectively, and which, despite interfertility,

obviously cannot interbreed under natural conditions (Stebbins, 1950, p. 199).

Environmental or ecological isolation results from differentiation between species in ecological preferences. This differentiation leads to species occupying different habitats where they coexist in the same geographical area. If the species hybridize, and if their respective habitats are discrete, most of their hybrid progeny will be unable to survive because of the absence of intermediate or recombination-type environments (Anderson, 1948). Thus *Salvia apiana* and *S. mellifera*, which are ecologically differentiated but sympatric in southern California, and which are capable of interbreeding to a limited extent, do not hybridize in undisturbed natural communities, but do occasionally produce hybrid swarms in artificially disturbed areas (Epling, 1947; Anderson and Anderson, 1954; K. A. and V. Grant, 1964).

Reproductive isolation, finally, refers to blocks to gene exchange between populations which stem from genotypically controlled differences in their reproductive organs, reproductive habits, or fertility relationships. These blocks may be external, acting before cross-pollination, as in the case of mechanical, ethological, and seasonal isolating mechanisms. Or the blocks may be internal, operating after cross-pollination as in incompatibility, hybrid inviability, hybrid sterility, and hybrid breakdown.

It goes without saying that the three main forms of isolation may be combined in various ways in actual cases. The several kinds of reproductive isolating mechanisms also usually work in combination. In the following pages we will briefly survey the methods of reproductive isolation (see Grant, 1963, Ch. 13, for a more detailed account).

External Reproductive Isolation

Seasonal isolation exists between two plant species when their periods of flowering and pollination occur at different times of the year or of the day. The seasonal isolation may be partial if the peaks of the two flowering periods are well separated in time but the ranges overlap, or complete if the two periods have nonoverlapping ranges. Examples are found in Pinus, Phlox, Salvia, and many other plant groups.

Oenothera brevipes and *Oe. clavaeformis* (Onagraceae), for example, grow and bloom together in the western American desert, and are both pollinated by solitary bees, chiefly Andrenas, but form hybrids only rarely. Raven (1962) has found that the flowers of *Oe. brevipes* open before sunrise and are visited by early-morning bees, while the flowers of *Oe. clavaeformis* open in late afternoon and are visited by bees that fly late in the day.

Mechanical isolation results from structural differences between the flowers of two or more species which prevent or interfere with inter-specific cross-pollination. Floral differences of the type that leads to mechanical isolation are found chiefly in plant families with complex floral mechanisms, such as the Orchidaceae, Papilionaceae, Labiatae, Scrophulariaceae, and Asclepiadaceae.

Thus *Pedicularis groenlandica* and *P. attollens* (Scrophulariaceae) grow and flower together in the Sierra Nevada of California and are pollinated by the same kinds of bumblebees, but do not hybridize. This is due partly to the fact that the floral mechanism brings about venter pollination in *P. groenlandica* and head pollination in *P. attollens* (Sprague, 1962). Other cases of mechanical isolation have been worked out between species of Stanhopea, Ophrys, Aquilegia, Asclepias, Penstemon, Mimulus, and Salvia.

Ethological isolation due to psychological preferences of individuals for members of their own species in mating plays a very important role in the reproductive isolation of higher animals. This type of isolation was not expected to occur in plants at all. Yet, as a second-order effect of the species-specific flower-visiting behavior of some insects, it can occur in plant groups pollinated by such insects.

Bees, hawkmoths, and, to a lesser degree, some other classes of flower-visiting insects have sensory perceptions enabling them to distinguish between different species of flowers on the basis of odor, form, and color, and also possess instincts of flower constancy leading them to feed preferentially on one species of flower during a succession of feeding visits. This flower-constant behavior tends to channelize the cross-pollinations chiefly or entirely within the limits of single species in sympatric mixtures of plant species differing in their floral recognition signals (Grant, 1949, 1950b). Ethological isolation of related species due to preferential flower visits by bees has been observed in natural habitats in Pedicularis (Sprague, 1962), Stanhopea (Dodson and Frymire, 1961), and several other genera.

Internal Reproductive Isolation

Interspecific cross-pollination may occur but fail to lead to the development of a viable hybrid embryo as a result of any one of several incompatibility barriers. The pollen may fail to germinate on a foreign stigma, as in *Datura meteloides* X *stramonium*, or the pollen tubes may not grow successfully in the foreign style, as in *Iris tenax* X *tenuis* (Avery, Satina, and Rietsema, 1959, p. 239; Smith and Clarkson, 1956). Or, if fertilization takes place and the hybrid embryo is formed, it may die prematurely as a result of the degeneration of the hybrid endosperm, as in certain crosses in Datura, Iris, Gossypium, and other genera.

Interspecific F_1 hybrids are frequently vigorous; but they may also display various constitutional weaknesses, known collectively as hybrid inviability, which block gene exchange between species in the vegetative phase of the F_1 generation. Certain strains of *Crepis tectorum* when crossed to *C. capillaris* produce inviable F_1 progeny (Hollingshead, 1930). Red dwarf plants appear in the F_1 generation of certain interspecific crosses in Gossypium, Gilia, and Papaver (Gerstel, 1954; V. and A. Grant, 1954; Latimer, 1958; McNaughton and Harper, 1960a). Genetically determined tumors develop on F_1 hybrids derived from many interspecific crosses involving *N. langsdorffii* and other parental types in the genus Nicotiana (H. H. Smith, 1958).

Hybrid sterility, like incompatibility, embraces a wide variety of phenomena. They can be classified according to the developmental stage in which the block is expressed into diplontic and gametic sterility. Sterility phenomena can be classified again on the basis of the controlling factors into genic, chromosomal, and cytoplasmic sterility. Various combinations of the different stages and factors are found in actual cases.

The interactions between the genes of different plant species frequently result in the abortive development of the anthers or other essential organs in the hybrid. Disharmonious interactions of nuclear genes with a foreign cytoplasm bring about male sterility in certain hybrids in Zea, Epilobium, Streptocarpus, Nicotiana, Capsicum, Solanum, and other plant groups (Michaelis, 1933; Oehlkers, 1940; Caspari, 1948; Clayton, 1950; Jones, 1956; Edwardson, 1956; Peterson, 1958; Jain, 1959; Putrament, 1962; Grun, Aubertin, and Radlow, 1962).

These unfavorable gene interactions may also express themselves in the stages of meiosis and gametogenesis, upsetting those delicately coordinated processes in some way. Genically determined gametic sterility has been found in hybrids in Nicotiana, Geum, Gilia, Bromus, and other genera (Greenleaf, 1941; Gajewski, 1953; Grant, 1956a; M. S. Walters, 1957, 1960).

Plant species frequently differ by chromosomal rearrangements, as we noted in the preceding chapter. The rearrangements may interfere with normal bivalent formation and/or with normal anaphase distribution of chromosomes at meiosis in the hybrid. A proportion of the products of meiosis carry deficiencies and duplications for particular chromosome segments and consequently do not develop into viable pollen grains and embryo sacs. This proportion of inviable gametophytes rises rapidly with increase in the number of segmental rearrangements. A hybrid which is heterozygous for only six independent translocations, for example, will have an expected pollen fertility of 1.5%, which is comparable to the sterility level found in highly sterile species hybrids.

It is well known that chromosome behavior at meiosis is affected by genic factors as well as by structural differences (Darlington, 1932, Chs. 6 and 13; Rees, 1961; John and Lewis, 1965b). Both causes of aberrant meiosis and hence of gametic sterility are combined in many actual cases. Indeed, chromosomal sterility, which is present especially in certain life-form classes of plants, can be regarded as a condition superimposed on a more widespread and perhaps more basic condition of genic sterility.

If the hybrid produces any F_2 or B_1 progeny, finally, they are likely to include varying proportions of weak or defective individuals. As we noted in Chapter 5, about 42% of the later-generation progeny of *Gilia malior* X *modocensis* were subvital. Eighty percent of the F_2 progeny of *Layia gaillardioides* X *hieracioides* (Compositae) were subvital (J. Clausen, 1951, pp. 109–111).

Conclusions

The adaptively valuable gene combinations of each species can be broken up by hybridization. Hybrid breakdown is a manifestation of this disintegration. The other reproductive isolating mechanisms serve to protect the adaptive gene combinations from disintegration.

These reproductive isolating mechanisms may come into existence as by-products of evolutionary divergence. This is clearly the case in hybrid breakdown. Evolutionary divergence can be, and usually is, the direct cause of the various other forms of reproductive isolation. Thus the mechanical and seasonal differences between *Salvia mellifera* and *S. apiana* are related primarily to the differential characteristics of their normal pollinators—small spring-flying bees and large summer bees, respectively—and their effects on reproductive isolation are probably by-products of the differentiation.

But this is not the only mode of origin of reproductive isolation. As we shall see later, the mechanisms operating in the parental generation may be built up in certain cases by selection for isolation per se. The adaptive gene combinations of the species are then protected from disintegration by mechanisms arising as products of direct selection for reproductive isolation.

PART II

Divergence of Species

CHAPTER 7

Patterns of Species Relationships

Introduction · Woody Plants · Perennial Herbs with Prominent Species-to-Species Differences in Floral Mechanism · Perennial Herbs without Prominent Species-to-Species Differences in Floral Mechanism · Annual Herbs · Autogamous Annuals · General Patterns · General Correlations · Conclusions

Introduction

Plant hydridizers have long recognized that striking differences exist between plant groups in the ease of crossing species and in the fertility of the hybrids. In some genera, related species are difficult to cross and their hybrids, if obtained, are usually sterile. At the other extreme are plant groups in which wide crosses between morphologically different species, sections, or even genera are possible, and the hybrids derived from the wide crosses are fertile.

The results of different hybridization experiments are unpredictable to a large extent. However, some general patterns become apparent when we make a comparative study of a fairly large sample of plant groups.

In the main body of this chapter we review a series of plant groups, mainly from the north temperate zone, which have been studied taxogenetically by different workers. These are grouped by life form and breeding system. In the summary sections we delineate the general patterns and point out some correlations.

Woody Plants

CEANOTHUS

The North American genus Ceanothus (Rhamnaceae) ranges from southern Canada to Guatemala and has its center of distribution on the Pacific slope. Forty-four of the 55 species in the genus are endemic in California. The plants are woody shrubs. The small, fragrant flowers are pollinated by miscellaneous insects which bring about obligate outcrossing in the self-incompatible species and predominant outcrossing in the self-compatible species. The chromosome number is $2n = 24$ in all known species (Nobs, 1963).

The species fall into two well-marked sections, Euceanothus and Cerastes. The pattern of species relationships within the California members of Cerastes and between Cerastes and Euceanothus has been worked out by Nobs (1963) on the basis of combined field and experimental studies.

The species of section Cerastes occur in a wide variety of ecological habitats on the Pacific slope. They are differentiated morphologically as well as ecologically. In most combinations these species (or semispecies) are separated geographically. In other combinations they overlap in range and occur on different but adjacent soil types in the same area (Nobs, 1963).

Thus *Ceanothus cuneatus* overlaps in range with at least 5 other species of section Cerastes in the area north of San Francisco Bay,

SECTION CERASTES

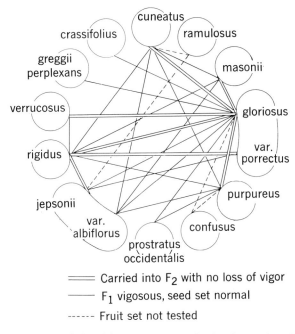

Figure 16 Fertility relationships between species in the section Cerastes of the genus Ceanothus. (Nobs, 1963.)

California. *Ceanothus cuneatus* occurs broadly on sandstone soils and comes into contact with two serpentine endemics, *C. jepsonii* and *C. pumilus*, in certain areas where serpentine crops out within the more widespread sandstone formations. In some of these zones of contact there is little or no natural hybridization, and in others there is considerable hybridization, between *C. cuneatus* and the serpentine-inhabiting species or semispecies (Nobs, 1963).

Nobs intercrossed 12 species of section Cerastes; the results are summarized in Figure 16. The species all cross easily to produce vigorous F_1 hybrids. The hybrids were fully fertile as to pollen and seeds with 88 to 97% good pollen. Meiosis was also normal in the hybrids, 12 bivalents being formed regularly (Nobs, 1963).

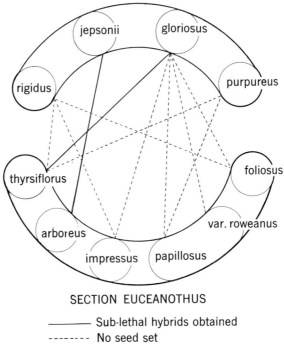

SECTION CERASTES

SECTION EUCEANOTHUS

———— Sub-lethal hybrids obtained

-------- No seed set

Figure 17 Fertility relationships between the two sections of the genus Ceanothus. (Nobs, 1963.)

Members of the sections Cerastes and Euceanothus are sympatric over a wide area, and species belonging to these two sections commonly grow and bloom together in the same plant community. Natural intersectional hybrids are very rare despite the large area of sympatry (Nobs, 1963).

It is of interest to learn, therefore, that the members of the two sections are reproductively isolated by strong internal barriers. Most of the intersectional crosses attempted by Nobs failed completely, no hybrid seeds being formed (Figure 17). In rare instances a few hybrid seeds and seedlings were obtained, but the hybrids were usually weak and stunted and died out in early stages. Still more rarely did the intersectional hybrids reach maturity. They were then highly sterile with

0 to 8 % good pollen and irregular meiosis. Chromosome pairing in a spontaneous hybrid of *C. masonii* X *foliosus* ranged from 2 to 11 bivalents and averaged 7 bivalents (Nobs, 1963).

QUERCUS

The large and widespread genus Quercus (Fagaceae) contains numerous species of trees and shrubs. They are diploid with $2n = 24$ chromosomes in the cases investigated cytologically. The plants are outcrossing and wind-pollinated. The species fall into two distinct subgenera, the white and black oaks, or Lepidobalanus and Erythrobalanus, respectively. The evidence concerning interspecific relationships in the oaks is considerable but scattered and fragmentary, and we shall follow recent reviews of the situation in selected geographical areas by Stebbins (1950, p. 61) and Benson (1962, pp. 53 ff.).

The species or semispecies belonging to the same subgenus are usually segregated into different but often interdigitating climatic or edaphic zones in the same area. Thus, among California white oaks, *Quercus lobata* occurs on valley floors and *Q. douglasii* on dry foothills; *Q. durata* on serpentine and *Q. dumosa* in surrounding nonserpentine country. In Texas, *Q. mohriana* occurs on limestone, *Q. havardii* on sand, and *Q. grisea* on igneous outcrops (C. H. Muller, 1952).

Where the species or semispecies of the same subgenus come into contact, they often hybridize naturally. Figure 18 shows the network of occasionally hybridizing semispecies of white oaks in California. This syngameon is composed of semispecies which are extremely different in morphology and ecology. *Quercus garryana* is a forest tree with large, deeply cleft leaves. *Quercus dumosa* and *Q. turbinella californica* are low xeric shrubs with small spiny leaves. Yet these extreme forms are connected directly by hybridization or indirectly through inter-mediary populations (Figure 18).

Similar syngameons are known in the white oaks of other regions: the southwest, eastern North America, and Europe; and again among the black oaks (Stebbins, 1950; Benson, 1962).

The natural interspecific hybrids within a subgenus appear to be more or less fertile and cytologically normal (Stebbins, 1950, p. 61). Tucker (1953) found that the hybrid of *Quercus garryana* X *durata*, a wide cross in morphological and ecological terms, had 93 to 97 % good pollen

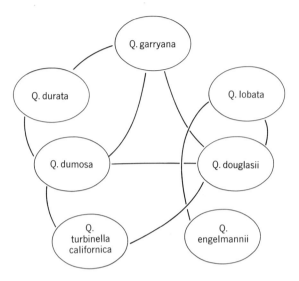

Figure 18 Natural hybridization in various combinations between semi-species of white oaks in California.

and regular meiosis, but was highly sterile as to seeds, and the hybrid of *Q. garryana* X *dumosa* was also seed-sterile.

White oaks and black oaks also come into contact rather frequently in nature. Nevertheless intersubgeneric hybrids have not been found in nature and have been produced only rarely in artificial crossings (Stebbins, 1950, p. 65).

RIBES, PINUS, EUCALYPTUS

The genus Ribes (Saxifragaceae) contains over 150 species of shrubs. They are diploids with $2n = 16$ in the known cases. The genus is sub-divided into four well-differentiated subgenera and numerous sections.

Numerous interspecific crosses have been made by Keep (1962) and others. When the sections are adjusted to conform to the crossing data, the following generalizations emerge. Species of the same section cross easily and produce fertile hybrids in every case tested. Crosses between members of different sections in the same subgenus may or may not succeed; and, when the intersectional hybrids are obtained, they are partially fertile (Keep, 1962).

The main breeding barrier in Ribes is between the subgenera, which are also well-differentiated morphologically. This is a strong incompatibility barrier. The F_1 hybrids which arise rarely as a result of breaching this barrier are then sterile (Keep, 1962).

A generally similar pattern occurs in Pinus ($2n = 24$) which is subdivided into two distinct subgenera, the yellow pines (Diploxylon) and white pines (Haploxylon). The species of yellow pines can be crossed successfully in many combinations, though not in others. The successful crosses are usually those within a species group, such as the *Pinus ponderosa* group or the *P. radiata* group. Some of the interspecific hybrids are known to be fertile. Likewise, within the assemblage of white pines, artificial crosses can be made in some hybrid combinations. But no crosses have been successful between the two subgenera, which are separated by a strong incompatibility barrier (Righter and Duffield, 1951; Duffield, 1952; Wright and Gabriel, 1958; Mirov, 1967).

This pattern of species relationships is found again in Eucalyptus (Myrtaceae) in Australia. The genus contains five subgenera and numerous species, most of which have $2n = 22$ chromosomes. Pryor (1959) reports that the species belonging to the same subgenus usually occupy different ecological habitats and are interfertile in a number of known cases. By contrast, species belonging to different subgenera often occur sympatrically, and are prevented from interbreeding by strong internal barriers (Pryor, 1959).

Perennial Herbs with Prominent Species-to-Species Differences in Floral Mechanism

AQUILEGIA

Aquilegia (Ranunculaceae) is a genus of perennial herbs with showy flowers and $2n = 14$ chromosomes. The genus has a widespread distribution in Eurasia and North America. Throughout this area there are numerous named species, many of which appear to be geographical races. The diversity within the genus can be classified naturally into the following four main species groups (Grant, 1952c; Prazmo, 1960).

(1) The most primitive species is *Aquilegia ecalcarata* of the Himalaya Mountains with spurless white or purple flowers. It approaches the

related genus, Isopyrum ($2n = 14$), which has similar flowers. (2) Extending across Eurasia and into northern North America are the many forms comprising the *A. vulgaris* group with nodding, blue or purple, long-spurred flowers. (3) Extending across temperate North America is the *A. canadensis* group which is characterized by nodding, red, stout-spurred flowers. (4) In western North America we find the members of the *A. chrysantha* group with their erect yellow flowers bearing long thin spurs.

In North America the three species groups occur sympatrically in various areas. They hybridize naturally in some localities but remain distinct in other areas of sympatric overlap (Grant, 1952c).

The early work on fertility relationships in Aquilegia by Anderson and Schafer (1931) and others has been extended in recent years by Prazmo (1960, 1961, 1965). These studies show that members of the four species groups cross readily in all combinations and yield fertile or semifertile F_1 hybrids with regular meiosis.

Thus, among the wider crosses within the genus, the F_1 of *Aquilegia ecalcarata* X *vulgaris* has a pollen fertility of 42% and seed fertility of 19% and produces numerous F_2 progeny (Prazmo, 1960). The F_1 of *A. ecalcarata* X *chrysantha* has 62% fertile pollen and 8% seed fertility (Prazmo, 1961). The F_1 of *A. ecalcarata* X *canadensis* is also semifertile as to pollen and seeds (Prazmo, 1965). And the *A. vulgaris, chrysantha,* and *canadensis* groups produce semifertile F_1s inter se.

The main barriers to interbreeding between the species groups must therefore be external. In the North American columbines these barriers are ecological, mechanical, and ethological isolation. The *Aquilegia vulgaris* group is normally pollinated by bumblebees, the *A. canadensis* group by hummingbirds, and the *A. chrysantha* group by hawkmoths (Grant, 1952c).

The results of crossings between Aquilegia and Isopyrum are entirely different from those within Aquilegia. Skalinska (1958) attempted the cross of *Aquilegia ecalcarata* ♀ X *Isopyrum thalictroides*. She encountered incompatibility blocks at various stages before and after fertilization. The pollen tubes of Isopyrum grow to a reduced extent in the styles of Aquilegia. Some ovules do become fertilized. But then, during the first two weeks of embryogeny, either the endosperm degenerates or the embryo itself fails to develop. The net result is failure to produce any hybrid seedlings (Skalinska, 1958).

ORCHIDS, ANTIRRHINUM, PENSTEMON, EPIPHYLLUM

Next we consider briefly four other groups of perennial herbs with showy flowers in which the species are interfertile within wide limits. The evidence in some of these groups has been derived from horticultural work, and is incomplete regarding the genetic and biosystematic aspects, but is nevertheless very suggestive.

The orchids are well known as a group in which very wide crosses can be made. Vanda has been crossed successfully with members of nine other genera. Ten genera in the subtribe Laelieae—Cattleya, Epidendrum, Laelia, and others—have been intercrossed in some twenty intergeneric combinations. Several genera of the subtribe Oncidieae have been crossed successfully with genera belonging to three other subtribes, the Cochliodeae, Comparettieae, and Trichocentreae (Lenz and Wimber, 1959).

The orchid genus Ophrys in the Mediterranean region contains a number of species which can be crossed artificially to produce pollen-fertile hybrids. These species occur in various sympatric combinations in nature without hybridizing to any great extent. Their isolation in nature is based primarily on floral differences which act as ethological and mechanical barriers to interspecific pollination by bees (Stebbins and Ferlan, 1956).

Interspecific hybrids in one section of the orchid genus Vanda have regular meiosis with complete chromosome pairing, but irregular meiosis occurs in hybrids in another section of the same genus (Kamemoto and Shindo, 1964). Meiosis is also normal in hybrids between certain species of Dendrobium (Shindo and Kamemoto, 1963).

In the genus Antirrhinum (Scrophulariaceae), most of the species in the section Antirrhinastrum ($2n = 16$) cross more or less easily with one another to produce fertile hybrids. These species are separated mainly by ecological and ethological barriers. But the species of section Antirrhinastrum are isolated by incompatibility barriers from *Antirrhinum orontium* and the section or segregate genus Asarina (Baur, 1914, Ch. 12; 1932; Mather, 1947).

Three species of Penstemon section Peltanthera (Scrophulariaceae) inhabit different ecological zones but meet sympatrically in various combinations in southern California. The species in question are *Penstemon grinnellii, spectabilis,* and *centranthifolius*. They are strikingly different in floral characters, *P. grinnellii* having two-lipped wide-throated flowers, *P. spectabilis* broad-tubular, bluish flowers, and *P. centranthifolius* red, trumpet-shaped flowers.

The species are interfertile in the experimental garden. In the wild they hybridize in some areas of sympatric contact but remain isolated in others. The isolation is due largely to mechanical and ethological factors. *Penstemon grinnellii* is normally pollinated by carpenter bees (Xylocopa), *P. spectabilis* by pseudomasarid wasps and medium-sized bees, and *P. centranthifolius* by hummingbirds (Straw, 1956).

The species of Epiphyllum (Cactaceae) also have very different types of showy flowers. Wide crosses are possible within Epiphyllum or between species which are sometimes placed in different segregate genera. Thus viable hybrids can be obtained between red hummingbird-flowered species and white moth-flowered forms (Haselton, 1951).

Perennial Herbs without Prominent Species-to-Species Differences in Floral Mechanism

GEUM

The fairly large genus Geum (Rosaceae), containing about 56 species, ranges widely throughout the northern hemisphere and occurs in temperate regions of the southern hemisphere. The plants are perennial herbs with an outcrossing breeding system. The basic chromosome number in the tribe to which Geum belongs is $x = 7$. There are no known diploid species in the genus Geum, however, which consists of polyploids at various levels from $4x$ to $12x$, with chromosome numbers ranging from $2n = 28$ to $2n = 84$ (Gajewski, 1957, 1959).

Gajewski (1957) classifies Geum into 11 subgenera. Most of the taxogenetic work has been carried out with two of these subgenera, namely, Oreogeum and Eugeum. Oreogeum consists of three species restricted to the mountains of Europe. The two best-known species of Oreogeum are *G. montanum* ($4x$) and *G. reptans* ($6x$). The larger and more widespread subgenus Eugeum contains 25 species, among which are the following (all $6x$): *G. rivale, urbanum, macrophyllum, molle, coccineum, hispidum*, and *canadense* (Gajewski, 1957).

In nature the species are isolated mainly by ecological and seasonal differences. Where these differences break down between two species, they hybridize naturally. *Geum rivale* and *G. urbanum* hybridize in many zones of contact throughout Europe (Gajewski, 1949, 1957).

In the experimental garden the species can be intercrossed with ease. Any two species of the subgenus Eugeum can be crossed regardless of differences in ploidy level. And Eugeum can be crossed successfully with Oreogeum and with four other subgenera. But at a higher phylogenetic level, as between Geum and the related genera Waldsteinia and Coluria, very strong incompatibility barriers come into play, and all attempts to make the intergeneric crosses have failed (Gajewski, 1957).

The viability of the artificial interspecific hybrids ranged from heterotic to sublethal in different combinations. Within Eugeum the F_1s showed normal vigor in crosses of *Geum macrophyllum* X *G. rivale, montanum*, or *canadense*, but had strong inhibitions of growth in crosses of *G. macrophyllum* X *G. urbanum, molle*, or *hispidum*. Most of the intersubgeneric F_1 hybrids were vigorous, but the F_1s of Eugeum X Woronowia never flowered (Gajewski, 1957).

Hybrid sterility also showed a wide range of conditions from nearly normal fertility to complete sterility. Within Eugeum, for example, the F_1 of *Geum rivale* X *urbanum* had a pollen fertility of 78% and seed fertility of 72%, and F_1s of *G. rivale* X *coccineum, molle*, and *hispidum* were also highly fertile. But the F_1 of *G. rivale* X *macrophyllum* was highly sterile with less than 1% good pollen and seed set, and many other hybrids within Eugeum were also quite sterile. In Oreogeum the hybrid of *G. montanum* X *reptans* was highly sterile as to pollen and completely sterile as to seeds (Gajewski, 1957).

Some intersubgeneric crosses between Eugeum and Oreogeum yielded semifertile F_1s, as, for example, *G. montanum* X *rivale* with 34% pollen fertility and 15% seed fertility. Other hybrid combinations of Eugeum and Oreogeum, like *G. rivale* X *reptans*, were highly sterile. The other intersubgeneric crosses were all highly sterile (Gajewski, 1957).

A high degree of chromosome pairing was found in hybrids in the species group including *Geum rivale, urbanum, molle*, and *hispidum*. These hybrids had 17 to 21 bivalents in different cells, and their chromosomes were homologous or nearly so. But the sterile hybrid of *G. macrophyllum* with *G. rivale* had low pairing with 0 to 7 bivalents and numerous univalents. There is good homology between two of the three genomes in *G. rivale* (6*x*, Eugeum) and the two genomes in *G. montanum* (4*x*, Oreogeum). The hybrids of Eugeum with other subgenera have reduced chromosome pairing (Gajewski, 1957).

The behavior of a spontaneous amphiploid of *Geum rivale* (6*x*) and *G. macrophyllum* (6*x*) suggests that these species differ with respect to chromosomal rearrangements. The F$_1$ hybrid had low pairing, with 0 to 7 bivalents, and was highly sterile as noted earlier. It produced two F$_2$ plants with a doubled chromosome number, hence 12*x*, and these amphiploids formed up to 42 bivalents at meiosis and were fertile (Gajewski, 1953).

However, gene-controlled disturbances of chromosome pairing are indicated by other facts in the case of *Geum rivale* X *macrophyllum*. When either *G. rivale* or *G. macrophyllum* is crossed with a third hexaploid species, *G. aleppicum*, full pairing in 21 bivalents is found, and therefore the chromosomes of *G. rivale* and *G. macrophyllum* must be more homologous than appears from the meiotic behavior of the *G. rivale* X *macrophyllum* hybrid (Gajewski, 1953). The sterility of this and other species hybrids in Geum is thus probably genic as well as chromosomal (Gajewski, 1953, 1957, 1959).

SOLANUM

Solanum section Tuberarium (Solanaceae) comprises a group of tuber-bearing, outcrossing species which form a polyploid series from 2*x* to 6*x* (*x* = 12). The species cross easily in most, though not all, combinations (Swaminathan and Hougas, 1954; Grun, 1961). Some interspecific hybrids are fertile and have full chromosome pairing, others are sterile with reduced pairing, but in general the species are closely related cytogenetically (Swaminathan and Hougas, 1954; Magoon, Cooper, and Hougas, 1958). There is evidence for small structural rearrangements between some of the species (Swaminathan and Howard, 1953).

IRIS

One of the natural groups within the beardless irises (Iris section Apogon) is the series Californicae or Pacific irises. This group consists of 11 semispecies ranging along the Pacific slope of North America from Washington to southern California. The plants are perennial herbs with showy outcrossing flowers and $2n = 40$ chromosomes (Lenz, 1958, 1959a).

These semispecies occur in different ecological zones in the overall distribution area of the group. For example, *Iris douglasiana* grows along the coastline, *I. macrosiphon* in the foothills back from the coast, and *I. hartwegii* in the yellow pine zone of the Sierra Nevada.

The habitats of these and other semispecies of Pacific irises are often contiguous, permitting marginal or neighboringly sympatric contacts between the semispecies themselves. Furthermore, the ecological barriers have been broken down in many places by logging, road building, and other human disturbances of the environment. The semispecies then engage in natural hybridization in numerous combinations (Figure 13, in Chapter 4) (Lenz, 1958, 1959a).

The Pacific irises have been intercrossed artificially in many combinations by Smith and Clarkson (1956) and Lenz (1959a). The semispecies cross easily. The F_1 hybrids are vigorous and fully fertile, with 84 to 99% good pollen. Meiosis in the hybrids is normal, with 20 bivalents (Smith and Clarkson, 1956; Lenz, 1959a).

The closest allies of the Pacific irises are a group of 40-chromosome Eurasian species belonging to the series Sibiricae. The two series can be crossed successfully to produce vigorous hybrids. But the hybrids are sterile (Lenz, 1959a).

A still more distant relative of the Pacific irises is *Iris tenuis* $(2n = 28)$ belonging to the series Evansia (Lenz, 1959b). *Iris tenuis* occurs in Oregon, where it is sympatric with members of the Californicae such as *I. tenax*. There is no natural hybridization between *I. tenuis* and *I. tenax*.

Smith and Clarkson (1956) found a strong incompatibility barrier between *Iris tenuis* and *I. tenax*. This barrier consists of blocks at several stages. The pollen tubes grow slowly and often burst in the foreign pistils. Fertilization, if it occurs at all, is delayed. And the hybrid embryos, if formed, fail to develop past the 12-celled stage. No artificial hybrids could be produced between *I. tenuis* and *I. tenax*.

SILENE

The genus Silene (Caryophyllaceae) contains numerous species of perennial herbs and some annuals. The basic chromosome number is

$x = 12$, and many diploid species ($2n = 24$) exist, especially in Europe, but polyploidy is also common in the genus.

In western North America there are 33 species of herbaceous perennials at various ploidy levels from $2x$ to $8x$. The majority of them are tetraploids. A few diploid, hexaploid, and octaploid species are also found here (Kruckeberg, 1954, 1960). The western American species are sharply delimited morphologically (Kruckeberg, 1961).

Kruckeberg (1955, 1961) found that the western American species could be crossed easily in most combinations to produce viable hybrids. But the interspecific hybrids, with few exceptions, were highly sterile. There was generally a low degree of chromosome pairing and other meiotic aberrations in the hybrids (Kruckeberg, 1955, 1961).

Annual Herbs

GILIA

One of the natural subdivisions of the heterogeneous genus Gilia (Polemoniaceae) is the section Eugilia or Leafy-stemmed Gilias. This section consists of ten species of annual herbs. We have already introduced two of these species, *Gilia capitata* and *G. tricolor*, in another connection in Chapter 2.

The Leafy-stemmed Gilias occur on the Pacific slope of North America and in temperate South America, where the species overlap in range in various combinations and frequently grow side by side in nature. The breeding system ranges from outcrossing to autogamous in different members of the group. The basic chromosome number of $2n = 18$ is found in 6 species, and the derived tetraploid condition of $2n = 36$ in 3 species. A tenth species in the section is unknown cytologically or genetically (Grant, 1954b, 1965).

The species are separated by strong incompatibility barriers in most, though not all, combinations. The F_1 hybrids which can be obtained are then highly sterile. Meiosis is irregular in the hybrids with a reduction of chromosome pairing to as few as 1 bivalent per cell in some cases (Grant, 1954b).

The low degree of pairing observed in the hybrids is due mainly but not entirely to structural rearrangements between the parental species, as indicated by the behavior of artificial amphiploids. Sterile and meiotically irregular species hybrids representing four combinations have doubled spontaneously in the experimental garden to produce amphiploid progeny with good bivalent pairing and medium to high fertility (Grant, 1954b, 1965). Segregation in the early amphiploid generations for both chromosome behavior and fertility suggests strongly that gene-controlled disturbances of meiosis are present in addition to structural rearrangements (Grant, 1952d).

The same pattern is found again in the related section Arachnion of Gilia. This section also consists of numerous species of annual herbs which frequently occur sympatrically in nature. The species are usually separated by strong incompatibility barriers, and their hybrids, when obtained, are usually highly sterile. The hybrid sterility is largely chromosomal and partly genic in nature (V. and A. Grant, 1960; Grant, 1964c; Day, 1965).

LAYIA

Layia (Compositae) is a genus of 15 species of spring-flowering annual herbs in the foothills and valleys of California. In this area close sympatric occurrences of 2 or 3 species are common. The species fall into three series on chromosome number. Six diploid species have $2n = 14$; eight diploid species have $2n = 16$; and there is one tetraploid species with $2n = 32$ (Keck, 1957).

The 7-paired species can be crossed successfully with one another. Most of the 8-paired species are intercrossable too. There are strong incompatibility barriers, however, between the 7-paired and the 8-paired species, and again between certain 8-paired species (Clausen, Keck, and Hiesey, 1945; J. Clausen, 1951).

Hybrid inviability is found in some crosses, such as *Layia gaillardioides* X *heterotricha*, and hybrid breakdown in others like *L. gaillardioides* X *hieracioides*. Hybrid sterility associated with irregular meiosis characterizes many of the interspecific combinations (J. Clausen, 1951; Keck, 1957).

The degree of chromosome pairing in the interspecific hybrids varies widely in the different hybrid combinations. There is complete or only slightly reduced pairing in F_1 hybrids of diploid species belonging to the same species group. Hybrids between members of different species groups, on the other hand, have a low degree of pairing. Thus *Layia heterotricha* ($n = 8$) X *glandulosa* ($n = 8$) has 0 to 2 bivalents, and *L.*

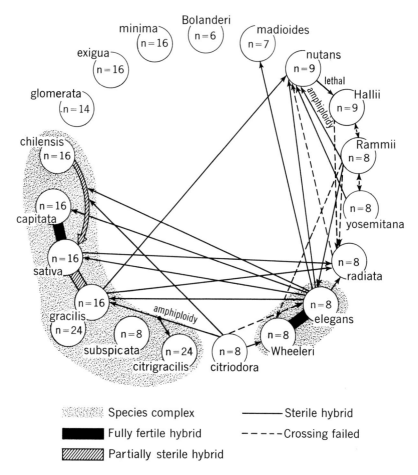

Figure 19 Fertility relationships between species of Madia. (From *Stages in the Evolution of Plant Species*, by J. Clausen, copyright 1951, Cornell University Press, Ithaca, New York, by permission.)

heterotricha ($n = 8$) X *platyglossa* ($n = 7$) has 0 to 3 bivalents (J. Clausen, 1951; Keck, 1957).

Clausen believes that the reduced chromosome pairing in species hybrids in Layia stems from genic disharmonies between the parental species but not from structural rearrangements in their respective genomes (J. Clausen, 1951, pp. 105, 132). However, the following evidence suggests that both chromosomal and genic sterility are involved in Layia.

The F_1 hybrid of *Layia platyglossa* ($n = 7$) X *pentachaeta* ($n = 8$) has 0 to 3 bivalents. It doubled to produce some tetraploid progeny with $2n = 30$ chromosomes. In the tetraploid plants the chromosomes paired regularly in 14 to 15 bivalents, and apparently no multivalents or secondary associations were found. The later stages of meiosis were abnormal, however, and the tetraploid plants were sterile (Clausen, Keck, and Hiesey, 1945, pp. 46 ff.; Keck, 1957).

CLARKIA

The genus Clarkia (Onagraceae) contains many species of annual herbs which have frequent sympatric contacts in nature. As in Gilia and Layia, the species are isolated by strong incompatibility barriers. Hybrid inviability occurs in some crosses. And the hybrids which can be grown to maturity are more or less sterile. The sterility is largely chromosomal (Lewis and Lewis, 1955).

Essentially the same pattern is found again in several other genera of annual plants. Three examples in the Compositae are Madia (Figure 19), Microseris, and Helianthus (J. Clausen, 1951, pp. 133 ff.; Chambers, 1955; Heiser, Martin, and Smith, 1962).

In the Cruciferae we find the pattern exemplified by Brassica (Yarnell, 1956). In the Scrophulariaceae we see the same pattern in Collinsia (Garber and Gorsic, 1956; Garber, 1957, 1960). In the Solanaceae we see it in several annual sections of Nicotiana (Avery, 1938; Goodspeed, 1954). And in the Gramineae it is displayed inter alia by Bromus section Ceratochloa (Stebbins and Tobgy, 1944; Stebbins, Tobgy, and Harlan, 1944; Hall, 1955).

Autogamous Annuals

GILIA INCONSPICUA GROUP

The *Gilia inconspicua* complex is a group of small-bodied, small-flowered, autogamous species belonging to Gilia section Arachnion (Polemoniaceae). There are 25 known species in the complex. They range widely throughout arid regions of western North America, where 23 species occur, and are found again in temperate South America. The

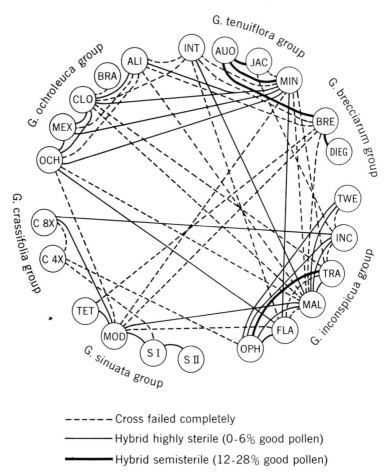

- - - - - - Cross failed completely

———————— Hybrid highly sterile (0-6% good pollen)

▬▬▬▬▬ Hybrid semisterile (12-28% good pollen)

Figure 20 Fertility relationships between sibling species of the *Gilia inconspicua* group. (From Grant, 1964a, in *Advances in Genetics*, copyright 1964, Academic Press, New York, by permission.)

Abbreviations of species, listed in clockwise order, are as follows. INT = *G. interior*. AUO = *G. austrooccidentalis*. JAC = *G. jacens*. MIN = *G. minor*. BRE = *G. brecciarum*. DIEG = *G. diegensis*. TWE = *G. tweedyi*. INC = *G. inconspicua* sens. str. TRA = *G. transmontana*. MAL = *G. malior*. FLA = *G. flavocincta*. OPH = *G. ophthalmoides*. S II = *G. sinuata* II. S I = *G. sinuata* I. MOD = *G. modocensis*. TET = *G. tetrabreccia*. C 4x = *G. crassifolia* 4x. C 8x = *G. crassifolia* 8x. OCH = *G. ochroleuca*. MEX = *G. mexicana*. CLO = *G. clokeyi*. BRA = *G. "bradbury."* ALI = *G. aliquanta*.

basic chromosome number is $x = 9$. Diploids and tetraploids are common, and one octoploid is known (Grant, 1964a).

The species are similar though not identical in their morphological characters and are difficult to identify by ordinary taxonomic procedures. In Chapter 2 we discussed and illustrated the similarities between five members of the *Gilia inconspicua* complex, namely, *G. transmontana, malior, minor, aliquanta*, and *clokeyi* (Figures 7 and 8). Furthermore, the plants frequently occur together in sympatric flocks of 2, 3, or 4 species. Therefore several early taxonomists logically concluded that they comprise a single variable species or two intergrading species. We now know from combined field and experimental studies that the *G. inconspicua* group is a complex of sibling species (Grant, 1964a).

Strong incompatibility barriers separate the species in nearly all combinations (Figure 20). When obtained, the interspecific hybrids are highly sterile in all but a few exceptional cases (Figure 20). The hybrids usually have a low degree of chromosome pairing. Several lines of evidence point to chromosomal rearrangements between the species, and hence to chromosomal sterility in the hybrids, while other evidence indicates that genic sterility is also present (V. and A. Grant, 1960; Grant, 1964a, 1964c; Day, 1965).

EROPHILA, VICIA, ELYMUS

The *Erophila verna* complex of the family Cruciferae is a group of morphologically similar, autogamous, annual herbs in Europe. The basic chromosome number is $x = 7$. Plants are found with diploid ($2n = 14$), polyploid, and various aneuploid numbers up to $2n = 64$ (Winge, 1940).

The whole group has traditionally been treated as a single taxonomic species, *Draba verna* or *Erophila verna*. Its taxogenetic composition has been analyzed by several workers, especially by Winge (1940) who recognizes four sibling species. These species occur sympatrically in various paired combinations in Europe (Winge, 1940).

Winge (1940) showed that the species of the *Erophila verna* group are separated by incompatibility and chromosomal sterility barriers.

The *Vicia sativa* group (Leguminosae) is another example of autogamous annual plants which comprise a complex of sibling species with different chromosome numbers (Hanelt and Mettin, 1966).

The pattern of sympatric sibling species separated by chromosomal sterility barriers is known from experimental evidence to occur in the autogamous perennial grass, *Elymus glaucus* (Snyder, 1950, 1951).

There is suggestive cytotaxonomic evidence pointing to the existence of this pattern in at least three other groups of autogamous annual plants in the California flora. They are the *Festuca microstachys* group (Gramineae), the *Mentzelia albicaulis* group (Loasaceae), and the *Oenothera micrantha* (or *Camissonia micrantha*) group (Onagraceae) (Kannenberg and Allard, 1967; Raven, 1969).

General Patterns

Our next task is to reduce the evidence presented in the main body of this chapter to a set of generalizations. The general patterns which can be recognized are as follows. The examples listed below are the genera or subdivisions of genera specifically discussed in the preceding sections.

1. The Ceanothus pattern. Woody plants with an outcrossing breeding system and more or less promiscuous flowers. Species inter-compatible, interfertile, and chromosomally homologous within wide limits. Species isolated or semi-isolated in nature by ecological and other external factors. Strong incompatibility barriers between sections or subgenera. Ceanothus, Quercus, Ribes, Pinus, Eucalyptus.

2. The Aquilegia pattern. Perennial herbs with an outcrossing breeding system and specialized flowers. Species intercompatible, inter-fertile, and chromosomally homologous within wide limits. Species isolated primarily by mechanical and ethological mechanisms, and secondarily by other external factors. Strong incompatibility barriers between sections, subgenera, or genera. Aquilegia, Antirrhinum, Penstemon, Epiphyllum, orchids.

3. The Geum pattern. Perennial herbs with an outcrossing breeding system and with floral mechanisms having only moderate species-to-species differences. Species intercompatible within wide limits; but strong incompatibility barriers developed between sections or sub-genera. Species belonging to the same species group are interfertile and

chromosomally homologous; those belonging to different species groups form hybrids with chromosomal and genic sterility. Related species isolated in nature by ecological and other external factors. Geum, Iris, Solanum, Silene.

4. The Madia pattern. Annual herbs with breeding systems ranging from outcrossing to inbreeding. Related species usually separated by incompatibility barriers and by chromosomal and genic sterility barriers. Madia, Layia, Helianthus, Microseris, Gilia, Clarkia, Brassica, Collinsia, Nicotiana, Bromus.

5. The *Gilia inconspicua* pattern. Annual herbs or, less commonly, perennials, with an autogamous breeding system. A single taxonomic species is composed of several or many sibling species isolated by incompatibility barriers and by chromosomal and genic sterility barriers. *Gilia inconspicua, Erophila verna, Vicia sativa, Elymus glaucus, Festuca microstachys, Mentzelia albicaulis, Oenothera micrantha.*

Most taxogenetic studies have been carried out in north temperate plant groups. Much less is known about the nature and distribution of sterility barriers in south temperate and tropical plants. These biases enter into the sampling of genera reviewed in this chapter. Consequently, the set of patterns recognized here is undoubtedly incomplete, and can be extended in the future on the basis of further taxogenetic work.

Some groups of woody plants in Australian and other south temperate floras, such as the Proteaceae and Podocarpus, have undergone a chromosomal structural differentiation comparable to that found in many herbaceous groups of northern floras (Smith-White, 1959; Hair, 1966). A tropical American genus of woody plants, Theobroma (Sterculiaceae), contains much interspecific incompatibility, like temperate annual genera, and much hybrid inviability (Addison and Tavares, 1952).

Considering the patterns identified and discussed in this chapter, it is remarkable to find a number of common characteristics in the genetic systems of phylogenetically unrelated genera. The similarities between Ceanothus, Quercus, and Pinus, or between Madia, Clarkia, and Gilia, and the collective differences between these sets of genera, are very striking. The patterns seem to reflect real convergences between members of different plant families.

General Correlations

In order to provide a broader base for our generalizations, we can take a larger sample of plant groups than that considered up to now in this chapter. Some years ago I surveyed the literature on 72 plant groups— genera, sections, or subtribes—which had been studied taxogenetically. I classified these plant groups into different categories on the basis of the presence or absence of incompatibility and sterility barriers between related species (Grant, 1958). The same phylads can also be grouped into three classes according to life form.

The correlation between life form and type of fertility relationships is shown in Table 1. Let us consider first the distribution of incompatibility barriers. Species belonging to the same section or species group usually cross freely in woody plants and perennial herbs, but are usually separated by incompatibility barriers in annual herbs (Table 1).

The distribution of hybrid sterility barriers shows a similar correlation with life form. F_1 hybrids between related species are usually fertile or semifertile in woody plants and perennial herbs, but are mostly sterile in annual herbs (Table 1).

Similar conclusions have been reached on the basis of independent surveys by Darlington (1956a; 1956b, pp. 103–104), Stebbins (1958a), and Ehrendorfer (1964).

Table 1 · Distribution of incompatibility and sterility barriers in plant groups with different life forms (Modified from Grant, 1958)

Life form	Number of phylads in which crosses between related species usually have the following results			
	Crossability		Hybrid sterility	
	High	Low	Fertile	Sterile
Trees and shrubs	19	2	11	2
Perennial herbs	28	0	15	6
Annual herbs	3	19	3	19

In a few cases it is possible to compare different sections of the same genus with respect to life form and strength of sterility barriers. In the genus Helianthus the perennial species comprising the *H. giganteus* group are interfertile, whereas the annual species of the *H. annuus* group are intersterile (Heiser, Martin, and Smith, 1962).

In Crepis (Compositae) three sections of annual herbs, namely, Hostia, Phytodesia, and Nemauchenes, have a preponderance of intersterile crosses. The perennial species of section Lepidoseris, on the other hand, are mostly interfertile; but another perennial section, Berinia, contains much interspecific sterility (Babcock, 1947).

In Knautia (Dipsacaceae), the species of the annual sections Tricheroides and Knautia have well-differentiated genomes, and are well isolated reproductively by both external and internal mechanisms. The perennial species of the section Trichera, by contrast, are genomically similar and weakly isolated reproductively (Ehrendorfer, 1962).

It happens that most of the trees, shrubs, and perennial herbs considered in our survey have an outcrossing breeding system, whereas many of the groups of annuals considered contain some partially or predominantly self-fertilizing species. It is possible, therefore, that the correlation shown in Table 1 between sterility and life form is a reflection of a more fundamental relationship between type of breeding system and the formation of sterility barriers.

Conclusions

We find a characteristic pattern of isolation in many groups of perennial plants in the north temperate zone. Related species are generally interfertile, and their isolation in nature depends largely on ecological and other external factors. But species belonging to different circles of affinity—sections, subgenera, or genera—are isolated internally by various blocks to crossing or to successful growth of the hybrid embryos.

These gene-controlled incompatibility and inviability barriers appear suddenly in the taxonomic structure of the plant group. The internal channels of gene exchange between species are open within relatively wide taxonomic limits, then beyond these limits become closed.

A different pattern of isolation is found in many annual and some perennial herbaceous groups of north temperate regions. Here related species are usually isolated internally by incompatibility and hybrid sterility. The type of sterility barrier which predominates in these groups, furthermore, is chromosomal sterility. External isolating mechanisms and genic sterility are often present too, but play a subordinate role.

The interfertile species groups and the intersterile species groups can be seen to have some basic modes of isolation in common. Ecological isolation, external reproductive isolation, and various gene-controlled blocks are developed in both types of plant groups. One of the main differences between them is the interpolation of chromosomal sterility within a circle of affinity in the one case and the absence of this feature in the other (Grant, 1958).

CHAPTER 8

Primary Speciation

Introduction

The term, primary speciation, is used here to describe the process of evolutionary divergence between populations to the species level. The term is intended to include the speciation phenomena involved in primary evolutionary divergence, and to exclude the phenomena of hybrid speciation and isolation reinforcement. The latter are discussed separately in later chapters in order to enable us to focus our attention on the former here.

Species differ, as we have seen in Chapter 5, in respect to adaptive traits controlled by complex gene combinations. Therefore primary speciation involves the development of new and different gene combinations in separate populations. This aspect of the speciation process is true for evolutionary species in general. In the case of biological species composed of cross-fertilizing organisms, speciation involves in addition the formation of reproductive isolating mechanisms.

It is convenient in discussing primary speciation to regard the foregoing developments—new gene combinations and isolating mechanisms—as goals of the speciation process. Then we can go on to consider the pathways by which these goals can be reached in plants.

It will be our task in this chapter to consider the pathways of primary speciation. Our discussion will enable us to explain some, but only some, of the results of the speciation process presented in the preceding chapter (Chapter 7).

Levels of Variation

The speciation process requires a supply of genetic variations as the starting condition. Such variations are of universal occurrence in the organic world. They occur on three levels which are believed to be significant for speciation. These levels are: polymorphism, local racial variation, and geographical variation.

Polymorphism is the segregational variation within a population. A gene or supergene, A, present in two or more allelic forms, a_1, a_2, etc., gives rise by segregation to the various homozygous and heterozygous genotypes. Since an essential aspect of speciation is the establishment of new gene combinations, the polymorphic condition of special relevance to the present discussion is that involving two or more genes or supergenes, A, B, C, ... The population could be polymorphic for A, B, and C simultaneously or sequentially in the line of descent leading to the new species.

As regards the realities of the situation, suffice it to say here that polymorphism is a characteristic feature of cross-fertilizing plants, as shown by progeny tests in many groups, and is common also, though not universal, in uniparental plants.

Local racial variation refers to genetic differentiation between populations separated by relatively short distances, for example, several miles or even a fraction of a mile. Such local racial differentiation may be an extension of polymorphism. Two neighboring populations may be alike, insofar as they are both polymorphic for A and B, and moreover contain the same polymorphic types (a_1, a_2, b_1, b_2), but differ statistically in the frequencies of these alleles.

Diplacus longiflorus (Scrophulariaceae) provides an example of this type of local racial variation. Most individuals of *D. longiflorus* have orange flowers, but red-flowered and yellow-flowered plants also occur in low frequencies in some populations (Beeks, 1962). In a transect through the San Gabriel Mountains of southern California, Beeks found two populations 1.5 miles apart which possessed all three color forms in different frequencies, the first population having more red-flowered and fewer orange-flowered individuals than the second, while a third population about 0.3 mile from the second differed from the latter in having more yellow-flowered plants (Beeks, 1962).

Or the local racial variation may be expressed as qualitative differences between neighboring monomorphic populations. One population might contain only a_1 and b_1 in its gene pool, and the next population only a_2 and b_2. This type of variation pattern occurs more frequently in plants with a colonial population structure than in those with large continuous populations. Although polymorphism is not found in the colonies in question, their racial differences presuppose an ancestral polymorphic condition, at least an ephemeral one, followed by a sorting out of the original variants differentially into the separate colonies.

For example, most natural populations of *Avena barbata* in an area sampled in central California are monomorphic. But different marker genes are often fixed in different populations (Jain and Marshall, 1967). A similar variation pattern has been described in Gutierrezia (Compositae) (Solbrig, 1960) and in other plant genera.

Geographical variation is an extension of local racial variation into a larger geographical arena. This extension often brings with it, as would be expected, an accumulation of greater racial differences (see Figures 5 and 6 in Chapter 2).

Neighboring colonies of Diplacus on a 2-mile transect in the San Gabriel Mountains differ in the frequencies of red, orange, and yellow flower forms, as we have noted above. This 2-mile transect, however, is a segment of a larger belt 50 to 100 miles wide in southern California in which the Diplacus populations occur (Beeks, 1962). On a longer transect across this belt we observe greater differences of racial or semispecific magnitude, ranging from tall red-flowered plants near the coast (*D. puniceus*), through bushy orange-flowered plants in the foothills (*D. longiflorus*), to low yellow-flowered plants in the arid interior (*D. calycinus*) (Figure 11 in Chapter 4).

Juniperus virginiana in the eastern United States varies in biochemical constituents as well as in external morphology. Flake, Rudloff, and Turner (1969) have analyzed a series of populations spaced about 150 miles apart on a transect from Washington, D.C., to Texas for their terpenoids, using gas chromatography combined with special numerical classification methods. They find that the populations differ quantitatively in various terpenoid constituents, and that these terpenoids, moreover, vary gradually or clinally along the northeast-southwest transect (Flake, Rudloff, and Turner, 1969).

Races

Races which inhabit different areas also live in different environments insofar as environmental factors vary along geographical transects. And, since environmental selection is a most important force in molding racial characteristics, it follows that races are usually and generally adapted to their particular respective environments (for further discussion see Grant, 1963, pp. 432 ff.).

Long ago Turesson (1922) demonstrated the adaptive nature of many observable racial characteristics in plants on the basis of three lines of evidence. In the first place, the distinctive racial traits are often specializations to the dominant environmental conditions in the habitats of the races. Thus the race of *Atriplex litorale* on the sheltered south coast of Sweden is tall and erect, while the race inhabiting the windy west coast is low and spreading (Turesson, 1922). Secondly, parallel racial differences occur in other species which occur in the same range of habitats, such as *Atriplex sarcophyllum* and *A. praecox*. And, finally, one race of *A. litorale*, when grown in the environment of the other race, develops the characteristics of the latter as a result of phenotypic modification (Turesson, 1922, 1925).

The environment of a plant has various aspects, all of which participate in shaping the racial characteristics. The aspect of the environment which has received the greatest amount of attention in this respect is climate. The vast majority of the ecotypes studied by Turesson (1922, 1925, 1930) in Atriplex, Geum, Caltha, Achillea, and many other genera are climatic ecotypes. The classical studies of Clausen, Keck, and Hiesey (1940, 1948) on *Potentilla glandulosa*, *Achillea millefolium*, and other species also deal with climatic races.

The selective effect of climate is clearly seen in racial differences in earliness of flowering. The growing season is of course shorter in northern Sweden than in southern Sweden. Corresponding to these climatic differences, the northern Swedish strains of *Fragaria vesca*, *Campanula rotundifolia*, *Geum rivale*, and *Caltha palustris* come into flower earlier than the southern Swedish strains of the same species when cultivated together in a uniform transplant garden (Turesson, 1922, 1930). Similarly, the races of *Potentilla glandulosa* and of the *Achillea millefolium* complex from the alpine zone in the Sierra Nevada of California bloom earlier in a uniform garden than the mid-altitude races from the same mountains (Clausen, Keck, and Hiesey, 1940, 1948).

In certain cases it has been possible to demonstrate the existence of racial differences with respect to edaphic conditions within a single climatic race. Populations comprising a single climatic race in *Achillea borealis*, *Gilia capitata capitata*, and *Streptanthus glandulosus* occur on both serpentine and nonserpentine soils in California. Kruckeberg (1951) showed that the serpentine-inhabiting populations and the nonserpentine populations differ racially in their physiological tolerance of this soil type. Edaphic racial differentiation is superimposed on climatic racial differentiation in many plant species (Kruckeberg, 1951).

Biotic races have been described in *Brassica nigra* by Sinskaja (1931). In Asia Minor the black mustard occurs both in open fields and in cultivated fields of yellow mustard (*B. campestris*). The two races of *B. nigra* differ in a combination of vegetative, floral, and seed characters. The more specialized race is that restricted to yellow mustard fields, and its distinctive characters are such as to fit it to grow among yellow mustard plants and ripen and disperse its seeds with those of the host crop plant (Sinskaja, 1931).

More recently we have obtained evidence for racial differentiation in respect to pollinating animals in several species of Polemoniaceae (V. and A. Grant, 1965). The floral differences between races of *Gilia splendens* are correlated with pollination predominantly by the beefly, Bombylius, by the long-tongued cyrtid fly, Eulonchus, and by humming-birds, respectively. Both *Eriastrum densifolium* and *Gilia achilleaefolia* have broad-throated races pollinated by large bees near the California coastline, and small-throated races pollinated by small bees in the interior mountains. An array of pollination races adds to the network of racial differentiation in some, or perhaps in many, plant species (V. and K. Grant, 1965).

The basin of ancient Lake Bonneville in Utah has changed from a lake and glacier-covered area to dry land during Pleistocene and recent times. These changes can be dated approximately. *Mimulus guttatus* (Scrophulariaceae) has invaded many of the newly exposed land areas. The ages of its populations can also be estimated from the known ages of their habitats. On this basis, Lindsay and Vickery (1967) estimate that a normal amount of racial differentiation in quantitative characters and crossability has taken place in 4000 years (Lindsay and Vickery, 1967).

Pathways

We have considered three levels of variation which are relevant to our discussion of primary speciation. The levels in question are polymorphic variation, local racial variation, and geographical racial variation. The relationship between these levels describes the pathway of primary speciation. The possible pathways are shown in Figure 21.

The best-known and best-documented pathway is that running from polymorphic variation through local and then geographical races to

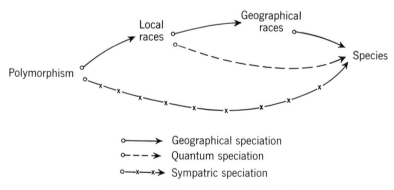

Figure 21 Possible pathways of speciation.

species (Figure 21). This is the pathway delineated by the geographical theory of speciation, which is discussed more fully in the next section.

The geographical theory of speciation has been challenged by various students. The proposed alternative short-cut routes to new species are shown by the broken arrows in Figure 21.

The route which bypasses geographical race formation on the way to species formation has been called quantum speciation. It is discussed later in this chapter.

The pathway of sympatric speciation, or, to be more precise, biotically sympatric speciation (see Grant, 1963, pp. 305, 463–69), bypasses local races as well as geographical races (Figure 21). Now there is good evidence for sympatric speciation in the case of hybrid speciation in plants, as is shown in a later chapter. Primary speciation in a biotically sympatric field, on the other hand, comprises a controversial problem, which we attempt to deal with later in this chapter.

The problem of the pathways of primary speciation is controversial and unsettled because the phenomenon in question, like other large-scale evolutionary changes, has not been subject to direct observation. The evidence bearing on the course of primary speciation is indirect and suggestive but inconclusive. This is the background for our discussion of various pathways in the next three sections.

Geographical Speciation

The geographical theory of speciation has had a long history of development. The basic concepts have been stated and restated with clarifications by a succession of zoologists. Prominent among them are M. Wagner (1889), K. Jordan (1896), D. S. Jordan (1905), Rensch (1929), Dobzhansky (1937b, 1951), and Mayr (1942). A historical review of the subject is given by Mayr (1963, pp. 482 ff.).

The theory in its modern form holds that geographical races are the precursors of species in a continuous process of evolutionary divergence (Figure 10 in Chapter 4). The spatial isolation of populations at the racial stage of divergence enables the separate populations to develop and maintain the gene combinations determining their distinctive morphological and physiological characters. The beginnings of reproductive isolation set in at the same stage. Species formation is then an extension of these processes. The reproductive isolating mechanisms which develop during a period of spatial isolation of two or more populations permit these populations to coexist without interbreeding if and when their range extensions bring them together in the same area.

The evidence for this theory can be summarized briefly, following Mayr (1942, pp. 162 ff.). (1) There is no qualitative difference between the types of characters which differentiate geographical races and those which differentiate species. (2) Population systems which are intermediate between geographical races and species, i.e., semispecies in our terminology, are common in many groups. (3) Cases are known in which a series of intergrading races replace one another geographically in one part of the species area, but overlap sympatrically in another part of the area. (4) Partial reproductive isolation is often found between geographical races of the same species.

The intermediate stages between geographical races and species, referred to in arguments 2 and 3 above, were exemplified in Chapter 4. Later in the present chapter we discuss argument 4 in more detail. Here we describe a case in which all of the stages of divergence demanded by the geographical theory of speciation are found within the limits of a single natural plant group.

Our example involves a group of interrelated races and species of diploid annual plants in the southwestern American deserts and mountains belonging to the more inclusive assemblage of the Cobwebby Gilias (Gilia section Arachnion). The morphological, ecological, geographical, and cytogenetic relationships between these races and species have been worked out by A. and V. Grant (1956) and V. and A. Grant (1960).

The geographical races of *Gilia latiflora* are well differentiated morphologically in their extreme forms. These races are interfertile and contiguous, with the result that they are connected by a series of intergrades in nature (Figure 22A). The related species *G. leptantha* comprises a series of geographical races which are also morphologically different and interfertile, but occupy disjunct areas, and consequently do not intergrade continuously (Figure 22B).

Gilia latiflora, *G. leptantha*, and a third related semispecies, *G. tenuiflora*, have largely allopatric distribution areas with overlapping boundaries, as shown in Figure 22C. The morphological and ecological differentiation of these three semispecies is greater than that between races within any one of them. Crossability and hybrid fertility are also

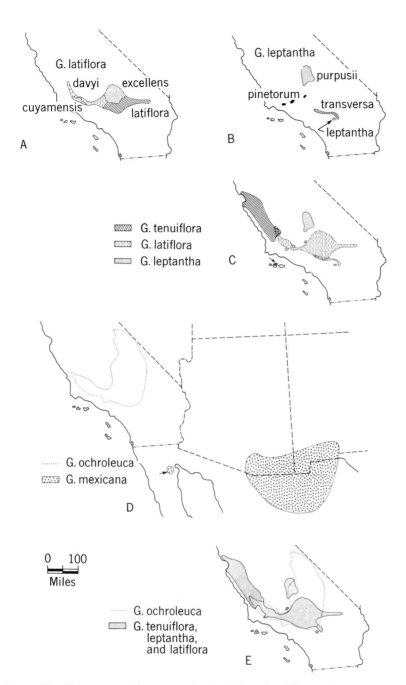

Figure 22 Five stages of divergence in the Cobwebby Gillas in the American southwest. (A) Contiguous geographical races. (B) Disjunct geographical races. (C) Marginally sympatric semispecies. (D) Allopatric species. (E) Sympatric species.

reduced in the inter-semispecific combinations as compared with the race crosses. Nevertheless, the channels of gene exchange remain open. In some regions of sympatric contact the semispecies hybridize; in other sympatric regions they maintain themselves distinct.

Gilia ochroleuca and *G. mexicana* have completely allopatric areas, as shown in Figure 22D. These two species are separated by a strong sterility barrier and a prominent morphological discontinuity.

Gilia ochroleuca is sympatric with the *G. tenuiflora-leptantha-latiflora* syngameon in southern California, and their individuals grow and bloom together in many places (Figure 22E). The morphological discontinuity between the two population systems remains intact throughout the extensive area of sympatry. The crossability and sterility barriers between the two species are very strong.

Quantum Speciation

The theory of quantum speciation represents a synthesis of several diverse concepts. The basic concepts involved are: haphazard local racial variation in species containing semi-isolated populations (Gulick, 1905); genetic drift (Wright, 1931); aberrant characteristics in peripheral populations (Mayr, 1942); quantum evolution (Simpson, 1944); and genetic revolutions (Mayr, 1954).

We can define quantum speciation as the budding off of a new and very different daughter species from a semi-isolated peripheral population of the ancestral species in a cross-fertilizing organism. This mode of speciation was described and its population genetic basis was identified by Mayr (1954) in his extension of an earlier discussion by Simpson (1944, pp. 206 ff.). The term quantum speciation, being the historically most appropriate, was introduced and applied to the process in my later review of the problem (Grant, 1963, pp. 456 ff., 566).

As compared with geographical speciation, which is a gradual and conservative process, quantum speciation is rapid and radical in its phenotypic or genotypic effects or both. The argument is that gene flow in the more or less continuous central populations of a cross-fertilizing species has a swamping effect on new variations. Furthermore, these conditions favor genes which have good combining ability with alien genes brought in on the continual stream of gene flow.

Now the isolation of a few such individuals from the normal stream of gene flow, by locating them in a semi-isolated peripheral population, perhaps a newly founded colony, leads to inbreeding with its accompanying drastic genetic and phenotypic effects. Most of these drastic changes can be expected to have a low adaptive value and suffer extinction. But, in a large sample of isolated peripheral populations, one or a few may undergo radical changes to a new well-adapted condition, becoming the progenitors of a deviant daughter species (Mayr, 1954, 1963; see also review in Grant, 1963, pp. 437 ff. and 456 ff.).

The theory of quantum speciation is supported by the observation of the following variation pattern in many species or species groups. The main central body of the species exhibits geographical variation of a clinal and conservative type. But races and species with deviant characteristics occur in geographically isolated positions on the periphery of the ancestral species. Mayr cites the examples of the East Indies kingfishers (*Tanysiptera galatea* and *T. hydrocharis*) and drongos (*Dicrurus hottentottus*) in this connection (Mayr, 1942, 1954, 1963). Carson, White, and others have provided additional examples in Drosophila and grasshoppers (Carson, 1959; White, 1959).

Quantum Speciation in Plants

It will be noted that the various ideas included in the theory of quantum speciation have been put forward by zoologists, and the problem has received relatively little attention from botanists. Similar ideas have been advanced, however, by H. Lewis and collaborators in regard to plant speciation (Lewis and Raven, 1958; Lewis, 1962, 1966). Extension of the theory to plants inevitably calls for a further modification of concepts, since in many plant groups the deviant phenotypic effects of inbreeding can be attained by self-fertilization as well as by crossing in isolated small populations.

The biosystematic and cytogenetic studies of the annual genus Clarkia (Onagraceae) by Lewis and his collaborators have revealed a number of species pairs in which the relationship is evidently that of parent and offspring. The phylogenetic relationship between *Clarkia biloba* and *C. lingulata* is quite clearly one of ancestral and daughter species respectively (Lewis and Roberts, 1956). The same is true of *C. rubicunda* and *C. franciscana; C. unguiculata* and *C. exilis; C. mildrediae* and *C. virgata;* and *C. mildrediae* and *C. stellata* (Lewis, 1962, 1966; Vasek, 1968).

The daughter species deviates from the ancestral species in morphological characters, ecological preferences, chromosome segmental arrangements, or in all three respects simultaneously, and often occupies a restricted geographical area lying off the periphery of the ancestral species area (Lewis and Raven, 1958; Lewis, 1962).

Lewis (1962) points out that peripheral populations in ecologically marginal habitats of a species are subject to extinction or near extinction in climatically unfavorable years. Near extinction followed by regeneration of the population from one or a few genotypically aberrant survivors sometimes leads directly to the rapid formation of a deviant daughter species (Lewis, 1962). In other words, rapid speciation sometimes results from drastic fluctuations in population size in peripheral areas. Lewis (1962) attributes the rapid evolutionary changes to extremely strong selection, or "catastrophic selection" as he calls it, but in my opinion it is more appropriate to regard these changes as products of the joint action of selection and drift.

It is fairly common to find peripheral populations with deviant morphological and chromosomal characters within a single species of Clarkia. Such semi-isolated, peripheral, local races exemplify the type of population which may give rise occasionally to a new daughter species with deviant characteristics (Lewis, 1962, 1966; Vasek, 1968).

The types of chromosomal rearrangements which differ between species of Clarkia are unlike those which differ between geographical races of one species. And this is further suggestive evidence for the view that many new species of Clarkia have developed, not out of geographical races, but from deviant peripheral populations (Lewis, 1962).

The case of *Clarkia biloba* and *C. lingulata* has been thoroughly analyzed by Lewis and Roberts (1956). These two California species of self-compatible annual herbs are closely related in morphology and in genome, but are separated by a chromosomal sterility barrier. The main morphological difference involves petal lobing, the petals of *C. biloba* being heart-shaped and those of *C. lingulata* tongue-shaped. The chromosome number is $2n = 16$ in *C. biloba* and $2n = 18$ in *C. lingulata* (Lewis and Roberts, 1956).

The phylogenetic relationships between the two species are indicated by their comparative chromosome numbers and ecological preferences. *Clarkia biloba*, with $n = 8$, is close to the basic number of $x = 7$ in the genus Clarkia, whereas *C. lingulata*, with $n = 9$, is more advanced. *Clarkia biloba* occupies moister habitats than does *C. lingulata*, and thus comes closer to the ancestral condition also in ecology. The cytotaxonomic and ecological evidence suggests strongly that *C. lingulata* is derived from *C. biloba* (Lewis and Roberts, 1956).

The geographical relationships between these species point to further conclusions regarding the mode of phylogenetic derivation. *Clarkia biloba* ranges through a large area in the northern Sierra Nevada and neighboring regions in California. *Clarkia lingulata*, by contrast, occupies a very small area in one river canyon on the southern periphery of the range of *C. biloba*. The restricted and peripheral distribution of *C. lingulata* suggests that it arose from the ancestral *C. biloba* fairly rapidly and fairly recently (Lewis and Roberts, 1956).

Cytogenetic analysis of the interspecific hybrids reveals the nature and extent of the chromosomal repatterning involved in the differentiation of *Clarkia lingulata* from *C. biloba*. Four of the chromosomes in the complements of the two species are homologous, and pair in bivalents at meiosis. Two other chromosomes in each complement form a ring of four in the hybrid, indicating that the species differ by a reciprocal translocation on these particular chromosomes (Lewis and Roberts, 1956).

The remaining chromosomes form a chain of five in the hybrid. The extra, ninth, chromosome of *Clarkia lingulata* occupies the middle position in this chain, indicating that it is a tertiary trisomic. This means that a second and independent translocation must have occurred in the ancestry of *C. lingulata*, and that one of the translocation chromosomes became established in homozygous condition, that is, as a tertiary tetrasomic, to produce the nine-chromosome complement of modern *C. lingulata*. Furthermore, the extra ninth chromosome of *C. lingulata* appears to carry the genes determining the distinctive petal shape of that species (Lewis and Roberts, 1956).

Two separate translocations thus play a basic role in the genetic differentiation of *Clarkia biloba* and *C. lingulata*. Together they account for most of the sterility barrier between the two species. The one which became established as a tertiary tetrasomic brought about the aneuploid

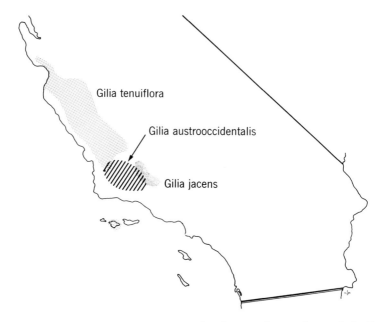

Figure 23 Geographical distribution of *Gilia tenuiflora* and two derivative species in southern California.

increase from $n = 8$ in *C. biloba* to $n = 9$ in *C. lingulata*. And the extra ninth chromosome in *C. lingulata* is associated with the main morphological character difference between it and the ancestral *C. biloba* (Lewis and Roberts, 1956).

An example in Gilia which is best explained on the basis of quantum speciation involves *Gilia tenuiflora*, *G. austrooccidentalis*, and *G. jacens*. These three diploid species occur in the south Coast Range of California (Figure 23). They are closely related but morphologically quite distinct. They form semifertile F_1s with slightly reduced chromosome pairing in the three possible hybrid combinations (V. and A. Grant, 1960).

As compared with the predominantly outcrossing *Gilia tenuiflora*, the other two species are reduced in breeding system and morphology, being small autogamous annuals. Furthermore, they live in more arid habitats than those occupied by *G. tenuiflora*. Therefore the probable direction of evolution has been from an ancestral form like *G. tenuiflora* to *G. austrooccidentalis* and *G. jacens* as derivative forms.

The geographical relationship between the three species is shown in Figure 23. It will be seen that *Gilia austrooccidentalis* and *G. jacens* occur in restricted and more or less isolated areas to the south of the more wide-ranging *G. tenuiflora*. When all features are taken into consideration, the pattern suggests that *G. austrooccidentalis* and *G. jacens* diverged from the ancestral *G. tenuiflora* via some geographically and ecologically marginal populations of the latter as species adapted to more arid conditions.

Sympatric Speciation

The problem before us here is whether primary divergence to the species level can take place within the limits of a single breeding population. This is a long-standing, much debated, and unsettled question.

Dobzhansky (1950) and Mather (1955) pointed out that polymorphism and sympatric species are alternative responses of living organisms to an environment which contains diverse niches within a common territory. Mather, but not Dobzhansky, then went on to suggest that species may grow directly out of polymorphic types under certain specified conditions of selection. Thoday and others have developed this view further and have obtained some experimental evidence for it (Thoday and Boam, 1959; Thoday and Gibson, 1962; Streams and Pimentel, 1961; J. M. Smith, 1966). Mayr, on the other hand, has argued strongly against the feasibility of nongeographical speciation (Mayr, 1942; 1963, pp. 449–480).

Obviously we cannot hope to settle the question of sympatric speciation here. But we can analyze and discuss the issues. Let us begin our analysis by stating and defining the problems.

Primary sympatric speciation follows the short-cut pathway from polymorphism to species, a pathway which does not include an intermediate stage of geographical or microgeographical isolation (Figure 21). The first question to be asked is whether this evolutionary pathway is theoretically possible. Secondly, is it theoretically likely? And, thirdly, is it an actual pathway in nature? To a large extent, Thoday and his co-workers and followers have been dealing with the first two questions, and Mayr with the third.

The theoretical difficulties of primary sympatric speciation are greatest in the case of a foundation population composed of cross-fertilizing individuals. Here the process of divergence is constantly opposed by interbreeding between the divergent polymorphic types. Under conditions of strict self-fertilization the difficulties largely disappear. Therefore discussions of the theoretical feasibility of primary speciation must first face the problem as it exists in a cross-fertilizing breeding system.

Divergence between different polymorphic types in a cross-fertilizing population will be promoted by disruptive selection (Mather, 1955; Thoday and Boam, 1959; Thoday and Gibson, 1962; Streams and Pimentel, 1961; J. M. Smith, 1966). This is the type of selection which operates when one polymorphic type is favored in one facet of the heterogeneous environment and another polymorphic type is favored in a different facet. Several experiments with *Drosophila melanogaster* show that strong disruptive selection will lead to divergence in spite of random mating (Thoday and Boam, 1959; Millicent and Thoday, 1961; Streams and Pimentel, 1961; Thoday and Gibson, 1962). A recent experiment with the housefly *Musca domestica* yields similar results (Pimentel, Smith, and Soans, 1967).

In one significant experiment, Thoday and Gibson (1962) carried out disruptive selection for number of body bristles under conditions of random mating. This is a polygenic character in Drosophila. A marked divergence developed gradually between the high line and the low line during 12 generations of selection. Furthermore, the frequency of hybrids between the high and low lines declined in the later generations of the experiment. Evidently some sort of partial reproductive isolation, perhaps ethological isolation or hybrid inviability, had also developed between the high and low lines (Thoday and Gibson, 1962).

Alternatively, in a cross-fertilizing but nonpanmictic population, the effects of disruptive selection could be perpetuated by a system of assortative mating. A tendency of genetically like individuals to mate would reinforce disruptive selection.

Strong selective pressures are required for disruptive selection to have pronounced effects under random mating. The selective pressures employed in the Drosophila experiments have ranged from very strong

(Streams and Pimentel, 1961) to lower values (Gibson and Thoday, 1964) which would still have to be considered strong in natural situations.

In order to relate the disruptive selection experiments to panmictic situations in natural populations, we have to know whether strong selection also exists in nature. There is in fact evidence for high selective values of 20% or more in natural populations (Ford, 1964). Ford concludes that such high selective values are common and usual (Ford, 1964, p. 296).

It seems to me that this conclusion is based on a series of case histories chosen for special study by virtue of the conspicuousness of their responses to natural selection. High selective values are more likely to be the exception than the rule in nature. Be this as it may, strong selection prevails in at least some natural populations, and disruptive selection could bring about its effects in such cases despite random mating.

We have mentioned assortative mating as a possible contributing factor. Assortative mating is potentially important because it would enable disruptive selection to produce an evolutionary divergence at lower selective intensities than are required under conditions of panmixia. The combination of assortative mating and disruptive selection would open up a larger range of possibilities for sympatric divergence in nature.

Perhaps we should approach the problem of sympatric speciation from another standpoint, however. We can question the assumption that it is useful to extrapolate from polymorphism to speciation. Certainly the two forms of sympatric genetic differentiation have some properties in common. But they also differ qualitatively.

Polymorphism is segregational variation, whereas species differences are differences in complex gene combinations. Interbreeding between two well-adapted polymorphic forms is apt to yield a certain proportion of less well-adapted progeny. But interbreeding between two well-adapted species yields an enormously higher proportion of ill-adapted recombination types. It follows that the cost of disruptive selection, in terms of genetic deaths, would have to be much higher for true sympatric speciation than for maintaining and enhancing a polymorphic system, and might well be intolerably high in most natural populations.

In summary, biogeographic evidence for primary sympatric speciation, unlike that for geographical and quantum speciation, is lacking. And the theoretical difficulties of speciation by disruptive selection within a cross-fertilizing population are very great. These difficulties are much less with self-fertilization. Primary sympatric speciation is a possibility which has neither been confirmed nor ruled out, and more evidence is desirable.

Reproductive Isolation as a By-product of Divergence

We would expect the formation of different gene combinations determining different adaptive character combinations in divergent populations to have the incidental effect, in many cases, of reducing the possibilities of interbreeding between such populations. In short, reproductive isolation can be expected to originate as a by-product of evolutionary divergence. This mode of origin of reproductive isolation, which was suggested by Darwin (1859, 1868) and by many later authors, is supported by much evidence in both plants and animals.

Partial reproductive isolation is commonly present between races of the same species. Earlier in this chapter we considered examples of racial differentiation in relation to climate, pollinating agents, and other environmental factors. Partial seasonal isolation will often exist between different climatic races of the same plant species. And partial mechanical and ethological isolation may develop between divergent pollination races.

Style length relative to pollen tube length is an important factor in cross-compatibility in flowering plants. Within a given race, pollen tube growth is coordinated with the normal style length. But two races differing in flower size are often separated in addition by a partial incompatibility barrier. Thus artificial cross-pollinations between two large-flowered races of *Gilia ochroleuca* yield abundant capsules and seeds; but a definite barrier to successful crossing exists between the large- and small-flowered races of this species. The large-flowered and small-flowered races of *G. ochroleuca* also differ in ease of crossing with a foreign species, *G. cana* (V. and A. Grant, 1960).

Let us consider the gene combinations underlying the character differences between races. These have been selected for their harmonious and balanced effects on development within each race. But many of the products of interracial hybridization will inherit genotypes which lack internal balance. Some degree of hybrid inviability, hybrid semi-sterility, and hybrid breakdown is in fact found between races in many plant species.

Kruckeberg (1957) compared the fertility of F_1 hybrids in 334 inter-population combinations within *Streptanthus glandulosus* (Cruciferae). He found a statistically significant inverse correlation between hybrid fertility and distance apart of the parental populations. Hybrid fertility drops with increasing distance between the parental races and, after distances of 50 to 75 miles, the hybrid fertility falls off abruptly (Kruckeberg, 1957).

All of the modes of reproductive and environmental isolation which exist between good biological species are found also, in an incipient form at least, between races of the same species (see Grant, 1963, pp. 439–41). All that is needed to strengthen the degree of isolation from one typical of races to one typical of species is a continuation of the process of divergence.

Salvia mellifera and *S. apiana* are distinct sympatric species in southern California. Mechanical isolation plays a central role in their ability to maintain their distinctive characteristics in areas of coexistence. The very dissimilar floral mechanisms of the two plant species are adapted to dissimilar kinds of bees, *S. mellifera* to medium-small bees such as Anthophora and Chloralictus, and *S. apiana* to large carpenter bees of the genus Xylocopa (K. A. and V. Grant, 1964).

Seasonal isolation occurs as a probable by-product of these differences in mode of pollination. *Salvia mellifera* blooms in early spring when its bee pollinators have their peak of activity, and *S. apiana* in late spring and early summer when the carpenter bees are on the wing. Ethological isolation is another side effect of the same floral differences, since the bees learn to restrict their visits mainly to the floral mechanisms which they can work successfully. Finally, there is some indication of a partial incompatibility barrier, which might also be associated with the floral differences between the two species (K. A. and V. Grant, 1964).

The species of Picea can be crossed successfully in many hybrid combinations, but in many other cases the crosses fail. J. W. Wright

(1955) finds that the unsuccessful crosses are mostly those involving parental species which are widely separated geographically and well differentiated morphologically. Conversely, species of Picea with neighboring distribution areas can be intercrossed fairly easily in most cases. This correlation indicates that the incompatibility barriers have probably developed in the course of divergence and during periods of geographical isolation (J. W. Wright, 1955).

Conclusions

We surveyed some results of the speciation process in Chapter 7. In the present chapter we have dealt with a phase of the process itself, the phase designated as primary speciation. Let us now attempt to relate our discussion in the present chapter to that in the preceding chapter. Which groups of speciation phenomena, among those described in Chapter 7, can be explained as results of primary speciation?

Woody plants with promiscuous pollination systems, as exemplified by Ceanothus, Quercus, Pinus, and other genera, display a pattern of relationships which points to gradual divergence in relation to secular ecological conditions. This divergence produces races, semispecies, and eventually internally isolated species groups treated taxonomically as sections or subgenera. Natural hybridization often occurs at early and intermediate stages of divergence. There is no need to postulate any processes other than those of primary speciation—and simple natural hybridization regarded as a reversal of primary divergence—in order to account for the Ceanothus pattern of species relationships.

Perennial herbs with specialized pollination systems, like Aquilegia, Antirrhinum, Penstemon (in part), and many orchids, display a generally similar pattern of relationships. Here too we see evidences of gradual divergence through a series of stages to that of complete internal isolation, with natural hybridization often reversing the trend locally in the early and middle stages. The divergence is related to pollinating agents as well as to secular ecological conditions. The Aquilegia pattern, like the Ceanothus pattern, can be explained satisfactorily as a result of primary speciation.

There are some similar features again in perennial herbs without prominent interspecific differences in pollination system, like Geum and Iris, but also some dissimilar features such as chromosomal sterility between species groups. It is logical to attribute those features which the Geum pattern has in common with the Aquilegia and Ceanothus patterns to a common cause in the processes of primary speciation. The distinctive phenomena in the Geum pattern, particularly the occurrence of chromosomal rearrangements between species groups, are not subsumed by this same explanation and will require further discussion in another chapter.

The Madia pattern found in many genera of annual herbs differs still more from the Ceanothus and Aquilegia patterns, in that chromosomal sterility and incompatibility barriers occur between related species. These aspects are not fully explained by the forces promoting divergence per se as discussed in the present chapter. There are other basic features in the Madia pattern, however, which do point to the operation of the processes of primary speciation.

CHAPTER *9*

Chromosome Repatterning

---

*Introduction · The Pattern Effect Theory · The Gene Linkage
Theory · Fixation of New Structural Homozygotes · Conclusions*

Introduction

Chromosomal structural differentiation between species in plants and
in insects has been investigated from diverse standpoints during the
course of this century.

   In the early decades of the century the condition of interspecific
chromosome repatterning was described in various plant genera such
as Nicotiana, Triticum, and Viola. The types of rearrangements and
their pairing relationships were worked out. The meiotic behavior of
structural heterozygotes was related to the sterility of species hybrids
and to the genetic behavior of allopolyploids and other types of hybrid

derivatives. The achievements of this phase are summarized in Darlington's *Recent Advances in Cytology* (1932, 1937a).

In the next phase it was possible to ask questions concerning the meaning of the cytogenetic facts discovered in the earlier phase. What is the functional role of the species-specific karyotype in the life of the species itself? The answer to this question was sought along different lines. The structural rearrangements are related to favorable position effects (Goldschmidt, 1940), to favorable gene linkage systems (Darlington and Mather, 1949, pp. 322–23; Stebbins, 1950, pp. 244 ff.; 1958b), or both (Grant, 1956b) in the species carrying them.

With the extension of cytogenetic studies to a larger sample of genera, it also became possible to outline the systematic distribution of chromosome repatterning. In the higher plants, some general trends and correlations were pointed out by Darlington (Darlington and Mather, 1949, p. 324; Darlington, 1956a, 1956b), Stebbins (1950, 1958a), Grant (1958), Ehrendorfer (1964), and H. Lewis (1966). The main conclusions have been reviewed in Chapter 7.

In recent years, interest has focused on the speciation processes which lead to the formation of new species-specific karyotypes. This problem has been taken up by several schools of workers. It has been discussed with reference to plants by H. Lewis (Lewis and Raven, 1958; Lewis, 1962, 1966) and myself (Grant, 1964b, pp. 205 ff.); and with reference to insects by White (1957b, 1959, 1964), Carson (1959), and B. John and K. R. Lewis (1965a, 1966).

The literature discussions of the evolutionary aspects of chromosome repatterning vary widely in scope and depth. Many papers devote only a short passage to a single facet of the problem. Many authors appear to have worked in isolation from the previously published views of other authors. Nevertheless, a gradual progression of ideas is discernible in the literature.

The problem before us in this chapter is twofold: the functional role of a species-specific segmental arrangement, and its mode of establishment in speciation. To be sure, the first problem could be considered without reference to the second. And some authors have attempted to discuss the second problem with little or no reference to the first. But in the following discussion we adopt the position that the problem of establishment of a new segmental arrangement is inseparable from the problem of its functional role.

The Pattern Effect Theory

The phenotypic expression of a given gene is often affected by its spatial relations with neighboring genes in the chromosome. This is the well-known phenomenon of position effect.

Goldschmidt (1938, 1940, 1955) extended the concept of position effect to the particular segmental arrangement in the genome as a whole. Each segmental arrangement has its own individual "pattern effect" and produces its own characteristic phenotype (Goldschmidt, 1940, 1955). It follows that chromosome repatterning is a source of new pattern effects and hence a means of producing new character combinations (Goldschmidt, 1940, 1955).

Goldschmidt then correlated the differences between species in physiological and morphological traits with their differences in chromosome segmental arrangement. There is an evolutionary connection between the two types of interspecific differentiation. The formation of a new species starts with a large-scale mutational event in the chromosomes, a "systematic mutation," which scrambles and rearranges the segments. The product of chromosome repatterning, when viable, is a new species (Goldschmidt, 1940, 1955).

Serious objections to Goldschmidt's theory of speciation by systemic mutations have been raised by a number of evolutionists (reviewed in Grant, 1963, pp. 454, 496 ff.). The systemic mutation would first arise in heterozygous condition, and the individual carrying it, being a complete structural heterozygote, would be sterile. Nor would the systemic mutant, in homozygous or heterozygous condition, be likely to be viable. Even if the systemic mutant could get past the initial sterility and inviability bottlenecks, and this is doubtful, how does it become established in a population composed of nonmutant individuals?

Goldschmidt's views have been widely rejected by evolutionists. These views have been adopted as stated by a few paleontologists and with important revisions by a few students of plant speciation. Among the latter are Lewis and Raven (Lewis and Raven, 1958; H. Lewis, 1962, 1966).

The combined evidence of geographical distribution and cytogenetic relationship in Clarkia (Onagraceae) suggests that new species have budded off rapidly from older parental species and that a rapid repatterning of the chromosomes has accompanied this process (Lewis and Raven, 1958; H. Lewis, 1962, 1966). The daughter species has probably arisen from a small and ephemeral local race of the parental species. The pathway, in other words, is that of quantum speciation, as discussed in Chapter 8.

Lewis and Raven's specification of the pathway is not only plausible in itself, but also successfully meets one of the serious objections to Goldschmidt's original theory. Furthermore, these authors postulate a series of chromosomal rearrangements that are not so numerous as to cause complete sterility in the heterozygotes, or so drastic in their phenotypic effects as to cause inviability, and these postulations remove the other main objections to Goldschmidt's theory (Lewis and Raven, 1958; H. Lewis, 1962).

Lewis and Raven suggest that some factor—perhaps mutator genes (Lewis and Raven, 1958; H. Lewis, 1962) or forced inbreeding (H. Lewis, 1966)—may induce chromosome breakage and rearrangements during certain periods in the history of a plant population. Some of the new genomes may have pattern effects which engender new physiological properties and adapt their carriers to new ecological habitats (Lewis and Raven, 1958). Such adaptively valuable genomes could become fixed in homozygous condition in a small daughter population and then go on to expand as a new daughter species (Lewis and Raven, 1958; H. Lewis, 1966).

The concept of pattern effect has much merit. To be sure, position effects are very rare in plants, and most new chromosomal rearrangements in these organisms have produced no detectable phenotypic changes. Nevertheless, considerable evidence indicates that functionally related genes frequently occupy neighboring sites on plant chromosomes (review in Grant, 1964b, Chs. 6 and 8).

The pattern effect theory has a fatal weakness, however, as a complete or even primary explanation of karyotype differences between plant species, as I have pointed out elsewhere (Grant, 1964b, pp. 212 ff.).

Divergent species have different adaptive character combinations in all life-form classes of plants. If these interspecific differences in adaptive properties were determined by pattern effect, we would expect to find correlated interspecific differences in karyotype in all groups of plants.

This expectation is not realized. Chromosome repatterning is associated with speciation in some taxonomic groups and life-form classes of plants and not in others, as we saw in Chapter 7. Therefore the primary evolutionary role of chromosome repatterning must be sought in some function other than direct pattern effects.

The Gene Linkage Theory

The second main explanation of the role of chromosome repatterning in speciation holds that the rearrangements involved serve a function of gene linkage. This idea was suggested in the first place by the known cytogenetic fact that inversions and translocations in heterozygous condition are devices for locking up genes from effective recombination. It was logical to attribute a similar function of linkage to rearrangements fixed in homozygous condition during speciation, as was done by Darlington (1940), Darlington and Mather (1949, pp. 322–23), Stebbins (1950, pp. 244 ff.), and Grant (1956b).

The gene linkage theory of chromosome repatterning was strengthened by the observation that this process accompanies speciation most regularly in those plant groups, especially annuals and some perennials, in which strongly restricted recombination would be expected to have a high selective value on other grounds.

It had been pointed out by Stebbins that herbaceous plants of pioneering habitats, particularly annuals, require a relatively high degree of uniformity in seed reproduction in order to multiply rapidly any given genotype of proven adaptive worth; and that this requirement for reproductive constancy is commonly met in annuals by low chromosome numbers or low chiasma frequency as well as by self-pollination (Stebbins, 1950, pp. 176 ff., 445 ff.; 1957a). The list of recombination-restricting factors in annuals was later extended to include also interspecific differences in segmental arrangement (Grant, 1956b, 1958; Stebbins, 1958b).

The chromosome segments which differ by rearrangements between plant species were believed, on the basis of some hybridization experiments, to be independent of the morphological and physiological character differences between the parental species (Stebbins, 1950, pp. 228 ff., 324). On this premise it was difficult to see any genetic connection between chromosome repatterning and the adaptive differences between species. Recently, however, new evidence has been obtained pointing to the location of adaptively valuable blocks of morphology- and physiology-determining genes on the rearranged segments in several plant groups (Grant, 1966a, 1966b) (see Chapter 5). This evidence removes the older theoretical difficulty.

Still another line of evidence in favor of the gene linkage theory of repatterning emerged from the studies of phylogenetic reduction in basic chromosome number in Crepis by Babcock and his school. Babcock and co-workers were able to show from comparative morphological and cytological studies that a prevailing phylogenetic trend in the genus Crepis was from perennial herbs with a basic number of $x = 6$ to annuals with $x = 3$ (Babcock, Stebbins, and Jenkins, 1937; Babcock and Jenkins, 1943; Babcock, 1947; Stebbins, Jenkins, and Walters, 1953). Stebbins suggested, plausibly, that the adaptive value of the low basic numbers was an increase in the amount of gene linkage in the derived annual species (Stebbins, 1950, p. 458).

It was possible to demonstrate further that the mechanism of chromosome number decrease in Crepis involved unequal reciprocal translocations (Navashin, 1932; Tobgy, 1943; Babcock, 1947). And this demonstration brought chromosomal repatterning into the picture in combination with altered linkage relations.

Later studies by other workers revealed parallel trends from high-number perennials to low-number annuals in other plant groups. The cases conforming to this pattern which are known to me are listed below. We find high-number perennials and derived low-number annuals in Phacelia (Hydrophyllaceae), with $x = 11$ and $x = 7$, respectively (Cave and Constance, 1947). Eriophyllum and related genera (Compositae) exhibit a series from perennials with $x = 8$ to reduced annuals with $x = 3$ (Carlquist, 1956). Oncidium (Orchidaceae) has diploid perennials with $x = 14$ or more chromosomes and at the other extreme some derived facultative annuals with $x = 5$ (Dodson, 1956). Polemonium and related genera (Polemoniaceae) exhibit a trend from 9-paired perennials to 6-paired annuals (Grant, 1959). A series in the Aster tribe appears to run from primitive perennials with $x = 9$ to derived annuals with $x = 4$ (Huziwara, 1959; Solbrig, Anderson,

Kyhos, Raven, and Ruedenberg, 1964). However, Turner and Horne (1964) read this series in the opposite direction. In Knautia and related genera (Dipsacaceae), finally, a trend is apparent from 9-paired perennials to 5-paired annuals (Ehrendorfer, 1964).

The correlation of a derived annual life form with a reduced basic chromosome number in several independent plant groups strengthens the thesis that the two conditions are causally related.

Fixation of New Structural Homozygotes

The fixation of chromosomal rearrangements in structural homozygous condition in a derivative population brings with it, as a by-product, the formation of chromosomal sterility barriers between the new and old populations. This process is thus a special case of primary speciation. It is an aspect of speciation which is still poorly understood.

On the basis of the gene linkage theory of repatterning, it is a logical necessity to postulate an ancestral condition of genic and associated chromosomal polymorphism in the parental population. The genic and structural polymorphism could be either balanced and long-lasting or transient and ephemeral.

Another postulate is that one of the possible homozygous gene combinations, which occurs as a variant in the polymorphic parental population, has a superior adaptive value in some new habitat accessible to that population. Then homozygous segregates for the favored gene combination can colonize and multiply in the new habitat. The daughter population will differ genically and also structurally from the bulk of the parental population (Grant, 1964b, pp. 207 ff.).

Of course, this process can go on independently in different lines of divergence. Two or more different new available habitats could be the birthplace of two or more different genic and structural homozygotes derived from a common variable parental population. The result is adaptive differences and associated chromosomal repatterning between three or more populations.

In Drosophila, in grasshoppers, and in Clarkia (Onagraceae), the types of chromosomal rearrangements which differentiate related species are unlike those which occur as common polymorphic or racial variants within species (B. Wallace, 1959; Lewis and Raven, 1958;

H. Lewis, 1962; White, 1964; Stalker, 1966). In Drosophila again, chromosomal polymorphism is best developed in the central populations of a species, whereas the peripheral populations are often monomorphic for gene arrangement (Carson, 1955, 1959; Carson and Heed, 1964).

These facts point to a previous chromosomally monomorphic condition in the history of the species now exhibiting widespread chromosomal polymorphism (Carson, 1959; Carson and Heed, 1964; White, 1959, 1964; Stalker, 1966; John and Lewis, 1966).

The same evidence also suggests that the new species arose from a monomorphic peripheral population of the ancestral species (Lewis and Raven, 1958; H. Lewis, 1962; White, 1964). As far as we can judge from the available indirect indications, primary speciation involving chromosome repatterning has probably often taken place by the rapid quantum route rather than by the gradual geographical route.

Conclusions

We saw in Chapter 7 that chromosomal sterility barriers between related species are a prominent feature in annual plants included in the Madia pattern and the *Gilia inconspicua* pattern. The processes of primary speciation on the basis of genic variation as reviewed in Chapter 8 provided no explanation of this feature. But chromosome repatterning taken as a concomitant of speciation in certain types of plants does bring this additional group of facts concerning species relationships within the realm of explanation.

The Geum pattern, as we also saw in Chapter 7, is intermediate between the Madia and *Gilia inconspicua* patterns on the one hand and the Ceanothus and Aquilegia patterns on the other. Related species of perennial herbs belonging to the Geum pattern are differentiated genically and are reproductively isolated by external mechanisms, as in the Ceanothus and Aquilegia patterns, and these common features find a common explanation in primary divergence on the basis of genic variations.

But chromosomal sterility is found between species groups in genera belonging to the Geum pattern. Chromosome repatterning thus enters the picture here, but at a higher taxonomic level of divergence than in genera with the Madia and *Gilia inconspicua* patterns of relationship.

It must be concluded that the process of divergence in respect to chromosome segmental arrangement, which is taking place in the recent history of rapidly evolving annual herbs, has occurred in some earlier cycle of speciation in perennial plants belonging to the Geum pattern. In Geum, Iris, and similar genera, in other words, the most recent phase of speciation without repatterning has been preceded by a period of speciation with repatterning. Perhaps in that earlier period these perennial genera played a pioneering role in their communities comparable to that of annuals in certain recent communities.

Woody plants of the north temperate zone can be inferred to have undergone chromosome repatterning at still earlier periods in their evolutionary history. Whole genera or families are cytologically stable now, with a single basic number in all known species, but differ from related genera or families in basic number. Thus the Betulaceae have $x = 14$, the Fagaceae $x = 12$, and the Corylaceae $x = 14, 11,$ and 8 (Darlington and Wylie, 1955). Likewise, in the Australian flora there is evidence for an ancient chromosomal repatterning in some woody groups like the Proteaceae and Myrtaceae (Smith-White, 1959a).

In conclusion it should be emphasized that, although we have learned much about chromosome repatterning and its role in speciation, many problems remain unsolved.

CHAPTER *10*

The Wallace Effect

*Introduction · Special Conditions · Selection Experiments ·
Geographical Distribution of Ethological Barriers within Species
· Comparisons between Sympatric Species and Allopatric Species
· Incompatibility Barriers in Gilia · Recognition Marks in
Flowers · Ecological Aspects of Coexistence · Conclusions*

Introduction

Reproductive isolation in its various forms can and does develop as a
by-product of evolutionary divergence between populations, as we saw
in Chapter 8. Superimposed on this basic process is another complementary mode of origin of isolation. Certain types of reproductive
isolating mechanisms can be built up by direct selection for isolation
per se, when this condition is advantageous for the populations concerned.

The process of direct selection for reproductive isolation has been called the Wallace effect, after A. R. Wallace who first proposed it in 1889 (Grant, 1966d). Wallace opens his discussion with the following statement (Wallace, 1889, p. 173).

> It will occur to many persons that, as the infertility or sterility of incipient species would be useful to them when occupying the same or adjacent areas, by neutralizing the effects of intercrossing, this infertility might have been increased by the action of natural selection; and this will be thought the more probable if we admit, as we have seen reason to do, that variations in fertility occur, perhaps as frequently as other variations.

He goes on to say (page 175):

> It must particularly be noted that this effect would result, not by the preservation of the infertile variations on account of their infertility, but by the inferiority of the hybrid offspring, both as being fewer in numbers, less able to continue their race, and less adapted to the conditions of existence than either of the pure forms. It is this inferiority of the hybrid offspring that is the essential point; and as the number of these hybrids will be permanently less where the infertility is greatest, therefore those portions of the two forms in which infertility is greatest will have the advantage, and will ultimately survive in the struggle for existence.

Wallace's hypothesis (1889, pp. 173 ff.) was ignored in his time, probably because it was not understood, and it was eventually forgotten. Essentially the same idea was put forward again in the early modern period of evolutionary biology by Fisher (1930, pp. 130 ff.), Dobzhansky (1941, 1951, pp. 208 ff.), and Huxley (1942, pp. 288 ff.).

The early modern formulations of the problem were, of course, couched in more precise genetical terms than was Wallace's statement, and referred chiefly to ethological barriers in animals, but were nevertheless quite brief and general. Since 1950 a number of students have developed the hypothesis further and have sought evidence for it, as we shall see in subsequent sections of this chapter. Recent critical reviews of the problem are provided by Mayr (1963, pp. 548 ff.) and Grant (1963, Ch. 17; 1966d).

Stated in modern terms, the Wallace hypothesis argues that those individuals in two sympatric populations which produce inviable or sterile hybrids will contribute fewer offspring to future generations than sister individuals in the same parental populations which do not hybridize. Consequently, the genetic factors determining some block or aversion to interspecific hybridization will tend to increase in frequency within each species over the course of generations. Natural selection along such lines will then lead to reinforcement of the reproductive isolation which had previously developed as a by-product of divergence.

Special Conditions

The ability of selection to reinforce an incipient degree of reproductive isolation is affected by various conditions. Some of these conditions are essential, and others are favorable, for the Wallace effect.

It is obvious, in the first place, that the species must be sympatric, at least marginally, in order for the Wallace effect to occur. Factors determining sympatry are discussed in a later section of this chapter.

Secondly, the effects of hybridization must be deleterious. Some or all of the hybrids in the F_1 or later generations must be sterile or inviable. This condition is generally realized in interspecific hybridization in plants, as we have noted in previous chapters. The condition does not necessarily hold true for all products of hybridization in plants, however.

A particular hybrid genotype in F_1 or later generation may be sterile but possess a degree of vegetative vigor which is superior to that of either parental species in the habitat in which it occurs. If the well-adapted hybrid plant is capable of vegetative or agamospermous reproduction, it can then spread and increase in numbers relative to the parental species in its environment. Actual cases of hybrids outnumbering the parent forms in particular habitats have been observed by Kerner (1894–1895) and later botanists (i.e., Stebbins, 1959) in Rhododendron, Nuphar, grasses, and other plant groups. The Wallace effect occurs in the common complementary situation where interspecific hybridization does reduce significantly the output of well-adapted progeny by the hybridizing parental individuals.

Furthermore, selection for reproductive isolation does not necessarily lead to a complete prevention of hybridization, as Ehrman (1962) has pointed out. Reproductive isolation will be reinforced up to the level necessary to safeguard the populations from the net deleterious effects of hybridization. This level is relative to the tolerance of the populations for selective elimination. As Ehrman (1962) puts it, the minimum required level of reproductive isolation is that where "the gene exchange between the species population[s does] not exceed the rate at which the diffused genes can be eliminated by natural selection."

In the next place, the Wallace effect applies exclusively to reproductive isolating mechanisms which act in the parental generation. Ethological and incompatibility barriers are especially well suited to be reinforced by the Wallace effect. As between these two, moreover, a premating barrier like ethological isolation will conserve the reproductive potential of the parents more effectively than will a postmating barrier like incompatibility.

The salient role of ethological isolation in animals can be attributed partly or even largely to the Wallace process. In plants, however, ethological isolation is usually insufficient in itself to prevent hybridization, because the flower constancy of pollinating insects is subject to periodical lapses, whereas incompatibility can be very effective. Therefore selection for isolation can be expected to build up incompatibility as well as ethological barriers in plants.

Another condition for the effective action of selection for reproductive isolation is that the loss of reproductive potential involved in hybridization is significantly disadvantageous to the organisms concerned. The selective disadvantage of a loss of reproductive potential varies widely in different classes of organisms.

At one extreme are such short-lived organisms as ephemeral flies and annual plants, the individuals of which have only a single season in which to reproduce. At the opposite extreme are long-lived woody or herbaceous plants of closed communities which normally produce a great excess of seeds year after year and decade after decade. Needless to say, many plants and animals occupy intermediate positions between these extremes.

The selective disadvantage of a loss of reproductive potential is high in the case of annual plants and other ephemeral organisms, but is likely to be negligible in long-lived plants of closed communities (Stebbins, 1950, pp. 183 ff.). It follows that selection for reproductive isolation may be particularly strong in annuals and relatively ineffective in long-lived perennial plants (Stebbins, 1958b; Dobzhansky, 1958). Some perennial herbs probably approach annuals in their responsiveness to selection for isolation (Levin and Kerster, 1967b).

Let us now return to the question of incompatibility barriers with special reference to annual plants. From an embryological standpoint, cross-incompatibility is a complex of prefertilization and postfertilization blocks (Avery, Satina, and Rietsema, 1959). The selective value of a prefertilization block lies in the conservation of a given number of ovules for intraspecific matings. A postfertilization block which prevents the formation of mature hybrid seeds also has a definite selective advantage over no such block in an annual plant, because it conserves a limited supply of food reserves for storage in nonhybrid seeds (Grant, 1966d).

Selection Experiments

The Wallace hypothesis has been tested experimentally by artificial selection in Drosophila. Using mixed cultures containing two species or two strains of the same species, several workers have selected, during a succession of generations, against hybrids and hence for parental genotypes which do not hybridize freely. The artificial selection has been effective.

Koopman (1950) used a mixture of *Drosophila pseudoobscura* and *D. persimilis* in laboratory cages. These two species of flies are ethologically isolated in nature but interbreed freely at certain temperatures in the laboratory. The mixed species populations belonging to the initial generation produced from 22 to 49% hybrid progeny in replicate cages. Artificial selection against the hybrids led to a rapid decline in hybrid formation in subsequent generations. After five generations of selection, only 5% of the progeny in the mixed cultures were hybrids. Ethological barriers had evidently been enhanced by the artificial selection (Koopman, 1950).

B. Wallace (1954) and Knight, Robertson, and Waddington (1956) started with combinations of mutant strains of *Drosophila melanogaster*. They also obtained responses to artificial selection for reproductive isolation. Ethological barriers between the strains increased in strength during successive generations of selection.

The laboratory experiments show that the Wallace effect actually takes place as predicted under certain conditions. The problem then becomes one of assessing the role of this process in nature.

Geographical Distribution of Ethological Barriers within Species

In a pair of related species with overlapping distribution areas, it is possible to compare sympatric populations and allopatric populations of the two species with respect to degree of ethological isolation.

This comparison has been made in several species pairs in Drosophila and frogs by Dobzhansky, Blair, and others (review with references in Grant, 1966d). In each case it was found that ethological isolation is stronger between sympatric races of a species pair than between allopatric races of the same two species.

The findings can be interpreted as evidence for the Wallace effect. But contrary cases in which isolation is not stronger in the sympatric representatives of two species are also common in Drosophila and amphibians, as several critics pointed out. The apparent impasse stems from the fact that in the earlier studies there were too few points of comparison to justify generalizations for or against the Wallace hypothesis (Grant, 1966d).

The recent studies of Ehrman on the *Drosophila paulistorum* group overcome the above difficulty by taking and considering a larger sample of populations (Ehrman, 1965). The six semispecies comprising the *Drosophila paulistorum* superspecies are ethologically isolated to varying degrees. The strength of the ethological isolation can be measured and expressed quantitatively. The isolation coefficient ranges from 1 to 0, where 1 is complete isolation and 0 is random-mating.

The six semispecies also overlap distributionally in various combinations throughout South America. Furthermore, the flies exhibit racial variation in degree of ethological isolation. Ehrman had eight paired combinations of semispecies in which it was possible to compare ethological isolation between sympatric representatives with that between allopatric populations. The ethological isolation was consistently stronger between sympatric populations of a semispecies in seven of the eight paired combinations. The average isolation coefficient for all sympatric populations was 0.85 as compared with an average value of 0.67 for all allopatric combinations (Ehrman, 1965).

Comparisons between Sympatric Species and Allopatric Species

The reality of the Wallace effect in nature can be assessed in another way. This is by comparing sympatric species and allopatric species belonging to the same group in respect to the degree of reproductive isolation in the parental generation.

Comparisons of this sort have been made by several schools of workers in several groups of animals. The cases studied are the *Drosophila guarani* and the *D. willistoni* groups in tropical America, the *Erebia tyndarus* group of satyrid butterflies in the Alps, and the *Gambusia affinis* group of fishes in south-temperate North America (King, 1947; Dobzhansky, Ehrman, and Pavlovsky, 1957; Lorković, 1958; Hubbs and Delco, 1960).

In these cases the geographical relationships between the species are of the type shown in Figure 24A. Two or more mutually sympatric species are compared with an outlying and geographically isolated species as to degree of isolation. The isolating mechanism thus compared has been ethological isolation.

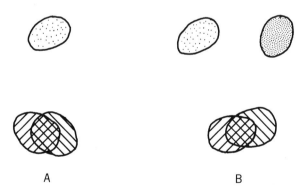

A   B

Figure 24 Two types of distribution pattern in a species group containing both allopatric species and sympatric species. It is possible to compare different classes of interspecific crosses with respect to degree of isolation in the two patterns, as follows. (A) Sympatric species inter se vs. sympatric X allopatric species. (B) Sympatric species inter se vs. allopatric species inter se. (Grant, 1966d.)

Similar results were obtained in each of the four groups. The ethological barriers are relatively strong between the sympatric species, and relatively weak between a geographically isolated species and the central species. These results point again to the efficacy of the Wallace process in nature.

In the cases mentioned above, the barriers between sympatric species are compared with those between an allopatric species and a sympatric species (see Figure 24A). In such comparisons at least one species in each class of interspecific combination has been exposed to sympatry and hence to selection for enhanced isolation; this complicates the story. Better evidence could be obtained from comparisons of the barriers between two or more sympatric species with those between two or more allopatric species, as diagrammed in Figure 24B (Grant, 1966d).

It is also desirable to extend the search for evidence in favor of the Wallace effect to organisms other than animals and to barriers other than ethological isolation. In the next section we present evidence for the selective origin of incompatibility barriers in a group of plants having the distribution pattern diagrammed in Figure 24B.

Incompatibility Barriers in Gilia

Favorable materials for demonstrating the Wallace effect with respect to incompatibility barriers are found in the annual Gilias (Polemoniaceae). Incompatibility barriers can and do arise as by-products of evolutionary divergence in these plants, as indicated by the presence of definite but weak incompatibility between races of the same species in many cases (see Chapter 8). But this is not the whole story.

Let us consider again the Leafy-stemmed Gilias, which we discussed in another context in Chapter 7. The ten species comprising this natural group occur on the Pacific slope of North America and again in temperate South America. Nine of the ten species have been studied cytotaxonomically and genetically as well as ecologically. The interspecific hybrids which have been produced are all highly sterile, with irregular meiosis. Indirect evidence indicates that the nine species would be intersterile in all possible combinations. Incompatibility barriers are strong between some of the species but weak between others (Grant, 1954b, 1965).

Foothill and Valley Species

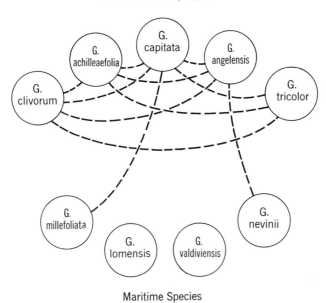

Maritime Species

Figure 25 Combinations of species of Leafy-stemmed Gilias which are known to grow side by side in nature. (Grant, 1966d.)

The plants are annual herbs which bloom in the spring and die as individuals when they go to seed in late spring or early summer. The loss in seed output in the production of sterile hybrids must therefore have a high selective disadvantage in these plants. Polymorphism for crossability with a foreign species, on which the selective process could operate, is known to occur in some Gilia populations (Grant, 1966d).

The nine genetically known species of Leafy-stemmed Gilia can be divided into two classes on the basis of their geographical relationships:

Five species—*G. capitata, G. tricolor,* and others—occur in the foothills and valleys of California and adjoining regions. The distribution areas of these species overlap extensively (Figure 2). Individuals belonging to two or more species in this subgroup grow and bloom side by side in many natural localities (Figure 25).

Four other species—*Gilia millefoliata, G. nevinii,* and others—occur on the coastal strand in four widely separated areas of North America and South America. These maritime species are completely allopatric in relation to one another. Two of them have rare sympatric contacts with some foothill-and-valley species, and two others lie wholly outside the distribution range of the latter (Figure 25).

The two geographical classes of Leafy-stemmed Gilias differ markedly in strength of incompatibility barriers. The strength of this barrier can be expressed quantitatively as the average number of hybrid seeds produced per flower cross-pollinated by a foreign species under controlled conditions.

The sympatric foothill-and-valley species are separated by very strong incompatibility barriers. The seed output ranged from 0.0 to 1.2 hybrid seeds per flower and averaged 0.2 for numerous crosses between different foothill species. The maritime species, by contrast, can be crossed inter se with the greatest of ease. Crosses between different maritime species yielded from 7.7 to 24.8 hybrid seeds per flower; and the average value for all interspecific combinations in this class of crosses was 18.1 seeds per flower (Grant, 1966d).

We expect some degree of cross-incompatibility to develop as a by-product of divergence, as noted earlier. But the amount of divergence between species is about the same in the two geographical classes of Leafy-stemmed Gilias. The observed differences in crossability between the two geographical classes of species cannot be accounted for on the basis of evolutionary divergence alone. The high degree of cross-incompatibility between the sympatric foothill-and-valley species is best explained as a product of the Wallace process (Grant, 1966d).

Recognition Marks in Flowers

Bees and various other flower-visiting insects possess instincts of flower constancy, impelling them to visit preferentially one species of flower at a time. This instinct is the basis of ethological isolation between different plant species which are pollinated by bees or other flower-constant insects, as we noted in Chapter 6.

Frisch's classical experiments demonstrated that honeybees can distinguish between certain colors, forms, and odors which differentiate species of flowers in many plant groups (Frisch, 1914, 1919). Similar perceptions are found in other bees and other kinds of flower-visiting insects (see Kugler, 1955; Faegri and van der Pijl, 1966). The insect is guided in its round of flower-constant visitations by the distinctive color, color pattern, form, and/or odor of a species of flower.

Long ago A. R. Wallace (1889) suggested that different sympatric plant species which are pollinated by flower-constant insects profit from having different recognition marks in their flowers, for such floral differences reduce the frequency of interspecific pollinations. He states (Wallace, 1889, pp. 318–319):

> It is probably to assist the insects in keeping to one flower at a time, which is of vital importance to the perpetuation of the species, that the flowers which bloom intermingled at the same season are usually very distinct both in form and colour. . . .
>
> This peculiarity of local distribution of colour in flowers may be compared, as regards its purpose, with the recognition colours of animals. Just as these latter colours enable the sexes to recognise each other, and thus avoid sterile unions of distinct species, so the distinctive form and colour of each species of flower, as compared with those that usually grow around it, enables the fertilising insects to avoid carrying the pollen of one flower to the stigma of a distinct species.

A survey of numerous genera of angiosperms was made some years ago in order to obtain estimates of the relative importance of floral characters in the taxonomy of plants with different pollination systems. It was found that 38% of the characters used by taxonomists to distinguish related species of bee-pollinated plants pertain to the petals and other inner floral parts. Similar percentages of floral characters are found in groups of species pollinated by other types of flower-constant insects. But only 15% of the character differences between related species which are pollinated promiscuously by many kinds of unspecialized and inconstant insects are floral characters (Grant, 1949).

These figures do not discriminate between floral differences between species related to mechanical isolation and those related to ethological isolation. However, it can safely be assumed that some part of the great preponderance of floral characters in nonpromiscuous angiosperms functions as recognition marks for flower-constant pollinators. This assumption can be confirmed by more detailed studies of specific cases.

For example, *Gilia capitata capitata* and *G. c. chamissonis* differ, among other features, in the shape of the petal lobes and the odor of the flowers. In an experimental plot where the two types of Gilia are intermixed and are both being visited by honeybees, the various individual bees generally remain constant to one type or the other. The behavior of the bees suggests that they are guided in their visitations by petal shape, floral scent, or both (Grant, 1949, 1950b).

Further evidence suggesting that floral recognition marks may be adaptively valuable in relation to ethological isolation has been obtained recently in Phlox by Levin and Kerster (1967b).

In the eastern United States, *Phlox pilosa* and *P. glaberrima* overlap in range and are biotically sympatric in many localities. The two species are incompatible, but are not well isolated mechanically or seasonally. They are pollinated by the same species of butterflies and other Lepidoptera. These agents carry out a considerable amount of inter-specific pollination in certain mixed colonies (Levin and Kerster, 1967a, 1967b).

The flowers are usually pink in both *Phlox pilosa* and *P. glaberrima*. White-flowered forms also occur as polymorphic variants in *P. pilosa*. In most polymorphic populations of *P. pilosa* the white-flowered form is rare, but in some populations it predominates.

The geographical distribution of the predominantly white-flowered populations of *Phlox pilosa* is very interesting. These populations always occur sympatrically with pink-flowered *P. glaberrima*. This correlation suggests that the white flower colors in such mixed stands serve as an aid to species discrimination by the Lepidopteran pollinators (Levin and Kerster, 1967b).

Butterflies going from *Phlox glaberrima* to *P. pilosa* in mixed colonies strongly prefer the pink form of *P. pilosa* and discriminate against the white form. Counts of *glaberrima* pollen grains on *pilosa* stigmas show that the pollinators transport five times as many pollen grains from *P. glaberrima* to the pink flowers of *P. pilosa* as to equal numbers of white *pilosa* flowers (Levin and Kerster, 1967b). Thus the white-flowered form of *Phlox pilosa* appears to have a selective advantage over the pink form in the mixed colonies. The high frequency attained by the white-flowered form of *P. pilosa* in the area of overlap with *P. glaberrima* may be due to this mechanism (Levin and Kerster, 1967b).

Ecological Aspects of Coexistence

The coexistence of two or more species has a secular ecological aspect as well as a reproductive aspect. An interaction exists between these two aspects, as I have attempted to show elsewhere (Grant, 1963, pp. 512 ff.).

Related species living in the same territory are likely to come into competition for common essential resources—raw materials, food, space, and the like—whenever these resources are available in limited amounts. Common species of large-bodied organisms will, in general, make heavier demands on environmental resources than common species of small organisms. Consequently, the interspecific competition for critical limiting resources will generally be keener in the large organisms than in small organisms.

The interspecific competition is likely to run its course to species replacement in the larger animals and plants. A certain amount of empirical evidence is in agreement with this expectation. Related species of large-bodied organisms are often segregated into different ecological zones or habitats (see Grant, 1963, pp. 414 ff.).

In plants of the California flora, for example, the different species in arboreal genera like Pinus and Quercus, and in perennial herbaceous genera like Polemonium and Calochortus, characteristically occupy different ecological zones on transects from coastline to interior mountains and deserts. Sympatric contacts between related species are usually confined to the contiguous or overlapping margins of their respective areas (Grant, 1963, p. 416).

In small-bodied plants and animals, which make lighter demands on environmental resources, interspecific competition is less severe and does not necessarily run full course to species replacement within a given habitat. The attainment of biotic sympatry is not strongly opposed by the forces of interspecific competition often prevailing in such organisms.

In the California flora again, biotically sympatric flocks of species are found commonly and almost exclusively among annual herbs. Clarkia, Mentzelia, and Gilia furnish good examples. Two or three species of Gilia frequently grow side by side in nature, and it is not unusual to find six or seven species of annual Gilias in the same natural habitat (Grant, 1963, p. 417; 1964a).

Sympatric contacts between related species set up a challenge which elicits different types of response. The response in some plant groups is ecological segregation and differentiation or, in other words, reinforcement of environmental isolation. In other plant groups the response is reinforcement of reproductive isolation in a sympatric field. In many cases a combination of both types of response probably occurs. And the magnitude of the requirements for environmental resources on the part of the plants is a factor determining the type of response.

Conclusions

In Chapter 7 we described the salient features in several patterns of species relationships. Most of these features can be explained as by-products of primary speciation, as noted in Chapter 8; and some others are consequences of chromosome repatterning, as discussed in Chapter 9.

The processes discussed in the present chapter put us in a position to understand still other features of certain species relationship patterns. We are referring in the first place to the high frequency of strong incompatibility barriers between related species in annual herbaceous plants.

The situation in annual plants is propitious for the Wallace effect. Ecological interactions are often such as to permit frequent sympatric contacts between related species. This exposes the species populations to the challenge of interspecific hybridization. The loss of reproductive potential involved in the formation of sterile or inviable hybrids is particularly disadvantageous in annual plants which have only one season in which to produce progeny. This threat to successful seed reproduction is evidently opposed by natural selection for incompatibility barriers.

Floral characters which serve as recognition marks for flower-feeding insects and birds form a prominent component of the species-to-species differences in angiosperms with specialized nonpromiscuous modes of pollination. It is probable that these floral features, in both annual and perennial plant groups, are also products of the Wallace effect.

PART III

Refusion and
Its Consequences

CHAPTER 11

Natural Hybridization

Introduction

The subject of natural hybridization in the plant and animal kingdoms has always held a special fascination for mankind. Greek mythology abounds in fanciful and farfetched examples of animal hybrids; the medieval herbalists described some actual hybrids in the European flora. Scientific inquiry into the subject dates from the middle of the 18th century.

As starting points of scientific hybridization studies in plants we may take the following works by the opposing schools of Kölreuter and

Linnaeus. Kölreuter's *Vorläufige Nachricht* appeared from 1761 to 1766. Linnaeus' *Disquisitio de sexu plantarum* was first published in 1760. The *Plantae Hybridae* of Haartman, a student of Linnaeus who reflects the views of the latter, is dated 1751.

After Kölreuter and Linnaeus, many authors have discussed natural hybridization in plants all through the 19th and 20th centuries. As a result, the literature on the subject is extremely voluminous. It is too voluminous to review completely and no attempt is made to do so here.

Much of the literature is descriptive in approach and consists of reports of hybrids in particular plant groups. Special knowledge of this sort finds its proper place in the archives of systematic biology. The evolutionary biologist is more interested in the general conclusions which are supported by specific evidence. Accordingly, the emphasis in our presentation of the subject is on general principles.

The various theoretical treatments of natural plant hybridization during the long history of this subject have dealt with two general problems. First, are natural hybrids freaks or are they normal phenomena? What is the frequency of their occurrence in different plant groups and floras? Second, does natural hybridization have any significant evolutionary effects? In other words, do new races or new species sometimes arise from natural hybrids? These problems have been discussed by many authors in successive generations. Different authors have often reached different conclusions from the same body of evidence. We will attempt to deal with the two broad questions stated above in this and the next two chapters.

The present chapter considers the formation of spontaneous hybrids in nature. Chapter 12 deals with the evolutionary products of such hybrids up to the racial level. In Chapter 13 we will go on to consider species formation as an outcome of hybridization.

It may be useful for many readers to list here some of the more important literature reviews. Roberts, in his *Plant Hybridization before Mendel*, brings together excerpts from the writings of the early students from Linnaeus and Kölreuter to Focke; this is a valuable source book (Roberts, 1929, 1965). Modern reviews of natural hybridization in plants with literature references are given by Heiser (1949a), Anderson (1949, 1953), and Stebbins (1950, Ch. 7; 1959). Natural hybridization in animals is reviewed by Mayr (1963, Ch. 6).

Hybridization as a Reversal of Divergence

The formation of two or more reproductively isolated populations from a common ancestral breeding population is a time-consuming process. The divergence between the separate lines may eventually reach a point of complete reproductive isolation when interbreeding is no longer possible at all. In higher plants this point of absolute reproductive isolation is probably set in nearly all cases by the attainment of complete cross-incompatibility, since there are ways out of the impasse of complete hybrid sterility, and the external isolating mechanisms are rarely, if ever, perfect in their effectiveness.

Between the initial and the ultimate stages of speciation there is a progressive decline in the amount of interbreeding between divergent populations and, conversely, there is a progressive development of environmental and reproductive isolation. Spontaneous interbreeding between populations which have undergone a previous history of divergence to the level of disjunct races, semispecies, or species, and which are separated by partial ecological or reproductive isolation or both, is natural hybridization.

The term hybrid has been applied by authors to a wide range of phenomena, including the products of crossing between different genotypes belonging to the same population and the products of grafting between specifically or generically different stocks and scions. Some terminological restriction is necessary in order to discuss natural hybridization usefully from an evolutionary point of view. Stebbins (1959) has defined hybridization as the "crossing between individuals belonging to separate populations which have different adaptive norms." A previous history of evolutionary divergence is implicit in Stebbins' definition and explicit in the one we have stated above.

Hybridization in this sense represents a reversal in the process of evolutionary divergence. The reversal may be local or widespread depending on the geographical extent of the hybridization. It may occur between sympatric species; or it may occur between races that were formerly disjunct and discrete, but which have reestablished contact along a new geographical front. The result in the latter case is a belt of variable intermediate types, known as a zone of secondary intergradation or hydrid zone (see Remington, 1968). The effects of natural hybridizaiton, finally, may be ephemeral or, as we shall see later, long-lasting.

Hybrids and Hybrid Swarms

Naturally occurring interspecific hybrids, presumably F_1s, are known in all major groups of plants and in all well-studied floras. Haartman described a number of natural as well as spontaneous garden hybrids in *Plantae Hybridae* in 1751. One of them was the natural hybrid of *Trifolium repens* and *T. pratense* which occurs in parts of Sweden (see Roberts, 1965, p. 28).

By 1891 Kerner was able to cite a list of wild hybrids in the European flora numbering over 1000 interspecific combinations and representing many groups from mosses, ferns, and horsetails to conifers, amentiferous trees, and herbaceous angiosperms (Kerner, 1894–1895, Vol. 2, pp. 582–86). More recent lists of natural hybrids are given by Allan (1949) and Heiser (1949a).

Botanists pay particularly close attention to the trees and shrubs of a region, and as a result hybrid individuals, even single hybrids, are readily detected and often reported in the larger woody plants. Kerner discovered a hybrid willow of the probable constitution *Salix incana* X *daphnoides* on the Danube in 1852 at a time when the existence of wild hybrids was still debated by botanists (Kerner, 1894–1895, Vol. 2, pp. 576–77). Natural willow hybrids belonging to scores of interspecific combinations have of course been found since Kerner's time.

Many authors have described natural hybrids of different parentage in oaks (i.e., Wolf, 1944; Palmer, 1948; Tucker, 1953; Cooperrider, 1957; Benson, 1962, Ch. 3). There are rare hybrids between *Pinus coulteri* and *P. jeffreyi* in California and between *Picea mariana* and *P. glauca* in Minnesota (Zobel, 1951; Little and Pauley, 1958). In Ceanothus, intrasectional hybrids are not uncommon and some intersectional hybrids are found in nature (McMinn, 1944; Nobs, 1963). The list of natural hybrids in woody and in herbaceous plants could be extended indefinitely.

The F_1 hybrid may be fertile, semisterile, highly sterile, or completely sterile. In all but the last case it can produce some later-generation progeny. The difference between a high degree of hybrid sterility and complete sterility may be small in terms of percentage points but is quite significant biologically, as many students have recognized.

A numerical but not unrealistic example is given by Stebbins (1959). Suppose that we have a highly sterile hybrid with a seed fertility of 0.001 % in a perennial grass such as Elymus which produces rhizomes. The hybrid individual could in time develop into a clone sending up 1000 flowering stems each year with 200 florets per stalk. This would give 200,000 florets and hence 200,000 potential ovules per year. The postulated seed fertility of 0.001 % would then mean that the hybrid clone could yield 2 seeds per year or 200 seeds per century (Stebbins, 1959).

The partially fertile F_1 hybrid may reproduce sexually by selfing, sib crossing with sister hybrid plants, or backcrossing to one or both parental species. The resulting second-generation progeny can then go on to cross with one another and with the original plants. The result is a hybrid swarm, an extremely variable mixture of species, hybrids, backcrosses, and later-generation recombination types.

Hybrid swarms are known in many plant groups. It suffices for our present purpose to list a few representative examples and refer to the corresponding special studies. In woody plants, hybrid swarms have been described inter alia in Juniperus, Eucalyptus, and Quercus (Fassett, 1944; Ross and Duncan, 1949; Clifford, 1954; Benson, Phillips, and Wilder, 1967). Comparable examples in herbaceous angiosperms are found in Carex, Viola, Helianthus, Geum, Aquilegia, and Diplacus (Drury, 1956; Brainerd, 1924; Russell, 1954; Heiser, 1949b; Marsden-Jones, 1930; Gajewski, 1957; Grant, 1952c; Beeks, 1962). An example in a fern is provided by Walker's (1958) study of Pteris.

A complete list of natural hybrids and hybrid swarms in plants would run from Achillea to Zinnia. Occasional or sporadic natural hybridization is normal in the plant kingdom when two or more related species come into contact. But it is not universal. Interest attaches to those exceptional cases in which two or more related species live in the same territory without hybridizing so far as can be determined from field studies.

Gilia tricolor does not hybridize with its sympatric relatives in California (Grant, 1952b). *Ranunculus acris, repens,* and *bulbosus* occur together frequently in Great Britain, but no hybrids are known

(Harper, 1957). *Papaver dubuim* and *P. lecoquii* also coexist in Britain without hybridizing (McNaughton and Harper, 1960b). *Lotus corniculatus* and related species remain quite distinct in California and northern Europe, but apparently hybridize in the Mediterranean region (W. F. Grant, Bullen, and Nettancourt, 1962). Seven species of Leavenworthia are sympatric in the southeastern United States but do not hybridize (Rollins, 1963). Natural hybridization is very rare in Sedum in central Mexico and in Silene in western North America (R. T. Clausen, 1959; Kruckeberg, 1961). *Juniperus virginiana* and *J. ashei* do not hybridize in various areas of sympatric contact in the Ozark plateau of the south-central United States, despite earlier reports to the contrary (Flake, Rudloff, and Turner, 1969).

Internal Control of Hybridization

The first important cause of nonhybridization between related sympatric species is incompatibility, construed broadly to include not only internal prefertilization blocks, but also early postfertilization blocks.

In the cases of nonhybridization cited in the preceding section, there is experimental evidence for very strong incompatibility barriers between species in the *Gilia tricolor* group, *Ranunculus acris* group, *Lotus corniculatus* group, and Leavenworthia (Grant, 1952b; Harper, 1957; W. F. Grant, Bullen, and Nettancourt, 1962; Rollins, 1963).

Natural intersectional hybrids are rare in Ceanothus and unknown in Quercus. And it is also known from experimental crossings that very strong incompatibility barriers exist between species belonging to different sections in these genera, as we noted in Chapter 7 (see Nobs, 1963).

But incompatibility barriers are only a part of the story. *Papaver dubium* and *P. lecoquii* can be crossed readily in the experimental garden, yet they do not hybridize naturally in Britain, as we have seen earlier (McNaughton and Harper, 1960b). Kruckeberg was able to make nearly every cross he attempted between 28 western American species of Silene, but these species remain distinct and form hybrids only rarely in nature (Kruckeberg, 1961).

Environmental Control of Hybridization

Many cases have been observed in which a given pair of related species of plants or animals does not hybridize in nature but forms hybrids freely in a botanic garden or zoo. Kölreuter referred to this common situation in 1761. He says (Kölreuter, 1761–1766; Roberts, 1929, pp. 41–42):

> In the orderly arrangement that nature has made in the plant kingdom, [it is improbable that] a hybrid plant should have arisen. Nature, which always, even in the greatest apparent disorder, adheres to the most beautiful order, has precluded this confusion. . . . [But] in the botanical gardens, where plants of all kinds and from all parts of the world, are together in a narrow space, hybrid plants will probably be able to originate, especially if one puts them together according to a systematic arrangement, and consequently those which have the greatest resemblance to one another. Man at least here gives to plants, in a certain manner, the opportunity which he gives to his animals brought from parts of the world lying far distant from one another, which he keeps confined, contrary to nature, in a zoological garden, or in a still narrower space.

Kerner later spelled out the role of the habitat in restricting natural hybridization in more concrete terms (Kerner, 1894–1895, Vol. 2, pp. 587–88).

> When a species thrives well at a particular place, is represented by a large number of individual plants, and renews itself in descendants which are in the main unchanged, it may be assumed that the organization of that species is suited to the soil and climate of the habitat in question. If there were no such harmonious relation there could be no question of the species flourishing, but on the contrary it would sooner or later die out. This suitability of the climate and soil to the organization manifested in the plant's external form must also exist in the case of the newly developed hybrid if the few individuals which spring up at any particular place are to survive in their original settlements, and to give rise to a numerous progeny. Sometimes such suitability does exist, but sometimes also it does not. In the latter case the hybrid is suppressed as soon as it sees the light. But even if its organization is adapted to the soil and climate of the place of origin, it has to enter upon a struggle with the species already established there, and especially with its own parent-species. If the latter grow luxuriantly and in large numbers at the spot, it is not easy for the new form to take possession of the ground.

Kerner concluded that hybrid plants can become established only if they are adaptively superior to the parental species in the ancestral habitat, or if they can find a suitable open habitat not occupied by the parental species. These conditions are fulfilled rarely. Kerner cites examples of Rhododendron, Salvia, and Nuphar where hybrids have successfully colonized new and distinctive habitats of their own (Kerner, 1894–95, Vol. 2, pp. 588 ff.).

In the modern period the importance of environmental factors in preventing or permitting natural hybridization has been emphasized by DuRietz (1930), Wiegand (1935), Epling (1947), Anderson (1948, 1949), Heiser (1949a, 1949b), and many other students.

Wiegand (1935) pointed out that natural hybridization between plant species is often restricted to certain localities in which the environment has been disturbed by man. In Newfoundland and Maine, for instance, hybrid types of Amelanchier and Rubus occur in disturbed habitats along railroad tracks, whereas the undisturbed forest or woods bordering the tracks harbor the parental species in uncontaminated form (Wiegand, 1935).

This observation has been extended to many other plant groups. In Iris, natural interspecific hybridization takes place chiefly in habitats disturbed by human activities—agriculture, road building, logging, etc.—in both the Mississipi delta region and the Pacific slope (Viosca, 1935; Riley, 1938; Anderson, 1949; Lenz, 1959a; Randolph, Nelson, and Plaisted, 1967). *Salvia apiana* and *S. mellifera*, which occur sympatrically over a vast territory in southern California, remain distinct where their sage scrub community is in its natural state, but hybridize locally in artificially disturbed places (Epling, 1947; Anderson and Anderson, 1954; K. A. and V. Grant, 1964).

The explanation of the correlation between hybridization and habitat disturbance which is favored by most students is an extension of the one proposed by Kerner in 1891. In a closed, stable community no habitat is available for such hybrid zygotes as are formed from time to time. Stabilizing selection eliminates them from the scene almost as soon as they arise. But, where the natural community has been broken into by road building, overgrazing, or the like, so that new open habitats are created, the hybrids can and do become established. In other words, environmental isolation operates to suppress hybridization between intercompatible species in a stable, closed community, but ceases to be fully effective in an open habitat.

Hybridization of the Habitat

Anderson (1948, 1949) carried the concept of environmental control beyond the F_1 to the second and later hybrid generations. He pointed out that the physiological differences between species segregate in the same way as the morphological differences. Whereas the F_1 generation is intermediate and also uniform in ecological preferences, the F_2 and later generations will contain a great diversity of recombination types with respect to physiological traits. If the parental species differ in several independent physiological characters, their later-generation hybrid progeny will require scores or hundreds of ecological niches. Anderson concludes that an intermediate habitat will accommodate the F_1 generation, but a varied array of ecological niches, a "hybridization of the habitat," must exist if any representative sample of the F_2 generation is to survive (Anderson, 1948; 1949, Ch. 2).

As Anderson puts it (1949, p. 17):

> It has been very generally recognized that if hybrids are to survive we must have intermediate habitats for them. It has not been emphasized, however, that, if anything beyond the first hybrid generation is to pull through, we must have habitats not only that are intermediate but also that present all possible recombinations of the contrasting differences of the original habitats. . . . Only by a hybridization of the habitat can the hybrid recombinations be preserved in nature.

Intermediate habitats are not too rare in nature. But habitats which are sufficiently hybridized to permit the relatively free reproduction of a plant species hybrid are much more uncommon. And such hybridized habitats, in the modern world, are produced mainly by human activities like road building, farming, and logging (Anderson, 1949, Ch. 2).

I wish to comment in passing that it is necessary to distinguish between hybridized habitats and open habitats. The F_2 and later-generation progeny of an interspecific hybrid might survive in either type of disturbed habitat. But they would do so for different reasons: in an open habitat because of the relaxation of interspecific competition and stabilizing selection, and in the hybridized habitat because of the diverse array of new niches. It is possible to imagine a hybrid swarm growing in either a uniform open habitat or in a scrambled but more or less closed habitat. Hybridization of the habitat is therefore not the only condition which will permit the establishment of segregating hybrid progenies.

The main effect of the human activities mentioned above is to clear away segments of natural vegetation and by so doing to create open habitats. Artificial hybridization of the habitat, by mixing soil types, by producing new gradients in shade or moisture, and so on, is a frequent but not a necessary concomitant of these activities. And where it does occur its permissive effects on hybrid reproduction are probably secondary to those resulting from the presence of open ground.

Open habitats, with or without habitat hybridization, are a common product of human activities in the modern world. But throughout earth history, up to and including the modern era, biotic communities have repeatedly been disturbed and new habitats opened up by purely natural agencies. Landslides, floods, volcanic eruptions, and lightning fires have created new open habitats on a local scale. On the grand scale, mountain building, land emergence from the sea, ice ages, and drought periods have all had their effects on natural communities. And biotic factors, like the immigration of new large herbivores or new disease organisms into a region, have also had far-reaching effects.

The climatic and geological changes promote plant migrations, bringing species into new sympatric contacts. At the same time these physical changes or the changed biotic factors or both may open up new uncolonized habitats which can serve as the breeding grounds for the natural hybrids, as Anderson (1948, 1953) and others have suggested.

Higher Plants Compared with Higher Animals

Natural hybrids have been recorded in the various major groups of higher animals. Compendious lists for birds are given by Cockrum (1952) and Gray (1958), and for mammals by Gray (1954). Natural hybridization in fishes is discussed by Hubbs (1955). The subject as a whole has been reviewed by Mayr (1963, Ch. 6).

The reader is referred to these sources and to their extensive bibliographies for discussions of, and details concerning, the complex subject of hybridization in animals. Here we are concerned, not with animal hybridization as such, but with broad comparisons between animals and plants as regards the frequency and the evolutionary role of hybridization. There does seem to be a difference between the two kingdoms in this regard.

Natural hybridization has been shown to affect the variation pattern in particular groups of animals, thus in certain butterflies, leaf hoppers, fishes, and birds (Hovanitz, 1949; Hubbs and Miller, 1943; Hubbs, 1955; Sibley, 1950, 1954; Ross, 1958; Remington, 1968). The list of examples could be extended. On the other hand, natural interspecific hybridization seems to be rare and inconsequential in many other animal groups, such as the Diptera and mammals.

But a multiplication of examples on one side or the other of the question does not advance our understanding of the role of hybridization in animal evolution very much. We have seen earlier in this chapter that natural hybridization occurs in some groups but not in others within the plant kingdom. Broad comparisons between the two kingdoms with regard to the importance of hybridization cannot as yet be placed on a satisfactory quantitative basis. Therefore other approaches to the problem must be sought.

Several generations of botanists have considered that hybridization plays an important role in plant evolution (i.e., Naudin, 1863; Kerner, 1894–1895; Lotsy, 1916; Ernst, 1918; Diels, 1921; DuRietz, 1930; Gustafsson, 1946–1947; Anderson, 1949, 1953; Heiser, 1949a; Stebbins, 1950, 1959). Several generations of zoologists have concluded that hybridization does not play an important general role in animal evolution (i.e., Weismann, 1913; Dobzhansky, 1937b, 1951; Mayr, 1942, 1963; Fisher, 1954; White, 1957a).

I am not citing the views of different authors here in order to decide a scientific question by a counting of noses or an appeal to authority. The point is that botanical authors and zoological authors express their views out of a store of background knowledge which should not be dismissed lightly.

A related comparison, on which there is also widespread agreement, concerns the relative frequency of good biological species in the two kingdoms. We noted in Chapter 4 that departures from an ideal standard of discrete biological species are very common in higher plants but much less so in higher animals. These departures from a biological species organization in the case of higher plants, moreover, are usually associated with natural hybridization.

Another approach is to compare the frequency of polyploidy in the two kingdoms. It is well known that about half of the species of higher plants are polyploid, whereas polyploidy is known in only a few groups of animals. Furthermore, the majority of naturally occurring polyploids in plants are allopolyploids (Clausen, Keck, and Hiesey, 1945). We have here a clear indication of the importance of hybridization in plant evolution and the absence of any comparable indication for the animal kingdom.

The ethological barriers which prevent hybridization are far more strongly developed in higher animals than in plants. In animals these mating barriers are primary and are highly effective except under abnormal circumstances. Such ethological barriers are lacking altogether in most segments of the plant kingdom. And where present, in the more advanced families of flowering plants, they are secondary, in being dependent on the behavior of flower-visiting insects, and are therefore less effective than the direct mating preferences expressed by animals.

Hubbs (1955) and Mayr (1963, p. 125) have pointed out that hybridization in animals is common mainly in the groups which have external fertilization, particularly fishes and some amphibians, and not in groups with internal fertilization, such as reptiles, birds, mammals, and insects. There are some exceptions to this generalization. The plant kingdom as a whole is comparable to that segment of the animal kingdom which has external fertilization, since cross-pollination is brought about by external agents in all plants. There is a correlated similarity between plants and externally fertilized animals in the frequency of hybridization.

CHAPTER *12*

Introgression

Introduction · Factors Promoting Introgression · Convergence as a Result of Introgression · Methods of Analysis · Examples · Ecological Segregation of Morphological Types · M-V Linkage in Relation to Introgression · Transgression of Chromosomal Sterility Barriers · Conclusions

Introduction

The reproduction of natural hybrids in a great many cases follows a definite pathway known as introgression. Introgressive hybridization, as defined by Anderson, is the repeated backcrossing of a natural hybrid to one or both parental populations. It results in the transfer of genes from one species or semispecies to another across a breeding barrier (Anderson and Hubricht, 1938; Anderson, 1949, 1953).

The early contribution of DuRietz (1930) to our understanding of introgression has been generally overlooked by later workers. DuRietz made population studies of species groups in Dracophyllum, Coprosma, and Salix in the New Zealand and Scandinavian floras. He observed that the members of a species pair hybridize in particular localities and, in addition, that the populations of the same parental species tend to converge in morphological characters in their areas of sympatric overlap. He attributed the convergence to hybridization followed by backcrossing. This process leads, as he put it, to one species becoming "infected" with particular genes from another species (DuRietz, 1930, pp. 376 ff., 411).

Factors Promoting Introgression

Consider a colony composed of one or two species and their F_1 hybrids. There are genetic factors and ecological factors which operate in such a colony to promote backcrossing and hence to steer hybrid reproduction along the pathway of introgression. Let us consider first the population-genetical factors.

The natural hybrid is usually represented by one or a few plants growing in the midst of larger populations of one or both parental species. Furthermore, the F_1 hybrids are usually at least semisterile and are often highly sterile. Under these conditions the hybrids contribute very few viable gametes to the gamete pool of the colony as compared with the contribution of the parental species. Consequently the vast majority of the effective fertilizations in which a hybrid takes part will involve a species plant as the other partner and hence will lead to the formation of backcross progeny (Stebbins, 1950, p. 261).

The ecological factors promoting introgression are a corollary of the habitat conditions described in the preceding chapter. Each parental species is well adapted to the ecological niche which it normally occupies. An open or intermediate habitat can accommodate F_1 hybrids and some intermediate F_2 types. An open habitat or a somewhat hybridized habitat can accommodate some of the recombination products of natural hybridization. But most of the opportunities for establishment of hybrid progeny in any normal meeting place between two species or semispecies will be for genotypes which approach one parent or the other in their physiological characters. The backcross progeny are most apt to have the requisite ecological preferences and are thus likely to be favored by the existing environmental conditions (Anderson and Hubricht, 1938; Anderson, 1949, Ch. 2).

Baetcke and Alston (1968) analyzed the composition of a hybrid colony of Baptisia in Texas. They used morphological characters and biochemical substances as criteria. These criteria point to the following composition of the colony:

Baptisia leucophaea, 470 plants
Baptisia sphaerocarpa, 561 plants
F_1 hybrids, 83 plants
Backcrosses to B. leucophaea, 37 plants
Backcrosses to B. sphaerocarpa, 19 plants

Thus under natural conditions the hybrid derivatives are seen to fall into the two reciprocal backcross classes (Baetcke and Alston, 1968).

Convergence as a Result of Introgression

The population-genetical factors and ecological factors described above continue to operate in successive hybrid generations and lead to repeated backcrossing. The result is the infiltration of certain genes or chromosome segments from one species into the genotype of another. The affected or introgressed populations of the recipient species thereupon converge toward the donor species in various morphological and physiological traits (DuRietz, 1930; Anderson and Hubricht, 1938; Anderson, 1949, 1953).

Introgressive hybridization is a force promoting convergence in morphological characters and ecological preferences in a zone of sympatric contact between two species or semispecies. Other forces work in the opposite direction.

Ecological coexistence and the resulting competition lead to character displacement. This phenomenon is well known in birds (Brown and Wilson, 1956; Mayr, 1963, pp. 82–86), but is either less common or less studied in plants. Levin and Kerster (1967b) have recently described a case in Phlox. A second force promoting divergence in a sympatric zone is selection for reproductive isolation (see Chapter 10). This process affects the external characters and internal conditions involved in the preliminary stages of reproductive isolation. It occurs in some plants, as we saw in Chapter 10, as well as in many animals.

Sympatry, then, is a challenge which evokes a variety of responses. Convergence in morphological and physiological traits happens to be the response we are interested in here. And it appears to be the predominant response in many plant groups.

Methods of Analysis

Multifactorial linkage, as will be recalled from our brief summary in Chapter 5, is the partial and interlocking linkage between different quantitative characters (see also Grant, 1964b, Ch. 7). Anderson showed that this condition of multifactorial linkage provides a basis for detecting and quantifying the effects of introgression (Anderson, 1939; 1949, Chs. 2 and 3; 1953).

The expected effects of backcrossing and introgression will be to produce a population which is variable and which approaches one parental species in its phenotypic characteristics. The hypothesis that this combination of variability and convergence is in fact due to introgression can be checked by the criterion of character association. This criterion is applicable in the case of developmentally unrelated characters like stem pubescence and flower color; it is not applicable to pleiotropic characters like leaf shape and sepal shape.

If some genes determining one character are introduced from a donor species into an introgressive population of another species, then, as a result of multifactorial linkage, other gene systems and their character expressions will be dragged into the recurrent parent with them. Therefore the introgressive population should resemble the recurrent parent but should vary in the direction of the donor species not only in one quantitative character but in many or all such characters (Anderson, 1949, 1953).

In practice one has to take a random sample of plants in a supposedly introgressive population, and measure each plant in the sample for many or all of the characters which exhibit individual variation. Correlations between separate characters can be detected and expressed from such data. The original method used by Anderson and his associates to express variability in character combinations was that of the hybrid index.

Take, for example, ten separate characters which differ between the two putative parental species. Assign an arbitrary score value of 0 to the extreme condition of each character found in species A, a score value of 2 to the contrasting character condition in the opposite species B, and a score value of 1 to the intermediate condition in known or putative F_1 hybrids. Alternatively, some morphological characters can be weighted more heavily than others in order to reflect their relative usefulness as indicators of the underlying genetic determinants.

In the unweighted example considered above, the total score, or hybrid index, of an individual plant belonging to species A, and measured for ten characters, will be 0, and the hybrid index value of a plant belonging to species B will be 20. These contrasting score values provide a standard of reference for assessing the intermediacy and variability of hybrid populations.

The next steps are to determine the hybrid index value for each individual plant in a sample taken from a hybrid colony, and to plot the frequency distribution of index values as a histogram. A colony consisting of parental species and F_1 hybrids will, of course, show up on the bar graph as a series of peaks in the extreme and intermediate ranges. A strongly introgressive population, on the other hand, will have a larger representation of hybrid index values in the range between the F_1 hybrids and a parental type. A slightly introgressive population, finally, reveals its nature by a perceptible shift of the mode away from one extreme and toward the intermediate condition accompanied by increased individual variation.

Two good examples of the application of the hybrid index method to actual cases of introgression are those of Riley (1938) in Iris and Heiser (1949b) in Helianthus. The Iris story was reviewed in Chapter 4, and the Helianthus story is presented in the next section of this chapter. The corresponding bar graphs are shown in Figures 12 (Chapter 4) and 26.

The hybrid index method pools the measurements of separate characters. Two plants can have the same intermediate score value as a result of different but compensating deviations from an extreme character combination. Much of the evidence concerning character correlations is therefore lost by this method of analysis. Consequently

Anderson later devised and introduced the method of pictorialized scatter diagrams, which has been widely adopted (Anderson, 1949; Anderson and Gage, 1952).

A conventional scatter diagram, in which two quantitative characters are represented by the coordinates and each individual plant by a dot, shows the variability for two characters in a population. In a pictorialized scatter diagram, additional characters are represented by the length of side whiskers attached to the dots. Such a diagram shows, almost at a glance, the total pattern of variation in a population which is varying simultaneously for three or more characters (see Figure 27).

One can then go on to draw a series of pictorialized scatter diagrams for different populations of the putative parental species and introgressive derivatives. Such a series enables one to compare the several populations directly with respect to the range and correlatedness of their variability (Figure 27).

A more recent development has been the introduction of biochemical criteria, as indicated by chromatographic and other methods, into the analysis of hybridizing or hybridized populations. Biochemical criteria and morphological criteria of hybridity are not always in good agreement, and therefore the use of both is desirable in many instances (Alston and Turner, 1962, 1963b; Baetcke and Alston, 1968).

As examples of the application of chemotaxonomic methods to studies of natural hybridization, we may mention the work of Turner and co-workers on Baptisia and of Levin on Phlox (Turner and Alston, 1959; Alston and Turner, 1962, 1963b; Baetcke and Alston, 1968; Levin, 1966, 1967). General reviews of the subject are given by Alston and Turner (1963a; Alston, Mabry, and Turner, 1963).

Examples

For presentation here we have selected, of the many good examples of introgression available, three which are illustrative of different situations. The three examples in question are drawn from Helianthus, Phlox, and Pinus; a fourth case in Iris was described in Chapter 4. Reviews containing references to many other studies are given by Heiser (1949a), Stebbins (1950), and Anderson (1953).

Our first case involves the sunflower species, *Helianthus annuus* and *H. bolanderi*, as worked out by Heiser (1949b). The common sunflower, *H. annuus*, occurs as a weed throughout temperate North America. In California, *H. annuus* is not a part of the native vegetation, and was probably introduced by the Indians in fairly recent times. *Helianthus bolanderi* occurs in both natural habitats and weedy places in California and Oregon. The native race of *H. bolanderi* grows on serpentine outcrops in the foothills of California and Oregon, while the weedy race is found in the central valley of California (Heiser, 1949b).

The two species differ in numerous morphological characters. Thus *Helianthus annuus* is a tall plant with ovate leaves, and *H. bolanderi* a low to medium-tall plant with lanceolate leaves. The pubescence, involucre, disk, rays, and achenes also differ between the species. These morphological character differences provided the markers for scoring individual plants and analyzing populations by the hybrid index method (Heiser, 1949b).

Experimental hybridizations show that *Helianthus annuus* and *H. bolanderi* will cross readily in either direction. The F_1 hybrid is morphologically intermediate, has irregular chromosome pairing at meiosis, and is highly sterile as to pollen and seeds (Heiser, 1949b).

Helianthus annuus and *H. bolanderi* come into contact in various places in California. Their blooming periods overlap and they hybridize locally. The variation pattern of a hybrid swarm in the Sacramento Valley of California is shown in Figure 26E. On the basis of morphological characters and correlated pollen fertilities, this hybrid colony is judged to consist of the two parental species, some F_1 hybrids, and the two reciprocal backcross types (Heiser, 1949b).

The characteristics of unintrogressed populations of *Helianthus annuus* from St. Louis, Missouri, and *H. bolanderi* from serpentine outcrops in northern California are shown in Figure 26, A and B. Weedy populations of the same two species from the central valley of California are shown for comparison in Figure 26, C and D. It will be seen that each of these latter populations varies in the direction of the opposite species. They lie in the range of the backcross types in the hybrid swarm, and are logically concluded to be products of introgressive hybridization (Heiser, 1949b).

The morphologically extreme race of *Helianthus bolanderi* is restricted to serpentine soils in the foothills, while the introgressive race of the

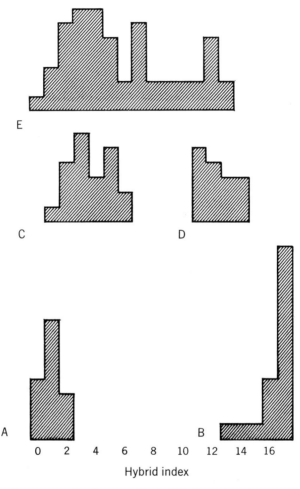

Figure 26 Frequency distribution of hybrid index values in five populations of *Helianthus annuus* and *H. bolanderi*. The populations are from California unless specified otherwise. (A) *H. bolanderi*, foothill serpentine race. (B) *H. annuus*, Missouri race. (C) *H. bolanderi*, valley weed race. (D) *H. annuus*, San Joaquin Valley. (E) Hybrid swarm, Sacramento Valley. (Redrawn from Heiser, 1949b.)

same species occurs as a weed in the central valley along with *H. annuus*. This ecogeographic distribution suggests strongly that the introduction of *H. annuus* into California, perhaps by Indians, was followed by hybridization between *H. annuus* and the original wild form of *H. bolanderi* (Heiser, 1949b).

Introgression from *Helianthus annuus* into *H. bolanderi* then led to the formation of the valley race of *H. bolanderi*, which was able to spread throughout the weedy habitats being opened up concurrently. It is likely that *H. annuus* in central California was affected reciprocally by introgression from *H. bolanderi*. The introgressive valley race of *H. bolanderi* may be only a few centuries old (Heiser, 1949b).

Helianthus annuus overlaps and hybridizes with other species of annual sunflowers in other regions of the United States, thus with *H. debilis* and *H. argophyllus* in Texas and with *H. petiolaris* in the Great Plains and Southwest. *Helianthus annuus* has given rise to introgressive races varying toward *H. debilis* in Texas and toward *H. petiolaris* in the central United States. Conversely there is an introgressive race of *H. petiolaris* (Heiser, 1947, 1951a, 1951b, 1954).

The second example involves *Phlox bifida* and *P. amoena* in Tennessee. This case was analyzed by the pictorialized scatter diagram method by Anderson and Gage (1952). The story begins with the discovery and analysis of a highly variable population of *P. bifida* near Nashville, Tennessee. The analysis revealed partial correlations between nine different vegetative and floral characters (Figure 27).

The trend of the correlated variations in this population terminated in two contrasting character combinations. At one pole were plants with long glabrous pedicels, long terminal internodes, small narrow leaves, deeply notched petals, four flowers per inflorescence unit, and deeply colored corolla tubes. These are the usual characters of *Phlox bifida* (Anderson and Gage, 1952).

The contrasting character combination found at the opposite pole in the pictorialized scatter diagram was the following. There were plants with heavy pubescence, short pedicels, short terminal internodes, large wide leaves, unnotched petals, three flowers per inflorescence unit, and light-colored corolla tubes (Figure 27). This character combination is unlike anything in normal *Phlox bifida* and was attributed to introgression from some second parental species (Anderson and Gage, 1952).

By extrapolation from the trends of variation seen in the Nashville

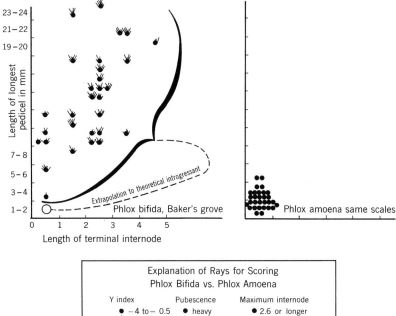

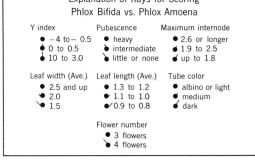

Figure 27. Pictorialized scatter diagram showing the trend of variation in nine characters in an introgressive population of *Phlox bifida* and in a normal population of *P. amoena*. (Redrawn from Anderson and Gage, 1952.)

population, the authors drew up a description of the putative second species. This species should possess the character combination described above in a somewhat more extreme form. No such species was found in the immediate vicinity of the Nashville population, nor was any such kind of Phlox known to the authors. They thereupon went through a taxonomic treatment of the genus Phlox and succeeded in finding an actual species description which matched their hypothetical description. The predicted second parental species was identified as *Phlox amoena*.

It was reported to occur in the same county as the introgressive Nashville population (Anderson and Gage, 1952).

When a population of *Phlox amoena* was later found in Tennessee and measured, it turned out to have the character combination which had been predicted by extrapolation from the trend of variation in the introgressive population (Figure 27).

Mason (1949) obtained paleobotanical evidence for natural hybridization in pines during a prehistoric period of climatic change and plant migrations in coastal California. In the Pleistocene the populations of *Pinus remorata* on Santa Cruz Island were morphologically quite distinct from the related *P. muricata*, which occurred on the neighboring California mainland at that time. In Late Pleistocene or Post-Pleistocene, *P. muricata* migrated to Santa Cruz Island. Natural hybridization between the two semispecies ensued. This hybridization has continued into Recent time and has altered the characteristics of some *P. remorata* populations (Mason, 1949). On the California mainland the fossil evidence indicates that introgression was occurring in the opposite direction, from *P. remorata* into *P. muricata*, in Pleistocene time, and this process is continuing today (Mason, 1949).

In many species groups, besides the *Pinus muricata* and *Helianthus annuus* groups, particular races have been produced by introgression. Races of introgressive origin are found in the following genera, among others: Dracophyllum (DuRietz, 1930), Coprosma (DuRietz, 1930), Abies (Mattfeld, 1930; review in Stebbins, 1950, pp. 280 ff.), Tradescantia (Anderson and Hubricht, 1938), Cistus (Dansereau, 1941), Gilia (Grant, 1950a; V. and A. Grant, 1960), Quercus (Cooperrider, 1957), and Purshia (Stebbins, 1959).

Ecological Segregation of Morphological Types

An introgressive population or a hybrid swarm often grows in a heterogeneous or somewhat hybridized habitat lying between the habitats of the parental species populations. In such a situation the microgeographical distribution of different backcross types is not at random. On the contrary, as might be expected, the individuals which approach one parental species in morphological characters tend to be clustered in the facies of the hybridized habitat which is most like the normal environment of that same species. Conversely, the opposite class of hybrid segregates and backcross types is found growing preponderantly in environmental niches approaching that of the opposite parental species.

I have observed this ecological segregation of different morphological forms in hybrid populations which I have examined in Aquilegia, Diplacus, Ipomopsis, and Gilia. No doubt, other botanists have made similar observations. The phenomenon has been documented by Benson, Phillips, and Wilder (1967) for a hybrid population of Quercus.

The parental semispecies in this case are *Quercus douglasii* and *Q. turbinella californica*. *Quercus douglasii* is a deciduous tree with glaucous blue leaves in the interior foothills of central and northern California. *Quercus turbinella californica* is an evergreen scrub oak of desert and semidesert slopes in central and southern California. The two semispecies come into contact and hybridize at various localities in the foothill zone of south-central California. The population analyzed in detail by Benson, Phillips, and Wilder (1967) is in Tejon Pass in this general zone of contact.

The Tejon population is spread over an ecologically diversified area with hill slopes exposed to different points of the compass. The hot dry slopes facing the south and southeast are most like the normal habitat of the desert scrub oak, *Quercus turbinella californica*. The slopes facing north and northeast, which are also dry but receive less insolation, approach the usual conditions of the more northern tree oak, *Q. douglasii*.

Benson, Phillips, and Wilder (1967) measured a sample of specimens in the Tejon population for 11 vegetative characters and scored them by the hybrid index method. This revealed the presence of numerous backcross types approaching *Quercus douglasii* and others approaching *Q. turbinella californica* in about equal proportions in the hybrid population as a whole. But, when the data were reanalyzed, taking into consideration the microgeographical position of the individual plants, a different picture emerged. The plants on the slopes facing north and northeast tend to resemble *Q. douglasii* in morphology. The slopes facing south and southwest, on the other hand, harbor mainly plants with hybrid index values approaching *Q. turbinella californica* (Benson, Phillips, and Wilder, 1967).

M-V Linkage in Relation to Introgression

Genes determining various morphological characters in higher plants, including those which are first expressed in the flowers, fruits, or late stages of vegetative development, are commonly linked with genes which affect growth and vigor in early developmental stages. And related plant species often differ allelically with respect to the linked morphological and viability genes. This is the *M-V* linkage which we discussed briefly in Chapter 5. There is some evidence, furthermore, for the existence of *M-V* linkage systems on the rearranged segments differentiating chromosomally intersterile plant species as well as on the homologous segments of interfertile races or species (see Chapter 5).

The *M-V* linkage is certainly a widespread and probably a general condition in plant species. Here we will discuss the bearing of this widespread type of linkage on the analysis of introgression. Suppose that we find a plant population exhibiting the following features. (1) The population is variable. (2) The plants closely approach a certain species, *A*, in a combination of morphological characters, but vary in the direction of another species, *B*. (3) They also approach species *A* in their ecological preferences. (4) They are interfertile with species *A* but form chromosomally sterile F_1 hybrids with species *B*.

By the standard methods of introgression analysis we would conclude that our hypothetical population is a product of introgression of particular genes from species *B* into species *A*. We would postulate a former event of hybridization of *A* ✕ *B*, followed by repeated backcrossing to *A*, with selection for favorable backcross types in the environment of *A*.

Undoubtedly, introgressive hybridization will produce the effects described above. But it must be pointed out that the same end results can also be attained by another and different mode of hybridization. If morphological genes are commonly linked with viability genes, and if plant species differ allelically in respect to these *M-V* linkage systems, then interspecific hybridization followed by straight inbreeding and selection of the inbred products will bring about the combination of effects listed in a preceding paragraph. That combination of features is therefore not absolutely diagnostic of introgression (Grant, 1967).

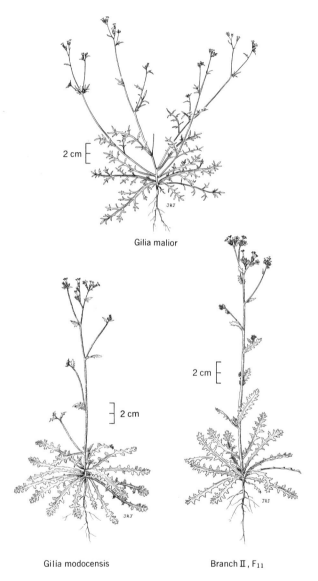

2 cm

Gilia malior

2 cm

2 cm

Gilia modocensis

Branch II, F_{11}

Figure 28 Two intersterile species of Gilia and one of their hybrid derivatives. The latter was derived by inbreeding and selection for viability. It is close to one parental species in morphology and fertility relationships. (Grant, 1967.)

We have an experimental case of introgressionlike effects without introgression in Gilia. The parental species involved in this case are *Gilia modocensis* and *G. malior*. They are autogamous and tetraploid species belonging to the *G. inconspicua* complex, which we introduced in Chapter 7. Our interest here centers on certain inbred lines derived from the F_1 hybrid of *G. malior* X *modocensis*, particularly the lines designated Branch II (Grant, 1966b, 1967). See Figure 28.

The F_1 hybrid of *Gilia malior* X *modocensis* is highly but not completely sterile and has a low degree of chromosome pairing. There is cytogenetic evidence apart from the observed pairing relationships in the hybrid to indicate that the parental species differ by numerous independent segmental rearrangements (Grant, 1964c, 1966c).

The F_1 hybrid produced some progeny without change in ploidy. These were propagated as a series of inbred lines. They were selected for fertility and for vigor in an environment favorable to the *modocensis*-like types. They were not selected on the basis of morphological characters as such. The later-generation selection products were vigorous and fertile (Grant, 1966b, 1966c).

One set of related lines derived from the same parental individual in F_2 and referred to collectively as Branch II turned out to resemble closely the *Gilia modocensis* parent in morphology (Figure 28). Branch II varied, however, in the direction of *G. malior*.

Fertile plants of Branch II were backcrossed to both parental species. The backcross hybrid between Branch II and *Gilia modocensis* was fertile with complete bivalent pairing, while the reciprocal backcross hybrid with *G. malior* was highly sterile with reduced pairing (Grant, 1966b).

Let us suppose now that Branch II were found growing in nature and were analyzed morphologically, ecologically, and cytogenetically. It would be routinely identified as a deviant race of *Gilia modocensis* containing particular genes and their morphological expressions introduced by introgression from *G. malior*. But we know the whole pedigree in this case and can positively rule out backcrossing in the ancestry of Branch II (Grant, 1966c, 1967).

It is apparent, therefore, from experimental studies as well as from theoretical considerations, that effects indistinguishable from those brought about by introgression can arise along an alternative pathway. The existing methods of introgression analysis do not discriminate between (a) backcrossing and selection and (b) selection in interaction with *M-V* linkage (Grant, 1967).

It becomes necessary under the circumstances to reexamine the criteria for identification of introgression and to reconsider some of the cases identified by these criteria. Perhaps the best criterion to use in the present stage of our understanding is the breeding system of the plants involved. In predominantly autogamous plants the probability of a hybrid reproducing by backcrossing is exceedingly low. To infer introgression from the observation of introgressionlike variations in such plants is unwarranted.

Introgression has in fact been inferred from an introgressionlike variation pattern in certain hybridizing autogamous plants, particularly the *Elymus glaucus* group and the *Aegilops variabilis* group (Snyder, 1951; Stebbins, 1956; Zohary and Feldman, 1962). But the interpretations in these cases should perhaps be reconsidered in the light of *M-V* linkage.

In outcrossing plant groups, on the other hand, backcrossing by hybrids is their most likely method of reproduction. This is a consequence, in the first place, of the breeding system in itself. It follows, moreover, from the relative numbers of viable gametes contributed to the gamete pool by the numerous fertile species plants and by the rare sterile hybrid plants, respectively, as discussed earlier in this chapter.

The inference of introgression from an introgressionlike variation pattern in outcrossing plants has a high probability of being correct, in line with the views of Anderson (1949, 1953), Stebbins (1950), and others. The examples of introgression selected for purposes of illustration in this chapter—Pinus, Quercus, Helianthus, Phlox, Iris—all involve outcrossing plants.

In plant groups possessing a breeding system intermediate between outcrossing and inbreeding, hybrid reproduction can be expected to follow a mixture of pathways. This may be the case in parts of the *Aegilops variabilis* group, to judge from recent work of Pazy and Zohary (1965).

Transgression of Chromosomal Sterility Barriers

Introgression between plant species possessing different structural karyotypes must involve the passage of genes through a chromosomal sterility barrier. The problem before us in this section is to consider the mechanism by which this transgression of chromosomal sterility barriers probably or necessarily takes place. The account given here is a revision and extension of earlier treatments (Grant, 1958; 1963, pp. 487 ff.).

The preconditions are the following. We have a hybrid population consisting of individuals of two species and their hybrids. The plants are outcrossing. The chromosomes of the parental species differ by segmental rearrangements, which lead to reduced pairing or unbalanced gametes or both in the hybrids, and the hybrids consequently exhibit chromosomal sterility or semisterility. Certain genes borne on these chromosomes in one parental species, when introduced into the genotype of the second parental species, form a recombination product which is adaptively superior in the environment of the second species. The foregoing conditions are probably often realized in nature.

The process of sterility transgression which is expected to take place under the preceding conditions is complicated. It involves the simultaneous operation of several separate subprocesses. In the first place, the plants constituting the hybrid population reproduce by outcrossing. Secondly, the homeologous chromosomes form bivalents and undergo crossing-over in at least some of the pollen and egg mother cells in the structural heterozygotes. Gamete selection is occurring so as to eliminate the deficiency-duplication products of meiosis and favor the genically balanced gametophytes. At the same time Darwinian selection between different individuals in the population is also going on. The latter selective process is favoring certain interspecific recombination genotypes in the environment of one parental species.

The process of sterility transgression will be considered first in relation to a particular model. We have a recipient species (R) and a donor species (D) which contributes some of its genes to the former by introgression. Recombination genotypes containing certain D genes in the genetic background of R have a high adaptive value in the environment of species R. These interspecific recombination types have a higher selective value in this environment than the corresponding allele combination found in unintrogressed species R.

The genes involved in the adaptively significant gene combination are borne on chromosomes which differ between the two species in structural rearrangements.

(A) Parental Types

Species R  $I_r I_r$

Species D $I_d I_d$

F_1 zygote $I_r I_d$

(B) Products of Meiosis in F_1 or Backcross Hybrid, with crossing-over between arms

Pairing 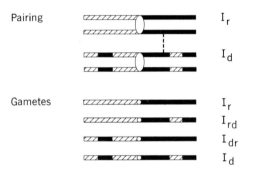 I_r

I_d

Gametes I_r

I_{rd}

I_{dr}

I_d

(C) Products of Meiosis in F_1 or Backcross Hybrid, with double crossing-over between rearranged segments

Pairing 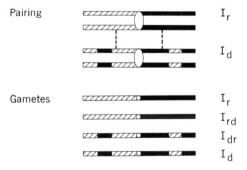 I_r

I_d

Gametes I_r

I_{rd}

I_{dr}

I_d

Figure 29 Cytogenetic behavior of a structurally differentiated pair of chromosomes in a hybrid population. The parental species R and D carry and introduce into the population two partially homologous forms of this chromosome, designated I_r and I_d respectively, which differ by a transposition. Further explanation in text.

In order to simplify the discussion, we focus our attention on one particular chromosome (I) present in two structural forms in the parental species (I_r and I_d). Chromosomes I_r and I_d differ with respect to a transposition, as shown in Figure 29A. A segment in the left arm has exchanged places with a segment in the right arm during the phylogenetic history of chromosome I_d, but this rearrangement has not occurred in chromosome I_r, and the two chromosome types therefore differ structurally in the manner indicated in Figure 29A.

The central or interstitial segments of chromosomes I_r and I_d are homologous (see Figure 29A again). We postulate that the genes of species D which form an adaptively valuable recombination with genes of R are borne on these interstitial segments of chromosome I_d. The R genes involved in the same valuable gene combination are on the corresponding homologous segments of I_r, or here and on other chromosomes of the complement too.

The $R \times D$ hybrids may have variable reduced pairing of the I chromosomes in consequence of the lack of complete homology. In the $I_r I_d$ bivalents, when such bivalents do form, crossing-over can be expected to occur fairly regularly in the homologous interstitial segments (Figure 29B). Half of the meiotic products of such crossing-over carry deficiencies and duplications for the transposed segments (Figure 29B), and yield inviable gametophytes, making the hybrid semisterile. The transposed segment, in other words, acts as a sterility factor when crossed over into the homeologous chromosome in such a way as to form a deficiency-duplication product.

The deficiency-duplication chromosomes will drop out of the gamete pool in each generation. And so will the D genes borne on their interstitial segments.

The only functioning gametes are those carrying the noncrossover I_r or I_d chromosomes (Figure 29B). Therefore the F_1 hybrid, after sib crossing, yields F_2 progeny segregating into the two parental and the hybrid classes ($I_r I_r$, $I_d I_d$, $I_r I_d$); and, on backcrossing to R, gives progeny segregating into one parental and one hybrid class ($I_r I_r$, $I_r I_d$). The $I_r I_d$ types are semisterile in F_2, B_1, and B_2, as they are in F_1. Conversely, the only fertile progeny in later generations are the parental types, $I_r I_r$ and $I_d I_d$.

With the usual mode of crossing-over in chromosome I, as diagrammed in Figure 29B, therefore, the hybrid population segregates into the parental types and the F_1 type for morphological and physiological traits determined by genes on this chromosome. Introgression of genes on chromosome I from species D into species R is blocked by the linkage of these genes with the sterility factors on the same chromosome.

But the genetic results of hybridization are very different if crossing-over takes place in the structural heterozygote at crossover points that separate the transposed segments from the homologous interstitial segments. Double crossing-over that will accomplish this separation is diagrammed in Figure 29C. This mode of crossing-over may or may not happen to occur in the F_1 generation; if not, it can take place in any structural heterozygote in any subsequent backcross generation.

Double crossing-over of this sort yields crossover chromosomes of the type labeled I_{rd} in Figure 29C. These I_{rd} chromosomes are genically balanced, and hence form viable gametes. They are structurally homologous with I_r chromosomes, so that they can form fertile heterozygotes of the constitution $I_r I_{rd}$, and thus they can be incorporated into the chromosome pool of species R.

These balanced I_{rd} chromosomes also carry large blocks of D genes on their central interstitial segments. The I_{rd} chromosomes which are at once balanced internally and homologous with I_r can therefore serve as a bridge for the passage of D genes into species R. Darwinian selection can now bring about an increase in frequency of the new I_{rd} chromosomes, or parts of them, in the environment of species R.

We have discussed the process of sterility transgression with reference to a simplified model involving a transposition. It can be shown that the process will work in a similar fashion with other types of rearrangements, particularly translocations and reinversions, under appropriate conditions of crossing-over. Furthermore, and significantly, it will work on larger numbers of rearrangements on separate chromosomes or chromosome arms, so as to circumvent strong chromosomal sterility barriers as well as weak ones.

Environmental selection for a new interspecific gene combination pulls the significant genes of the donor species through a sterility barrier composed of various types and numbers of chromosomal rearrangements. Such a selective process can build up a new race possessing various morphological and physiological characters like the donor parental species, but belonging to the fertility group of the recipient species.

It remains to consider the role of the genes on the rearranged segments in the process of introgression through a chromosomal sterility barrier. The convergence referred to above is limited by these genes.

The morphology-determining genes on the rearranged segments of species D will be weeded out along with the segments themselves by gamete selection in the hybrid population. If the morphological genes of D are linked with viability genes of D on the same segments, moreover, they will be discriminated against by Darwinian selection also in the environment of species R. The two types of linkage—that between the morphological genes and the chromosomal sterility factors and the M-V gene block on the rearranged segments—prevent the introgressive race from retaining too close a resemblance to the donor species in its morphological characters. The rearranged segments, in other words, restrict the introgression of alien genes, and act as a brake on the process of convergence in morphological and physiological traits.

Conclusions

Natural hybridization occurs fairly commonly between sympatric species or semispecies of plants where the ecological conditions are permissive. A large amount of evidence which is descriptive but not quantitative indicates that natural hybridization is more common in plants than in higher animals with internal fertilization.

The occurrence of hybridization in plants is not well correlated with the strength of internal reproductive barriers between the parental species, up to the limiting point of absolute cross-incompatibility. The occurrence of hybridization is, on the other hand, closely correlated with the availability of open habitats for the establishment of the hybrids and hybrid progeny. Such open habitats are formed on a wide scale by human activities in the modern world, but they have also been created by various natural processes throughout geological history.

The reproduction of natural hybrids in plant groups with an outcrossing breeding system commonly follows the pathway of backcrossing and introgression. The infiltration of genes from the donor species to the recipient species is much more strongly restricted where these parental species are differentiated in respect to chromosomal rearrangements than where they possess completely homologous chromosomes.

In autogamous plant groups, hybridization may lead to end results which are quite similar to those usually taken as diagnostic of introgression. The alternate pathway involves inbreeding (rather than backcrossing), segregation, and selection.

Introgressive hybridization plays a key role in race formation in many predominantly outcrossing plant groups. Races of introgressive origin are found in species groups in Pinus, Abies, Quercus, Purshia, Cistus, Coprosma, Dracophyllum, Helianthus, Gilia, Tradescantia, and other genera.

CHAPTER *13*

Hybrid Speciation

Introduction · Stabilization of Hybrid Reproduction · Historical Background · Amphiploidy · Introduction to Recombinational Speciation · Hybrid Speciation with External Barriers · Introduction to Apomictic Speciation · Conclusions

---

Introduction

One of the historical problems of plant evolution, a problem which was discussed by successive authors through the 18th and 19th centuries, is the role of hybridization in species formation. Is natural hybridization a mechanism for the production of new plant species?

This question was answered in the affirmative by several early authors. But, as we shall see, these same authors failed to consider some essential parts of the problem, especially segregation in the hybrid progeny, and their affirmative answers were therefore premature and prescientific.

The early discussions did, however, serve to keep the problem alive until it could be approached in the light of genetics by modern workers.

We saw in an earlier chapter that new races arise sometimes as products of natural hybridization. This is hybrid race formation but it is not necessarily hybrid speciation. The new race, once formed by processes involving hybridization, could go on to diverge to the species level in geographical isolation from the original ancestral race. This would be a special case of geographical speciation in which hybridization produced the necessary variations at an intermediate stage. We do not need to consider this case further here (see Chapter 8).

By hybrid speciation we mean the origin of a new species directly from a natural hybrid. This definition brings the problem into focus. It is clear that the same sexual process which produced the natural hybrid will also bring about the breakup of its gene combination by segregation in later generations. Therefore an essential part of the mechanism of hybrid speciation is the stabilization of the breeding behavior of the hybrids.

Stabilization of Hybrid Reproduction

Segregation in the F_1 or any subsequent generation derived from an interspecific cross can be completely stopped or greatly restricted by various mechanisms. The known methods of stabilization in hybrid reproduction are: (1) vegetative propagation; (2) agamospermy; (3) permanent translocation heterozygosity; (4) permanent odd polyploidy; (5) amphiploidy; (6) recombinational speciation; and (7) the segregation of a new type isolated by external barriers.

Methods 1 and 2, known collectively as apomixis, are introduced briefly in a later section of this chapter and discussed more fully in Chapters 18 and 22. Methods 3 and 4 are exemplified by the genetic systems in the *Oenothera biennis* group and the *Rosa canina* group, respectively, which are discussed in Chapter 23. Method 5, amphiploidy, or allopolyploidy, is introduced in this chapter and covered in more detail in Chapters 15 and 16.

Methods 6 and 7, in contrast to method 5, both involve recombination on the homoploid level. Method 6, designated recombinational speciation, is the formation of a new homozygous recombination type for chromosomal sterility factors. It is introduced in this chapter and discussed more extensively in Chapter 14. Method 7 is the formation of a new recombination type isolated by external barriers. We call this process hybrid speciation with external barriers, and discuss it briefly later in this chapter.

The various methods listed above permit the multiplication of any given genotype of hybrid origin. The genotype in question may be that of the F_1 hybrid itself or of a later-generation derivative, and it may be heterozygous or homozygous.

Under methods 6 and 7 the hybrid derivative attains a true-breeding condition by becoming homozygous, at least temporarily and at least partially. The hybrid derivative breeds true for a highly heterozygous condition under methods 1 to 4. Method 5 is a compromise solution combining features of both homozygosity and heterozygosity. The hybrid derivative under method 5 is homozygous as far as meiosis is concerned, and is therefore true-breeding but retains the opposite gene systems of two parental species.

In any case the particular gene combination of hybrid origin, if adaptively valuable in some available environment, is able to increase in numbers and form a population of its own in the area covered by the favorable environment. The derivative population may be a new species or microspecies of hybrid origin.

Now the method of stabilization of hybrid reproduction determines the type of species resulting from the hybridization. Vegetative propagation and agamospermy are asexual methods of reproduction. These means of uniparental multiplication can produce clonal or agamospermous microspecies. Permanent structural or numerical hybridity (methods 3 and 4) also produce uniparental microspecies. Amphiploidy, recombinational speciation, and hybrid speciation with external barriers, on the other hand, are methods of stabilization which take place in a context of sexual reproduction. They give rise, therefore, to new biological species.

The new species descended from hybrids or hybrid derivatives are evolutionary species in every case, according to the concepts discussed in Chapter 3. But only where sexuality and interbreeding persist throughout the critical formative stages can we speak of the hybrid origin of new biological species. The difficulties of hybrid speciation are greatest in the latter case where sexuality remains in force. The hybrid formation of a new biological species is theoretically the more difficult but also the more interesting process.

Historical Background

In the pre-Mendelian period the hybrid origin of new plant species was proposed by Linnaeus in 1744 and 1760, William Herbert in 1820, Charles Naudin in 1863, and Anton Kerner in 1891. The line of thought was continued in the early post-Mendelian period by Lotsy, Hayata, and others. The references as cited in our bibliography are: Linnaeus (1760, 1790), Herbert (1820), Naudin (1863), Kerner (1894–1895, Vol. 2, pp. 576 ff.), Lotsy (1916), and Hayata (1921): see also DuRietz (1930, pp. 400 ff.) for a review.

Linnaeus and his student, Rudberg, first suggested the hybrid origin of new plant species in 1744 in connection with Peloria, a form of Linaria with radially symmetrical flowers. Hybrid speciation is suggested again in the *Disquisitio de Sexu Plantarum* (1760). Here Linnaeus states that: "It is impossible to doubt that there are new species produced by hybrid generation."

Kerner (1894–1895) started with the known facts that crossing generates new forms and that natural interspecific hybrids occur in many plant groups, and concluded from these premises that "hybrids could originate new species." To be sure, many hybrids are formed in nature, but only a few of them succeed in becoming new species. The important limiting factor, according to Kerner, is the environment.

Only where there is a favorable and open habitat not occupied by the parental species can the hybrid plants establish themselves and increase in numbers. But such open habitats do occur here and there in nature. And so do species of hybrid origin. Kerner cites examples in Rhododendron, Salvia, and Nuphar in Europe.

Rhododendron intermedium, for example, is a product of hybridization between *R. ferrugineum* and *R. hirsutum*. In some places in the Alps, *R. intermedium* occurs as a rare associate of the parental species, but in other places it forms large populations of its own and outnumbers the parental types. Here *R. intermedium* has attained the status of a true species (Kerner, 1894–1895, Vol. 2, pp. 588–89).

Linnaeus and Kerner were the outstanding early exponents of the idea that natural hybrids can be the starting points of new species. Analysis of their discussions reveals that they both ignored two important aspects of the problem which any satisfactory hypothesis of hybrid speciation must take into consideration, namely, sterility and segregation.

Linnaeus and Kerner did not recognize the phenomenon of segregation. Indeed, Kerner postulates that: "hybrids transmit their form unchanged to their posterity . . . provided the pollen from other species is excluded." Linnaeus had done little experimental work on plant hybridization, and Kerner's experimental work dealt with other aspects of plant reproduction. Consequently neither author was prepared to realize that there is any problem of hereditary transmission in passing from natural hybrid to daughter species in fertile species crosses.

The list of early plant hybridizers who did know about segregation and did *not* discuss species formation through hybridization is a long one, ranging from Kölreuter (1761–1766) to Mendel, Darwin, and Focke (1881).

Naudin's (1863) paper is very interesting because it contains discussions of both segregation and hybrid speciation. Naudin treated the former as a fact and the latter as a possibility. He recognized that the fact of segregation created difficulties for the theory of hybrid speciation. However, he allowed that exceptions might occur in which hybrid characters become fixed in later generations, and such exceptional cases could lead to the formation of new species (Naudin, 1863).

The difficulty of sterility barriers was also set to one side in the early discussions of hybrid speciation. It was known that some species pairs are intersterile and others are interfertile. The possibility of hybrid speciation was restricted to cases in which the hybrid is fertile (Kerner, 1894–1895, Vol. 2, p. 587). This focused the attention arbitrarily on one special aspect of the whole question. It made the general problem of hybrid speciation appear to be simpler than it actually is.

With the rise of modern genetics in the early decades of this century, the old problem of hybrid speciation could at last be stated correctly. The problem could be seen to involve sterility barriers and segregation as well as morphological and physiological traits. In order to solve this problem it was necessary to find a mechanism by which a new, internally isolated, constant type could arise from a species hybrid without loss of sexuality.

The first major breakthrough was Winge's (1917, 1924) hypothesis of amphiploidy, to use a much later term, which postulates that a new constant species can arise from a hybrid between two preexisting species following chromosome number doubling. Winge's hypothesis was soon confirmed experimentally by other workers in Nicotiana, Raphanus-Brassica, Galeopsis, and other plant groups (Clausen and Goodspeed, 1925; Karpechenko, 1927; Müntzing, 1930b, 1932).

Amphiploidy is a mode of speciation which involves unidirectional increases in number of chromosome sets, whereas many plant species have obviously evolved on the diploid level, and the question therefore remained whether hybrid speciation could take place without change in ploidy.

Winge himself returned to this question in his later work on Trago-pogon and Erophila, and Lamprecht tackled it in Phaseolus (Winge, 1938, 1940; Lamprecht, 1941). The stated objective of the experiments in Tragopogon, Erophila, and Phaseolus was to determine whether new fertile types or microspecies could be produced by hybridization without the accompaniment of amphiploid doubling. These experiments paved the way for studies by other workers on a process of hybrid speciation, alternative to amphiploidy, which is now known as recom-binational speciation.

Amphiploidy

Chromosomal sterility stems from segmental rearrangements between the parental genomes which upset the course of meiosis, visibly or invisibly, in the hybrid. Either the homeologous chromosomes do not pair normally in bivalents and do not separate properly to the poles at anaphase, or they pair but segregate to yield daughter nuclei carrying deficiencies and duplications. In either case a proportion of the meiotic products are unbalanced and do not develop into functional spores and gametes. This proportion of inviable spores and, hence, the degree of sterility rises rapidly with increase in number of heterozygous rearrange-ments (see Chapters 5 and 6).

Let us suppose that the chromosomally sterile hybrid undergoes doubling of chromosome number. Then, disregarding other possible complicating factors for the present, it will become meiotically normal and gametically fertile.

This recovery of fertility on the new polyploid level is most clear-cut in the first case mentioned above where the chromosomes do not form bivalents regularly in the diploid hybrid. The structurally well-differentiated genomes of the parental species can be symbolized as A and B respectively. The genomic constitution of the two diploid species is AA and BB; and that of their hybrid is AB. The chromosomes belonging to the A set have no homologous partners to pair with at meiosis in the hybrid, and neither do the B chromosomes. But the situation is entirely different in the allotetraploid (or amphidiploid) derivative of this hybrid with the genomic constitution $AABB$. Now there exists a homologous pair of each chromosome type in each genome. Consequently meiosis and gamete formation can proceed normally.

In the second case mentioned earlier the homeologous chromosomes in the hybrid do form bivalents but segregate to produce genically unbalanced meiotic products. The diploid parents possess different subgenomes, A_s and A_t. Their F_1 hybrid has the genomic constitution A_sA_t and has chromosomal sterility associated with visibly normal meiosis. In this case also the tetraploid derivative, $A_sA_sA_tA_t$, is likely to recover fertility, or at least semifertility, as a result of what Darlington called differential affinity (Darlington, 1932). Each A_s chromosome and each A_t chromosome has a completely homologous partner in the tetraploid. These homologous chromosomes pair preferentially in the tetraploid and pass to opposite poles to yield segmentally and genically balanced products of meiosis.

The tetraploid derivative of the chromosomally sterile hybrid, whether it has the genomic constitution $AABB$ or $A_sA_sA_tA_t$, is not only fertile itself, but is also reproductively isolated from its diploid parents. Crosses between tetraploid and diploid plants frequently run into incompatibility barriers, and the hybrids, if any arise, are triploid and hence usually sterile.

The classical cases of *Primula kewensis*, Raphanobrassica, and *Galeopsis tetrahit* illustrate the various types of genomic constitution in experimental amphiploids.

The F_1 hybrid of *P. floribunda* ($2n = 18$) $\times$ *P. verticillata* ($2n = 18$), widely known by the name *P. kewensis*, is a diploid perennial herb like its parents. It showed normal bivalent pairing but was highly sterile.

In three different years the otherwise sterile hybrid plant spontaneously gave rise to fertile branches, which, when studied cytologically, turned out to be tetraploid ($2n = 36$) and to have predominantly bivalent pairing.

Evidently, preferential pairing of completely homologous chromosomes occurred in the tetraploid branches. These fertile shoots arose by somatic doubling in a bud or sector of a bud. They produced seeds which developed into fertile and fairly uniform F_2 progeny (Newton and Pellew, 1929; Upcott, 1939).

A contrasting condition was found in Raphanobrassica, the amphiploid derivative of *Raphanus sativus* X *Brassica oleracea*. The parental species are both diploid with $2n = 18$ chromosomes. The F_1 hybrid exhibits complete failure of chromosome pairing and is highly sterile. It produces some unreduced diploid gametes. Union of these gave rise to tetraploid ($2n = 36$) plants in F_2, which had normal meiosis with regular bivalent formation and were mostly quite fertile. They yielded a morphologically uniform F_3 generation. The new, fertile, true-breeding line is isolated by sterility barriers from the parental diploid species (Karpechenko, 1927).

Galeopsis tetrahit ($2n = 32$), unlike *Primula kewensis* and Raphanobrassica, is a naturally occurring tetraploid species in northern Europe and Asia. Müntzing proved that this annual herb is an amphiploid derived from two related diploid species in Europe, *G. pubescens* ($2n = 16$) and *G. speciosa* ($2n = 16$) (Müntzing, 1930b, 1932).

The artificial F_1 hybrid of *Galeopsis pubescens* X *speciosa* is meiotically irregular and fairly sterile with 8% good pollen and 5 to 8 bivalents. The F_1 hybrids, despite their sterility, produced an F_2 generation consisting of some sterile diploids and one sterile triploid plant. The latter arose from the union of one unreduced gamete and a reduced gamete in F_1. The triploid when backcrossed to *G. pubescens* yielded a single seed which gave a tetraploid plant in F_3. The tetraploid probably arose from the fertilization of an unreduced $3n$ egg by a normal $1n$ sperm. The tetraploid F_3 plant had 16 bivalents at metaphase of meiosis and produced 70% good pollen. It yielded fertile tetraploid F_4 progeny (Müntzing, 1930b, 1932).

The artificial allotetraploid resembled natural *Galeopsis tetrahit* in morphology as well as in chromosome number. Furthermore, it is isolated by an incompatibility barrier from *G. pubescens* and *G.*

speciosa, as wild *G. tetrahit* is. It could be considered to be a synthetic form of *G. tetrahit*. To test this assumption, Müntzing crossed the artificial allotetraploid with natural *G. tetrahit*. The artificial and natural tetraploids crossed easily to produce F_1s which were fertile with good chromosome pairing. This was the first experimental resynthesis of a naturally occurring amphiploid species (Müntzing, 1930b, 1932).

Introduction to Recombinational Speciation

Recombinational speciation is the term adopted to denote one of the lesser known forms of hybrid speciation (Grant, 1966e). The process takes place within a breeding system of sexual reproduction. It leads to the formation of a daughter species which is isolated from the parental species by a chromosomal sterility barrier, as in amphiploidy. But, in contrast to amphiploidy, the daughter species remains on the same ploidy level as its parents.

This result comes about by the formation and establishment in the later hybrid generations of a new homozygous recombination type for the various independent chromosomal rearrangements separating the parental species. The new homozygous recombination type is fertile within the line but intersterile with either parent.

The process of recombinational speciation can produce daughter species which are surrounded by sterility barriers of varying strength. The degree of isolation of the daughter species is directly proportional to the degree of intersterility between the parental species.

The leading idea for the hypothesis of recombinational speciation was provided by Müntzing (1930a, 1934, 1938). He proposed but did not develop the concept that separate pairs of chromosomal sterility factors can be recombined to give a new fertile type. Gerassimova (1939) confirmed this suggestion experimentally for translocations in Crepis. Stebbins then discussed recombination of sterility factors in relation to hybrid speciation in a general way (Stebbins, 1942; 1945; 1950, pp. 249, 286–89). Later Stebbins (1957a) and Grant (1958) presented a definite genetical model of recombinational speciation. Still later I reviewed the hypothesis and summarized the evidence for it (Grant, 1963, pp. 469 ff.).

The subject of recombinational speciation is introduced here so it can be compared with other modes of hybrid speciation. It is explained and discussed in more detail in the next chapter.

It may be noted in passing that the recombination process is just as capable of generating new genic sterility barriers out of preexisting ones as it is of compounding chromosomal sterility barriers. There is some evidence suggesting a recombinational formation of new genic sterility barriers in Phaseolus and new gene-controlled incompatibility barriers in Gilia (Lamprecht, 1941; Grant, 1954a). Too little is known about this aspect of the problem, however, to warrant a separate discussion in this book. In general the process would be essentially similar to the one, to be described later, involving chromosomal sterility factors.

Hybrid Speciation with External Barriers

Hybridization between two interfertile plant species may lead to the formation of diverse recombination types for morphological and physiological character differences between the parental forms. Some of the new character combinations may bring about external isolation of one sort or another between the hybrid derivatives and the parental species. Such externally isolated recombinants, if well adapted to some available environment, can then increase in numbers within the territory of the parent species.

Earlier in this chapter we mentioned a historical example of possible hybrid speciation in Rhododendron studied by Kerner (1894–1895). Kerner postulated that *R. intermedium* arose from a hybrid between *R. ferrugineum* and *R. hirsutum* and became successful in certain places in the Alps. The daughter species of hybrid origin differs from its parents in flower color and, to some extent, in soil preferences. Kerner's account of the case suggests that *R. intermedium* is partially isolated from *R. ferrugineum* and *R. hirsutum* by the pollinating behavior of bees and perhaps by edaphic factors. For a more recent description of the hybrid *R. intermedium* and its parental species see Hegi (1909–1931, Vol. 5, pp. 1627–44).

Delphinium recurvatum, D. hesperium, and *D. gypsophilum* form a group of interfertile diploid species in the foothills and valleys of California. On morphological and ecological grounds, *D. gypsophilum* is believed to have originated as a segregate in a hybrid swarm of *D. recurvatum* X *hesperium*. The former is intermediate between the putative parents in its ecological preferences and in its morphological characters. Some individuals in an artificial F_2 of *D. recurvatum* X *hesperium* resemble *D. gypsophilum* in morphology. In nature, *D. gypsophilum* is kept apart from the other two species mainly by ecological isolation (Lewis and Epling, 1959).

Three species of Penstemon in California, *P. centranthifolius*, *P. grinnellii*, and *P. spectabilis*, are interrelated, diploid ($2n = 16$), and somewhat interfertile. They differ in their floral mechanisms, mode of pollination, and vegetative ecology. *Penstemon centranthifolius* has red trumpet-shaped hummingbird flowers; *P. grinnellii* has flesh-colored, broad-throated, carpenter bee flowers; and *P. spectabilis* has flowers of an intermediate color and shape which are fitted for pseudomasarid wasps and anthophorid bees (Straw, 1955, 1956).

Morphological comparisons between the three species, ecological field studies, and the appearance of natural hybrids combine to suggest that *Penstemon spectabilis* arose as a product of hybridization between *P. centranthifolius* and *P. grinnellii*. The derivative species is kept distinct from the putative parental species in nature by mechanical, ethological, and ecological isolating mechanisms (Straw, 1955, 1956).

Stebbins and Ferlan (1956) postulated a hybrid origin of *Ophrys murbeckii* from *O. fusca* X *lutea* in the Mediterranean region. Ethological and ecological isolation are apparently the chief factors keeping the daughter and the parental species distinct in this case (Stebbins and Ferlan, 1956).

Tucker and Sauer (1958) have described a series of diploid populations of Amaranthus on the delta of the Sacramento River in California where diploid hybrid speciation may be in statu nascendi. Several well-established species are found in the region—*A. powellii*, *A. caudatus*, *A. retroflexus*, and along with them are aberrant local populations exhibiting different combinations of the characters of the aforementioned species. The aberrant populations, or some of them, may represent different diploid hybrid derivatives of the preexisting Amaranthus species (Tucker and Sauer, 1958).

In Alaska, *Carex rostrata* typically occurs in sweet-water lakes and *C. rotundata* in peat bogs. Hybrid swarms between these species are found in the transition zone between their respective habitats. In some areas a new, morphologically distinct, fertile population is segregating out of the hybrid swarms and is increasing in numbers relative to the parental species. This new form is regarded as a daughter species of hybrid origin and is named *C. paludivagans* (Drury, 1956).

We have mentioned a few possible examples of diploid hybrid speciation. The evidence is incomplete in one way or another in each case, but is very suggestive. Much more experimental and analytical work is needed on the process of formation of new external isolating mechanisms by homoploid hybrid speciation.

Introduction to Apomictic Speciation

An interspecific hybrid in the F_1 or some subsequent early generation may prove to be adaptively superior. Its highly heterozygous genotype may possess properties of heterosis or homeostasis or both. Such a hybrid is potentially worth perpetuating and multiplying. But the reproduction of the adaptively superior species hybrid is beset with two obvious difficulties. The first is hybrid sterility; the second is inability to breed true to type. However, there are also some ways around these difficulties, as Darlington pointed out long ago (1932, Ch. 16).

A heterozygote derived from a species cross can surmount its inherent reproductive difficulties of sterility and segregation, and can multiply its particular heterozygous genotype, by several of the methods of stabilization listed at the beginning of this chapter (methods 1 to 5). In this section we mention the first two methods, known collectively as apomixis, in relation to hybrid speciation. Apomictic speciation is introduced here and is discussed more fully later in Chapters 18, 21, and 22.

Opuntia fulgida and *O. spinosior* are two distinct species of cholla cactus in the southwestern American desert. They hybridize occasionally to produce fairly seed-sterile hybrids. Hybrid plants, despite their seed sterility, have become abundant in a local area along the Gila River in Arizona, owing to their capacity for vegetative propagation by the fallen stem joints (Benson, 1950, p. 30; Kearney and Peebles, 1964, p. 585).

The European blackberries belonging to the genus Rubus section Moriferus are a large complex of hybridizing sexual species and their apomictic derivatives. The hybrids possess both agamospermous and vegetative means of reproducing. Each hybrid type can multiply and spread by asexually produced seeds and by rooting of the stem tips, and many microspecies have arisen in this way (Gustafsson, 1943); see Chapters 18 and 21.

A hybrid with distinctive morphological characters and ecological preferences can thus spread apomictically and form a population of its own. The population composed of similar hybrid individuals of apomictic derivation is a clonal or agamospermous microspecies when it is poorly distinguished morphologically or narrowly endemic or both. If the hybrid derivatives possess a character combination which is recognizable in ordinary taxonomic practice, and if this character combination has been able to spread by apomictic means throughout a definite geographical area, the resulting population has attained the status of a taxonomic species.

Conclusions

It may be useful at this point to present a classification of the various modes of speciation which have been considered in this and earlier chapters.

1. Primary speciation; evolutionary divergence between breeding populations up to the level of reproductively isolated species, but not including the development of special reinforcement isolation.

A. Geographical speciation; the attainment of this level of divergence by a pathway running through an intermediate stage of geographical races (Chapter 8).

B. Quantum speciation; a more rapid process of primary speciation in which a daughter species arises directly from a local population of the ancestral species (Chapter 8).

C. Quantum speciation involving chromosome repatterning; a special case of B in which the daughter species becomes differentiated in chromosome segmental arrangement; also sometimes called speciation by saltation (H. Lewis, 1966) (Chapters 8 and 9).

D. Sympatric speciation; used here in a restricted but meaningful sense to refer to the disputed process of primary speciation within the spatial limits of a breeding population in a cross-fertilizing organism

(Chapter 8). Speciation is literally sympatric in the various forms of hybrid speciation, and it can be sympatric in uniparental organisms, but these cases present no great theoretical difficulties.

2. Hybrid speciation with sexual reproduction; the origin of a new biological species directly from an interspecific hybrid.

A. Hybrid speciation with external barriers; the formation of a new recombination type in the progeny of a hybrid which is separated from the parental species by external isolating mechanisms (Chapter 13).

B. Recombinational speciation; a chromosomally sterile or semi-sterile hybrid gives rise to homoploid or undoubled derivatives which possess a new homozygous recombination of the various independent chromosomal rearrangements separating the parental species; the new fertile line has its own character combination and is chromosomally intersterile with the parental species (Chapter 14).

C. Amphiploidy; a chromosomally sterile hybrid gives rise by doubling of the number of chromosome sets to a fertile polyploid derivative; the amphiploid derivative is separated from the parental species by a chromosomal sterility barrier (Chapters 13, 15).

3. Hybrid speciation with asexual or subsexual reproduction; the multiplication of a hybrid by mechanisms circumventing the normal sexual and chromosomal cycle, and the spread of this hybrid type throughout a definite geographical area, so that it attains the population size and composition of a uniparental microspecies or species.

A. Apomictic speciation; a new clonal or agamospermous micro-species develops from an interspecific hybrid by means of vegetative propagation or agamospermous seed formation, respectively (Chapters 18, 21, and 22).

B. Permanent structural or numerical hybridity; a translocation heterozygote multiplies by the *Oenothera biennis* genetic system, or an odd polyploid increases in numbers by the *Rosa canina* genetic system, to the population size of a heterogamic microspecies (Chapter 23).

4. The Wallace effect; the formation of supplementary reproductive isolation between sexual populations which have reached the level of elementary biological species by either primary speciation or hybrid speciation; this reinforcement of a preexisting degree of reproductive isolation takes place under the stimulus of sympatric contacts and by the action of selection for isolation per se (Chapter 10).

CHAPTER 14

Recombinational Speciation

Introduction

Recombinational speciation is the formation and establishment, in the progeny of a chromosomally sterile or semisterile hybrid, of a new homozygous recombination type for the two or more independent rearrangements differentiating the parental species. The hybrid derivative is fertile itself, and on the same ploidy level as the parental species, but is isolated from the latter by a chromosomal sterility barrier.

The process of recombinational speciation can produce daughter species which are surrounded by weak or by strong sterility barriers.

The hybrid origin of a new homoploid line which is fertile itself but highly intersterile with its congeners is clearly the more difficult case to explain. But, if we can find a satisfactory explanation of this case, we will have all the elements needed for understanding the theoretically easier cases involving weak sterility barriers.

In this chapter we describe the hypothetical mechanism of recombinational speciation, review the relevant experimental evidence, and inquire into the possible role of the process in nature.

The Hypothesis

A definite genetical model applicable to the theoretically interesting case of recombinational speciation with strong sterility barriers was presented by Stebbins (1957a) and Grant (1958). This model employs a chromosomal sterility barrier between two hybridizing parental species which is composed of separable segmental rearrangements. The model was based on earlier work by Müntzing and others as noted in our introduction to the subject in Chapter 13.

Suppose that two parental species are isolated by a chromosomal sterility barrier composed of two or more separable segmental rearrangements. Then their interspecific hybrid, which is chromosomally sterile but not completely so, can give rise to one or more new homozygous recombination types for the segmental rearrangements. These new types are fertile themselves but chromosomally intersterile with one another and with the parents (Stebbins, 1950, pp. 287–88; 1957a; Grant, 1958; 1963, pp. 469 ff.).

This process of recombination can be illustrated by a simple example. Let the two parental types differ with respect to two translocations on four chromosomes. Their chromosomal constitutions are $AA\ BB\ CC\ DD$ and $A_bA_b\ B_aB_a\ C_dC_d\ D_cD_c$ respectively (see Figure 30). Their F_1 hybrid, with the constitution $AA_b\ BB_a\ CC_d\ DD_c$, is heterozygous for two independent translocations and therefore has an expected gametic fertility of 25% (see Figure 30 again).

The functioning gametes include the two parental types, which do not interest us further here, and also two recombination types (ABC_dD_c and A_bB_aCD) which do interest us. For the latter can give rise to the new homozygous types in F_2, $AA\ BB\ C_dC_d\ D_cD_c$ and $A_bA_b\ B_aB_a\ CC\ DD$

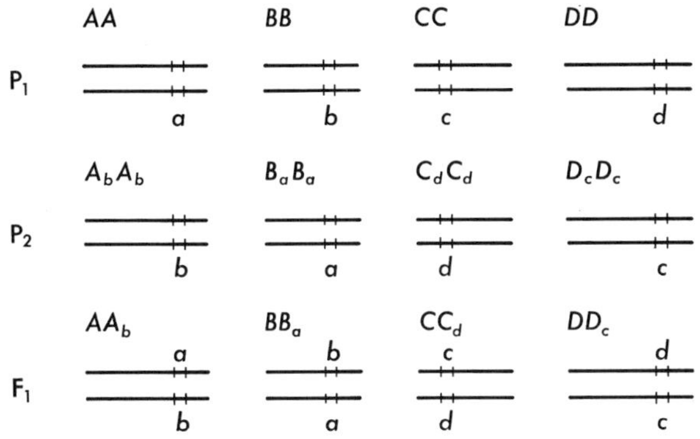

P_1

$AA \quad BB \quad CC \quad DD$

$a \qquad b \qquad c \qquad d$

P_2

$A_bA_b \quad B_aB_a \quad C_dC_d \quad D_cD_c$

$b \qquad a \qquad d \qquad c$

F_1

$AA_b \quad BB_a \quad CC_d \quad DD_c$

$a \qquad b \qquad c \qquad d$

$b \qquad a \qquad d \qquad c$

F_1 gametes

unbalanced

$ABCD_c$	A_bBCD
ABC_dD	A_bBCD_c
AB_aCD	A_bBC_dD
AB_aCD_c	$A_bBC_dD_c$
AB_aC_dD	$A_bB_aCD_c$
$AB_aC_dD_c$	$A_bB_aC_dD$

balanced

$ABCD$
ABC_dD_c
A_bB_aCD
$A_bB_aC_dD_c$

F_2 homozygotes

$AA \quad BB \quad CC \quad DD$

$AA \quad BB \quad C_dC_d \quad D_cD_c$

$A_bA_b \quad B_aB_a \quad CC \quad DD$

$A_bA_b \quad B_aB_a \quad C_dC_d \quad D_cD_c$

Figure 30 Breeding behavior of a hybrid heterozygous for two independent translocations. The first translocation involves the segments a and b in chromosomes A and B, and the second translocation involves segments c and d in chromosomes C and D. P_1 has the standard arrangement in all four chromosomes, while P_2 carries both translocations. The F_1 hybrid produces 16 classes of gametophytes as shown. In 12 of these classes there are deficiencies and duplications for the a-b segments, the c-d segments, or both pairs of segments. Four other classes are balanced and viable. The four classes of functioning gametes produce various types of translocation heterozygotes in F_2, which are not shown, along with the four types of structural homozygotes which are shown. Further explanation in text.

(Figure 30). These new homozygous recombination types are fully fertile within the line but are partially intersterile with one another and with the parents (Stebbins, 1957a; Grant, 1958).

A larger number of independent translocations, of course, produces a chromosomal sterility barrier of greater strength. Thus a heterozygote for six independent translocations would have an expected gametic fertility of 1.56%. Species hybrids in plants which are classified as highly sterile have outputs of good pollen in this range. In general, the proportion of good gametes produced by the translocation heterozygote is $1/2^n$, where n is the number of independent translocations.

As the chromosomal sterility barrier increases in strength, moreover, the number of new fertile homozygous types that can be produced by recombination in F_2 or later generations also increases. And these new structurally homozygous recombination types, or most of them, will be isolated from the parental species by strong sterility barriers.

Consider again the example of a hybrid heterozygous for six independent translocations. This hybrid can produce 62 new fertile homozygous recombination types in later generations. These recombination types, when backcrossed to the parental species, would then form hybrids with gametic fertilities ranging from 3 to 50%. And the most common class of recombination types would yield backcross hybrids with either parental species which have a gametic fertility of 12.5%.

Other kinds of segmental rearrangements besides translocations, particularly transpositions and reinversions, will also permit the multiplication of chromosomal sterility barriers by recombination. A transposition is a segmental interchange within a single chromosome as shown in Figure 15 (in Chapter 5). A reinversion is the result of two successive inversions in the same region, one of which is included within the other. If the gene order on the standard chromosome is ABCDEF, that of the reinversion chromosome is AECDBF. In the structural heterozygote, crossing-over between the transpositions or within the reinversion gives deficiency-duplication products. Multiple independent transpositions or reinversions in heterozygous condition can therefore bring about chromosomal sterility of a high degree, providing that crossing-over takes place regularly between the transpositions or within the reinversions. The functioning gametes are the noncrossover types for the transpositions or reinversions.

Among the functional gametes will be some in which the separate transpositions or reinversions borne on different chromosomes are recombined. These gametes can produce new fertile types of structural homozygotes in F_2 or later generations which will be isolated by sterility barriers from their parents. Transpositions and reinversions can be recombined like translocations to yield new fertile isolated lines.

An important factor in the process of recombinational speciation is the breeding system of the plants involved. The system of breeding in a hybrid population, whether by outcrossing or selfing, exercises a decisive influence on the production of new, fertile, internally isolated species by that hybrid population. Recombinational speciation is promoted by inbreeding (Grant, 1958).

The immediate end result of recombinational speciation is the formation of a new type of structural homozygote. Structural homozygosity is attained most rapidly in the progeny of a structural heterozygote under conditions of inbreeding. And the new homozygous type, once formed, can be fixed and maintained most readily by the same system of inbreeding. Outcrossing in a hybrid population, by contrast, greatly decreases the chances of either producing a new structurally homozygous type or of subsequently maintaining such a type in a true-breeding line of its own (Grant, 1958; 1963, pp. 478–79).

It is instructive to compare the ease of fixation of a new structural homozygote in a simple hypothetical case under contrasting breeding systems. Let the parental types differ by two separable rearrangements, as in Figure 30. Hybridization between these types gives rise to a mixed colony consisting of, for example, two F_1 plants and four P_1 plants. Reproduction of the hybrids produces an array of F_2 types which we will assume to be equally viable and equally fecund. What is the frequency of one particular new structural homozygote (i.e., *AA BB* C_dC_d D_cD_c in Figure 30) in the F_3 progeny of the hybrids as derived by self-fertilization and by random mating, respectively?

With self-fertilization this homozygous type comprises 11% of the plants in the F_3 generation. But the results are quite different if the plants have an outcrossing breeding system and if the F_1 hybrids mate at random with one another and with the P_1 individuals in the hybrid colony. Under these conditions the F_3 progeny of the hybrids will contain the same particular new homozygote in frequencies ranging from 0 to 5%, depending on which F_2 types reproduce, with a mode in the lower part of this range.

With outcrossing in a typical hybrid colony composed of intermixed F_1 plants and parental species plants, it is possible for a particular, new, structural homozygote to arise. But this genotype has a low probability of appearing and, once formed, moreover, it does not breed true. In a hybrid colony with the same composition, self-fertilization is many times more effective in producing and establishing the new species.

It remains to consider briefly the fate of the phenotypic character differences between the parental species during the process of recombinational speciation.

When the chromosome segments, both the homologous and the rearranged ones, become recombined, so do the genes borne on these segments. And, when the new recombination of chromosomal rearrangements is fixed in homozygous condition, so again are the constituent genes. As a result, the daughter species can inherit a new combination of the morphological and physiological traits of the parental species, and this new character combination will be true-breeding.

Experimental Approach

The hypothesis of recombinational speciation has been in process of formulation over a period of 30 years. During this time, hybridization experiments bearing on recombinational speciation have been carried out in a number of plant groups. The earliest experiments were those of Gerassimova (1939), Winge (1940), and Lamprecht (1941), and the most recent is that of Grant (1966e).

Consequently the early experiments were not designed to test the features in the hypothesis of recombinational speciation which were worked out in later years. In fact, some of the factual evidence obtained in the early period stems from studies which were not really experiments as far as the hypothesis of recombinational speciation is concerned. The relation between theory and experiment has not always followed the textbook method of scientific advance in this field. Nevertheless a fairly good body of experimental evidence has accumulated over the years, if we construe the term experiment broadly and if we follow the original data as well as the stated conclusions of the various workers.

The experimental evidence for recombinational speciation has come out of studies in Crepis (Gerassimova, 1939), Erophila (Winge, 1940),

Phaseolus (Lamprecht, 1941), Nicotiana (H. H. Smith, 1954; Smith and Daly, 1959), Elymus (Stebbins, 1957b), and Gilia (Grant, 1966e). All of these studies confirm at least one facet of the hypothesis of recombinational speciation. The more complete experiments, which test more aspects of the hypothesis simultaneously, are those in Nicotiana, Elymus, and Gilia. Most of the studies have dealt with weak sterility barriers. Only the experiments in Elymus and Gilia have attempted to test the hypothesis in terms of very strong chromosomal sterility barriers.

The Early Experiments

Gerassimova (1939) worked with four lines of *Crepis tectorum* ($2n = 8$) which were differentiated by homozygous translocations on one or both chromosome pairs. The normal line of *C. tectorum* has the standard arrangement on the four chromosomes: *A*, *B*, *C*, *D*. Mutant line I has a homozygous translocation involving the *A* and *D* chromosomes. Mutant line II carries a translocation in the *B* and *C* chromosomes. These two lines were fertile themselves but partially intersterile, as expected. The semisterile F_1 hybrid between lines I and II, which was heterozygous for the two independent translocations, yielded an F_2 generation containing an array of translocation heterozygotes and homozygotes.

One F_2 plant, designated line *nova*, contained both translocations in homozygous condition, and was fertile. Line *nova* differs from lines I and II by a single translocation and can be predicted to be partially intersterile with its immediate parents. Line *nova* differs from the normal form of *Crepis tectorum* by two independent translocations and is therefore expected to be even more intersterile with it. Gerassimova confirmed this expectation by producing the F_1 hybrid of *tectorum* X *nova* and finding that it was semisterile with irregular meiosis and 30% seed fertility (Gerassimova, 1939).

This was the first experimental production of a new isolated fertile type by recombination of separable chromosomal sterility factors. The sterility barrier was weak in this case, and the lines did not differ in phenotype.

The *Erophila verna* group ($x = 7$) is a complex of autogamous annual sibling species with various polyploid and aneuploid chromosome numbers (Winge, 1940), as we saw in Chapter 7. Winge crossed two forms of *E. verna* with the following names and chromosome numbers: *E. confertifolia* ($n = 15$) X *violacea-petiolata* ($n = 32$). The F_1 hybrids had much reduction in chromosome pairing and were fairly sterile as to pollen and seeds, the seed fertility being about 3% of normal (Winge, 1940).

The F_2 generation consisted of 451 plants which segregated for morphology, fertility, and chromosome number. Four F_2 plants were selected as the founders of as many F_3 families and of their inbred progeny to the F_9 generation. The lines continued to show segregation, becoming uniform variously from the F_5 to the F_9 generation. Eight lines which were analyzed cytologically became stable in F_7 to F_9 for new and different chromosome numbers. The cross of ($n = 15$) X ($n = 32$) produced derivative lines with $n = 22, 23, 25, 29, 29, 31$, and 34. These lines were fertile, morphologically uniform, and represented new aneuploid types (Winge, 1940).

Winge's experiment showed that hybridization between parents on different ploidy levels can yield new fertile constant aneuploid types. It is assumed, but not confirmed by the appropriate backcrosses, that the aneuploid derivatives are intersterile to some extent with the parental species. This experiment left open the question whether hybridization can also give rise to new microspecies on the homoploid level.

Another extensive hybridization experiment of the early period is that of Lamprecht (1941, 1944) with Phaseolus. *Phaseolus vulgaris* ($2n = 22$) and *P. multiflorus* ($= P. coccineus$) ($2n = 22$) are well differentiated morphologically in several plant parts. The F_1 hybrid has normal chromosome pairing but is fairly sterile. The pollen fertility in F_1 is about 10 to 30% and the seed set is much reduced. In the F_2 generation there is segregation for morphology, viability, and fertility (Lamprecht, 1941).

Lamprecht selected for and obtained some viable, fertile, and true-breeding lines in later generations. He was interested in producing fertile recombinations of the morphological characters of the parental species. This proved to be impossible for certain characters. He could not get any fertile lines in which a *multiflorus* stigma or cotyledon was recombined with a *vulgaris*-like condition for other characters of the flowers, inflorescence, or seed pods (Lamprecht, 1944). Some fertile recombination types were obtained, however, for other characters of the two parents, and they became constant by F_5 or later (Lamprecht, 1941, 1944).

Several lines of fertile recombinants were now intercrossed to determine their fertility relationships with one another and with the parental species. Some of the backcross hybrids were fertile, but others were not. Thus line MG-3 when backcrossed to *Phaseolus multiflorus* yielded an F_1 with 19% good pollen and reduced seed set. Other fertile recombination types, when crossed to the opposite parent, *P. vulgaris*, gave F_1 hybrids with 13 to 27% good pollen and considerable to high seed sterility. Interline crosses showed that some of the lines were isolated from others by sterility barriers of similar strength (Lamprecht, 1941).

Lamprecht succeeded in showing that interspecific hybridization and recombination are capable of increasing the number of intersterile forms. The results in Phaseolus are in agreement with the hypothesis of recombinational speciation as regards morphology and sterility. But cytological aspects of the problem are largely left untouched in this case. The sterility barriers are probably not chromosomal in nature here, and we do not know whether or not the derived fertile lines are homoploid.

The Later Experiments

The next experiments to be considered, in Nicotiana, Elymus, and Gilia, are in a different class from the early experiments in that they were performed when the hypothesis of recombinational speciation was in a more advanced stage of formulation. This made it possible to design these more recent experiments so as to put the hypothesis or important parts of it to a definite test.

Smith and his co-workers carried out a long-term selection experiment with homoploid inbred lines derived from an interspecific cross in Nicotiana (H. H. Smith, 1954; Smith and Daly, 1959). The parental species, *Nicotiana langsdorffii* and *N. sanderae*, are both diploid with $2n = 18$. They differ in flower size, *N. langsdorffii* having short-tubed and *N. sanderae* long-tubed corollas, and also in vegetative and other characters.

The F_1 hybrid was semifertile as to pollen, with an average of 55% good grains, and was fairly fertile as to seeds. Meiosis was somewhat disturbed. Most of the chromosomes paired in bivalents, but some of them formed a chain at metaphase or remained unpaired and lagged at anaphase (Smith and Daly, 1959). The observed reduction in pollen fertility is probably due primarily to these irregularities in chromosome pairing.

Three selection lines were started in F_2 and continued by self-pollination of single parental plants to F_{10}. The artificial selection was for corolla tube length. One line was selected for long corollas, like the *Nicotiana sanderae* parent; another for short corollas, like *N. langsdorffii;* and a third for flowers of intermediate size. The three lines responded to the selection for flower size, and became different also in several unselected floral and vegetative characters. They became differentiated morphologically from one another and from the parental lines (Smith and Daly, 1959).

The selection lines remained diploid like the parental species. The diploid derivatives with short corollas and associated characters are the most significant for us here, since they rose to nearly normal pollen fertility in the F_6 to F_9 generations, whereas the other selection lines remained or became semisterile in later generations. (Smith and Daly, 1959).

The small-flowered derivatives produced semisterile hybrids with both parental species. The pollen fertility was 46 to 58 % in backcross hybrids with *Nicotiana langsdorffii*, and 60 to 61 % in backcross hybrids to *N. sanderae*. Some meiotic aberrations occurred in both backcross hybrids. The small-flowered line is thus as intersterile with the two parental species as the latter are with one another (H. H. Smith, 1954; Smith and Daly, 1959).

The Nicotiana experiment provides a well-rounded confirmation of the hypothesis of recombinational speciation. The elements of chromosomal sterility, homoploidy, breeding system, and morphological differentiation are all present. This experiment confirms the hypothesis, however, in terms of a sterility barrier too weak to bring about much isolation of the derived line.

The first experiment on recombinational speciation utilizing a very strong chromosomal sterility barrier started with a grass hybrid in the *Elymus glaucus* complex. The parental species of this hybrid were *Elymus glaucus* and *Sitanion jubatum*, both with $2n = 28$ chromosomes. The experiment was performed by Stebbins and described in a brief report (Stebbins, 1957b).

The interspecific hybrid has a high degree of chromosome pairing, with 11 to 14 bivalents, but is highly sterile with less than 1% good pollen and about 1 seed per 10,000 florets. The cytogenetic behavior of the artificial allopolyploids derived from this hybrid indicates that the sterility is partly chromosomal, due to small rearrangements which do not interfere very much with bivalent formation, and partly genic (Stebbins and Vaarama, 1954; Stebbins, 1957b).

F_1 hybrids with the general parentage specified above were produced several times from crosses between different microspecies of the *Elymus glaucus* group and different strains of *Sitanion jubatum* or its relative, *S. hystrix*. The F_1 plants in the various hybrid combinations were then backcrossed to *E. glaucus* to obtain a very low proportion of B_1 seeds and individuals. Pooling the results of the several crosses, a total of 173,000 florets yielded 15 B_1 seeds and 11 undoubled backcross individuals. Most of these B_1 plants apparently represented reversions to *E. glaucus* in morphology and fertility balance, and consequently are of no further interest to us here (Stebbins, 1957b).

One backcross plant, however, possessed morphological characters of both the F_1 hybrid and the *Elymus glaucus* pollen parent, and was semifertile. It was self-pollinated to produce vigorous and fully fertile progeny in the I_1 and I_2 generations. The inbred families exhibited segregation for morphological characters, with some variations in the direction of *Sitanion jubatum*. (Stebbins, 1957b).

A plant in the I_2 generation was now crossed with the original parental strain of *Elymus glaucus* in order to determine its fertility relationships. The F_1 hybrids from this test cross were highly sterile with 0 to 3% good pollen (Stebbins, 1957b).

The Elymus study advances the experimental demonstration of recombinational speciation one step further. It confirms the hypothesis in the important case of a very strong sterility barrier which is at least partly chromosomal in nature. The isolated homoploid derivative was obtained by backcrossing in the F_1 generation, however, rather than by straight inbreeding. This design of the experiment reduces its applicability to those real situations in nature in which we would expect recombinational speciation to take place most readily.

The final experiment to be reviewed here was carried out with a pair of species and their hybrid derivatives in the *Gilia inconspicua* group (Grant, 1966e). The parental species, *G. malior* and *G. modocensis*, are autogamous annuals with $2n = 36$ chromosomes. Their F_1 hybrid is highly sterile and has about 2% good pollen and much less than 1% seed fertility. Chromosome pairing in the hybrid is reduced to 1 to 10 bivalents. There is independent evidence to indicate that the reduction in pairing is due largely to structural rearrangements between the parental species, and that the hybrid sterility is therefore mainly chromosomal (Grant, 1966c, 1966e).

The hybrid, despite its sterility, produced some F_2 and later-generation progeny on the homoploid level. Well over half of the plants in the F_2 to F_6 generations were subvital or sterile or both. A series of inbred lines were started in F_2 and selected in subsequent generations for vigor and for fertility associated with regular meiosis. This selection was successful in several lines which eventually became fully vigorous, fertile, and meiotically normal in the later generations without change in ploidy (Grant, 1966c, 1966e).

The fertile homoploid derivatives of the sterile species hybrid were next crossed with the parental species and with one another to determine their genetic relationships. One set of lines known as Branch II turned out to be a reconstituted form of *Gilia modocensis* in both its phenotypic characteristics and its fertility relationships. Branch II does not provide confirmation of the hypothesis of recombinational speciation.

Another fertile homoploid derivative, designated Branch III, inherited a new combination of the parental morphological characteristics. Branch III was like *Gilia modocensis* in four characters, like *G. malior* in three, and intermediate like the F_1 hybrid in four other traits (Grant, 1966e). Its branching pattern and leaf form are shown alongside those of the parental species in Figure 31.

Branch III proved to be chromosomally intersterile with *Gilia malior*, *G. modocensis*, and Branch II. The F_1 hybrid of *G. malior* X Branch III, for example, had reduced chromosome pairing with an average of 13 bivalents, and produced 4 to 18% good pollen. Its seed fertility was less than 1%. Branch III is thus isolated by a strong sterility barrier from its parents and siblings (Grant, 1966e).

The behavior of Branch III agrees with expectation on the hypothesis of recombinational speciation in all essential respects. It originated from a chromosomally highly sterile species hybrid. It was derived from this hybrid by straight inbreeding. Step by step through successive generations of selection, it achieved full fertility associated with regular

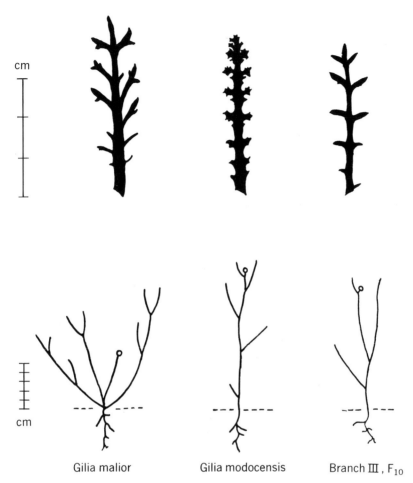

cm

cm

| Gilia malior | Gilia modocensis | Branch III, F_{10} |

Figure 31 Branching pattern and leaf form in two intersterile species of Gilia and a fertile isolated homoploid derivative of their hybrid. The particular hybrid derivative shown here is Branch III. See also Figure 28 for the same parental species and their Branch II derivative. (Grant, 1966e.)

meiosis without change in ploidy. It is intersterile with the parental species. It possesses its own particular combination of the parental morphological characters as well as of the parental chromosomal sterility factors.

The experimental evidence reviewed in this section is sufficient to confirm the hypothesis of recombinational speciation as such. This is not to say that the task of experimentation is completed in this field, however. We shall see that much more work is needed to enable us to assess the role of recombinational speciation in nature.

Block Inheritance in Relation to Recombinational Speciation

There is some experimental evidence in Gossypium, Clarkia, and Gilia to suggest that the rearranged segments in related species carry different alleles of genes affecting growth and vigor. It is possible to locate morphology-determining genes on the rearranged segments in species pairs belonging to these and other genera. The evidence for block inheritance of M-V gene systems has been presented elsewhere (Grant, 1966a) and was reviewed briefly in Chapter 5.

Our task here is to consider the effect on recombinational speciation of the M-V genes borne on the rearranged segments differentiating the parental species. Suppose that natural hybridization takes place between two species which differ in M-V gene blocks coextensive with segmental rearrangements. Suppose further that the hybrid progeny live and grow in the habitat of one of the parental species (P_1) and not in the ecologically different habitat of the other species (P_2).

In this situation environmental selection will obviously favor the hybrid segregates carrying the viability alleles of P_1. But, furthermore, we can expect selection to act in this hybrid population to favor the whole gene blocks, which is to say the rearranged segments, of P_1 over those of P_2. The homoploid hybrid derivatives in this situation are thus expected to revert to the genotype of one parental species (P_1) in their chromosomal fertility relationships, their critical ecological preferences, and the associated morphological characters.

This result was observed under controlled experimental conditions in the progeny of the chromosomally sterile hybrid of *Gilia malior* X *modocensis*. We are referring now to the homoploid derived lines designated as Branch II. These lines were selected for vigor in a *modocensis*-like environment, but not for their morphological resemblance or fertility relationships to *G. modocensis*. Nevertheless the selection products in Branch II turned out to be chromosomally homologous and interfertile with *G. modocensis*, and morphologically very similar to that same parental species (Grant, 1966b). The case of Branch II was discussed in Chapter 12 and illustrated in Figure 28.

The conclusion to be drawn is that the viability-affecting genes on rearranged segments are a potential brake or deterrent to the process of recombinational speciation. The potential brake is likely to become an actual one when the hybrid population grows in the habitat of one parental species.

The most suitable site for recombinational speciation is perhaps some new or altered or "recombination-type" habitat in the general vicinity of the hybrid population. Such an environment may be necessary in order to harbor and select for a new combination of rearranged segments, if these segments carry gene blocks which have adaptive properties. Recombinational speciation would thus be expected to occur mainly in habitats which have ecological characteristics unlike those of the parental species.

On the Role of Recombinational Speciation in Nature

The ultimate reason for conducting experiments in recombinational speciation is to determine the feasibility and the importance of this process in plant evolution. Let us now attempt to assess the role of recombinational speciation in nature in the light of the available experimental evidence.

In theory, recombinational speciation can be compared with the much better understood phenomenon of amphiploidy. The two processes represent alternative pathways from a chromosomally sterile hybrid to a new fertile, isolated daughter species. Now the amphiploid pathway is known to be very common in plant evolution on the basis of much experimental and cytotaxonomic evidence. It would be logical to suppose that recombinational speciation is likewise a common process in nature, perhaps about as common as amphiploidy, but is a process which, unlike amphiploidy, has left no cytotaxonomically detectable traces.

The Gilia experiment, which is the only one in which the derivatives descended by straight inbreeding from a highly sterile hybrid, can teach us a lesson in this regard. There was much hybrid breakdown in the early-generation progeny of *G. malior* X *modocensis*. The lines could scarcely be kept alive under favorable artificial conditions, and their survival under natural conditions would be even more unlikely (Grant, 1966c). Some of the lines which did come through the hybrid breakdown barrier, moreover, represented reversions to one parental species (Grant, 1966b). The only new daughter species to arise, conversely, had a long history of constitutional weakness under the environmental conditions of the experiment (Grant, 1966e). The Gilia experiment shows that it is extremely difficult to get past the barrier of hybrid breakdown in the case of hybridization between highly intersterile and well-differentiated species.

The force of the hybrid breakdown barrier was left out of consideration in the previous formulations of the hypothesis of recombinational speciation. Theoretical and experimental work on this process was conducted with one aspect of the problem, hybrid sterility, chiefly in mind. Amphiploidy is a way out of the impasse of chromosomal sterility in a species hybrid, and recombinational speciation could be considered to be an alternative way out of the same impasse. But what was not sufficiently recognized is that amphiploidy is also a means of circumventing the barrier of hybrid breakdown, whereas recombinational speciation is not.

The best interpretation which we can draw from the available experimental evidence is that recombinational speciation is a far less common mode of speciation in plants than is amphiploidy.

The experimental work does show, however, that recombinational speciation is a workable process under artificial conditions. This in turn suggests that it is a process which may occur occasionally in nature.

An example of a natural species which has probably originated by recombinational speciation is *Gilia achilleaefolia* ($2n = 18$). This diploid species occurs in the South Coast Range of California. It is believed to be of hybrid origin between some ancient members of the diploid *G. capitata* and *G. angelensis* groups (Grant, 1954a).

The evidence for this phylogenetic hypothesis is quite suggestive. In the first place, *Gilia achilleaefolia* is intermediate morphologically between the putative parental species in every plant part. It is also extraordinarily variable in its morphological characters. This variability is expressed in the form of local racial differentiation. Some races of *G. achilleaefolia* approach *G. capitata* in morphology, while other races approach *G. angelensis*. Indeed, the former races have been confused taxonomically with *G. capitata* in the past and the latter with *G. angelensis* (Grant, 1954a).

The two putative parental species hybridize naturally in certain localities in southern California today. It is significant that the introgressive products of this hybridization resemble *Gilia achilleaefolia* in morphology (Grant, 1954a).

The chromosomes of *Gilia achilleaefolia* are known to be partially homologous and partially nonhomologous with those of the *G. capitata* and *G. angelensis* groups (Grant, 1954b). Its genomic relationships are consistent with the hypothesis of a recombinational origin. Furthermore, *G. achilleaefolia* is self-compatible and, in some races, predominantly autogamous. Its breeding system is thus favorable for the establishment of a new recombination-type genome.

More combined taxonomic and genetic studies are needed in order to extend the list of probable examples. And more purely experimental work on recombinational speciation is also needed. We are still a long way from knowing how large a role recombinational speciation plays in plant evolution.

Conclusions

In Chapter 7 we described the salient features in several patterns of species relationships in higher plants. The processes of speciation considered in Chapters 8, 9, and 10 put us in a position to explain most of these features. Certain other features have not been accounted for as yet but can be considered now in the light of the processes reviewed in this and the preceding chapter.

We are referring to some features in the so-called *Gilia inconspicua* pattern. It will be recalled that this pattern is found in segments of Erophila, Elymus, Festuca, Mentzelia, Gilia, and other genera. The plants are usually autogamous annuals or, sometimes, autogamous perennial herbs. Related species are morphologically very similar, but are separated inter alia by chromosomal sterility barriers.

In plants belonging to the *Gilia inconspicua* pattern, a single taxonomic species turns out on taxogenetic analysis to be composed of several or many sibling species. The related sibling species in most such taxonomic species fall into two categories with regard to ploidy level. We find morphologically similar but chromosomally intersterile species on different ploidy levels, thus related diploids and tetraploids. We also find related sibling species on one and the same ploidy level, as exemplified by sets of diploid sibling species.

The two types of species relationships in the *Gilia inconspicua* pattern can be accounted for to a large extent by amphiploidy and by recombinational speciation respectively. These are different forms of hybrid speciation involving parental species which differ by chromosomal rearrangements. The daughter species of hybrid origin will resemble the parental species morphologically and will be isolated from them by second-order chromosomal sterility barriers.

The explanation of amphiploidy as applied to the polyploid fraction of sibling species in the *Gilia inconspicua* pattern is supported by a great deal of factual evidence. The parallel explanation of recombinational speciation as applied to the homoploid fraction of sibling species is still in the preliminary stages of verification. Recombinational speciation *could* produce the observed results but, as noted earlier in this chapter, we still do not know how often it actually does so.

PART IV

Derived Genetic Systems

Polyploidy

Introduction

The term polyploidy refers to a special arithmetic relationship between the chromosome numbers of related organisms which possess different numbers. The polyploid organism, in the simplest case, has a chromosome number which is twice that of some related form. Or the polyploid condition is some other multiple of the haploid number in a related organism, such as $3x$ or $6x$. Or, finally, the chromosome number of the polyploid is not the average but the sum of the numbers found in two related forms with different lower numbers, as in the series $2n = 10, 12, 22$.

All of these cases represent the formation of a higher chromosome number, in the polyploid organism, by the addition of extra whole chromosome sets present in one or more ancestral organisms. In short, polyploidy is the presence of three or more chromosome sets in an organism.

The phenomenon of polyploidy was discovered during the exploratory phase of plant cytogenetics and plant cytotaxonomy in the early years of this century. Winkler (1916), who introduced the term polyploidy, and Winge (1917), who offered a partial explanation, played key roles in the discovery.

Winkler (1916) was studying vegetative grafts and chimeras in Solanum. Callus tissue develops on the cut surface of the stem and can go on to regenerate a new shoot. Winkler observed that tetraploid plants of *S. nigrum* developed in this way from diploid stocks. This result must have come about through some kind of somatic doubling in the originally diploid tissue.

Winge (1917) compared related species belonging to the same genus in respect to chromosome number. He had some chromosome counts of his own in Chenopodium and previously published counts by Tahara (1915) in Chrysanthemum. Winge observed a regular arithmetic series in the chromosome numbers. Different species of Chenopodium have $2n = 18$ and $2n = 36$. Different species of Chrysanthemum have $2n = 18, 36, 54, 72$, and 90. Thus the chromosome numbers of related species are seen to be multiples of some common basic number.

In order to account for these observations, Winge (1917) proposed the fruitful hypothesis of chromosome number doubling in species hybrids. He suggested that polyploid series develop as a result of the following sequence of events. Interspecific hybridization gives rise to hybrid plants which are sterile owing to failure of chromosome pairing. Doubling of the chromosome number sometimes occurs spontaneously in the hybrid (or hybrid progeny) and converts it into a fertile type. The fertile polyploid plants may then multiply and become a new constant species. Thus two diploid species could give rise by hybridization to a new tetraploid species; hybridization between a diploid and a tetraploid could produce a hexaploid species; and other hybrid combinations could give rise to the higher polyploid types (Winge, 1917).

Winge's hypothesis was soon confirmed experimentally by artificial interspecific hybridizations in Nicotiana, Raphanus-Brassica, and Galeopsis (Clausen and Goodspeed, 1925; Karpechenko, 1927; Müntzing, 1930b, 1932).

It will be obvious that significant cytogenetic differences exist between the polyploid phenomena dealt with by Winkler and those dealt with by Winge and his followers. On the one hand we have intraspecific nonhybrid polyploidy and, on the other hand, we have polyploidy associated with interspecific hybridization. The first type of polyploidy involves the multiplication of one and the same chromosome set. The second type of polyploidy results from the doubling of the structurally dissimilar chromosome sets in a species hybrid.

These distinctions were clarified by the terminology—autopolyploidy and allopolyploidy—introduced by Kihara and Ono (1926). The terms autoploidy and amphiploidy, which are used for the most part in this book, are synonymous with autopolyploidy and allopolyploidy respectively (Clausen, Keck, and Hiesey, 1945).

Scope and Focus

Polyploidy research has expanded enormously since the early investigations, and has produced a voluminous literature. Polyploidy is important in relation to such diverse fields as cytogenetics, physiology, plant breeding, cytotaxonomy, and biogeography. It is obviously impossible to cover the whole subject of polyploidy in all its bearings here.

Good reviews of various aspects of polyploidy are given by Müntzing (1936; 1961, Ch. 26), Darlington (1937a, 1963), Clausen, Keck, and Hiesey (1945), Löve and Löve (1949), Stebbins (1950, Chs. 8, 9), Tischler (1953), and Rieger (1963). The reader will find much additional information in these works.

Our central focus in this book is on the evolutionary aspects of polyploidy in higher plants. In order to carry out our objective, we find that we need some, but only some, of the evidence from each of the separate fields concerned with polyploidy. Accordingly, we present selected groups of facts from cytotaxonomy, from cytogenetics, and from biogeography. The facts to be presented from the various fields are selected on the basis of their relevance to the question before us.

In the immediately following sections of this chapter we will review the evidence which shows that polyploidy has indeed played a very important role in the evolution of higher plants, particularly the angiosperms and pteridophytes. The evidence supporting this conclusion is twofold. A large proportion of the species of higher plants are polyploid. And high ploidy levels are reached in various genera.

The next step is to pose the obvious problem. What is the explanation of the observed facts? What are the reasons for the success of the polyploid condition in plant evolution? We attempt to deal with this question in the next chapter, Chapter 16.

Polyploid Series

For purposes of illustration we will list a few well-known examples of polyploidy in the angiosperms. The examples are grouped by type of polyploid series.

The simplest case is that of a genus, section, or species group containing diploids and tetraploids. Classical examples are Dahlia, Gossypium, and the *Nicotiana tabacum* group.

Several diploid species of Dahlia have $2n = 32$, whereas the garden dahlia (*D. variabilis*) is tetraploid with $2n = 64$ (for details and references see Darlington and Wylie, 1955). In Gossypium the diploid condition is found in numerous wild species and in the Old World cultivated cottons, *G. arboreum* and *G. herbaceum;* while the tetraploid condition occurs in the New World cultivated cottons, chiefly *G. hirsutum* and *G. barbadense.* The chromosome numbers are· $2n = 26$ and $2n = 52$, respectively, and the basic number is therefore $x = 13$ (Hutchinson and Stephens, 1947). The *Nicotiana tabacum* group ($x = 12$) contains diploid species like *N. glutinosa* and others ($2n = 24$) and the common tobacco, *N. tabacum*, which is tetraploid ($2n = 48$) (Goodspeed, 1954).

The next situation is a simple extension of the foregoing one. It is a polyploid series with even multiplies, thus $2x, 4x, 6x, 8x$. The species of Triticum, with chromosome numbers of $2n = 14$, 28, and 42, exhibit

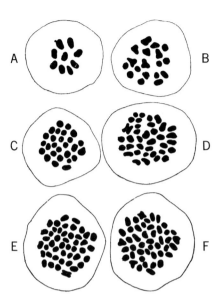

Figure 32 Polyploid series in Chrysanthemum ($x = 9$). Metaphase I in pollen mother cells. (A) *C. makinoi* ($2x$, $n = 9$). (B) *C. indicum* ($4x$, $n = 18$). (C) *C. japonense* ($6x$, $n = 27$). (D) *C. ornatum* ($8x$, $n = 36$). (E) *C. yezoensis* ($10x$, $n = 45$). (F) *C. pacificum* ($10x$, $n = 45$). (From Shimotomai, after Müntzing, 1961.)

such a series up to the hexaploid level on a basic number of $x = 7$. A longer polyploid series is found in Chrysanthemum ($x = 9$), as noted earlier. The $2x$, $4x$, $6x$, $8x$, and $10x$ levels are all represented in this genus along with some still higher numbers (Figure 32). A still longer series in Potentilla ($x = 7$) contains every even multiple from the diploid to the $16x$ level. (See Darlington and Wylie, 1955, for data and literature references.)

Polyploid series with both even and odd multiples are illustrated by Rosa and Crepis. The diploid roses have $2n = 14$. In the *Rosa canina* group there are, in addition, tetraploid, pentaploid, and hexaploid forms with $2n = 28$, 35, and 42 chromosomes (see Darlington and Wylie, 1955). The *Crepis occidentalis* group ($x = 11$) exhibits a longer series consisting of $2x$, $3x$, $4x$, $5x$, $7x$, and $8x$ forms (Babcock and Stebbins, 1938). The odd polyploid species in these and other such groups necessarily reproduce by asexual or subsexual methods.

Dibasic polyploids have chromosome numbers which are the sum of two different aneuploid diploid numbers (Darlington, 1956b, 1963). An example in Brassica consists of *B. oleracea* ($2n = 18$), *B. campestris* ($2n = 20$), and their tetraploid derivative, *B. napus* ($2n = 38$) (U, 1935; Clausen, Keck, and Hiesey, 1945). Another dibasic polyploid series is seen in a segment of the genus Clarkia containing the gametic numbers $n = 8$, 9, 17, and 26. *Clarkia prostrata* ($2n = 6x = 52$) is probably derived from *C. davyi* ($2n = 4x = 34$) and *C. speciosa* ($2n = 2x = 18$); and *C. davyi* ($2n = 34$) in turn is a tetraploid product of two diploid species with $2n = 16$ and $2n = 18$, respectively (Lewis and Lewis, 1955).

Polyploids contain much duplication of genetic material in their nuclei and can therefore often tolerate the loss of one or more chromosome pairs. This leads to modified polyploid series ending with what Darlington has called a polyploid drop. A modified series of this type is found in Hesperis (Cruciferae), where different species have gametic numbers of $n = 7$, 14, 13, and 12 (Darlington, 1956b, 1963; Darlington and Wylie, 1955).

In the ferns we find polyploid series based on $x = 37$ or $x = 41$ or other high basic numbers. The *Polypodium vulgare* group ($x = 37$), for example, has diploid, tetraploid, and hexaploid forms with $2n = 74$, 148, and 222. Another series in Polystichum is based on $x = 41$. In such cases it is probable that the basic number inferred from the existing polyploid series is not a truly basic number in the phylogenetic sense. The existing diploid members are probably old polyploids from an earlier cycle of evolution (Manton, 1950).

Similar considerations apply to some relatively high basic numbers in angiosperms. The number $x = 13$ in Gossypium, for example, may be a secondary basic number. The original basic number for this genus might well have been $x = 7$. This number is found in many other genera of Malvaceae. It could have given rise to the 13-paired complements of the existing diploid species of Gossypium by doubling and polyploid drop (Tischler, 1954).

Systematic Distribution

Polyploidy is widespread in the angiosperms. It is common in the Rosaceae, Rubiaceae, Compositae, Iridaceae, Gramineae, and many other families. It is relatively uncommon but definitely present in other families such as the Caesalpinaceae, Passifloraceae, and Polemoniaceae. In still other families, for instance the Fagaceae and Berberidaceae, the condition exists but is rare.

Stebbins (1950) has called attention to intergeneric differences in the frequency of polyploidy in several angiosperm families. Thus, in the Salicaceae, polyploidy is common in Salix but rare in Populus. In the Betulaceae, polyploidy is common in Betula but relatively uncommon in Alnus. The condition is common in Thalictrum but rare in Aquilegia (Ranunculaceae), and common in Tulipa but rare in Lilium (Liliaceae) (Stebbins, 1950, p. 301).

Polyploidy is common again in the Gnetales. Tetraploids are found in all three genera: Ephedra, Welwitschia, and Gnetum (Tischler, 1953). According to Khoshoo (1959), about 40% of the species of this group are polyploid.

But, in the gymnosperms exclusive of the Gnetales, polyploidy is a rare condition (Tischler, 1953; Khoshoo, 1959). Polyploidy is unknown in the cycads and in ginkgo. Among conifers, some species of Pseudolarix and Podocarpus with high chromosome numbers, which were formerly believed to be polyploids, have been reanalyzed and reinterpreted as products of ascending aneuploidy (Khoshoo, 1959). One of the very few authentic naturally occurring polyploids in the conifers is *Sequoia sempervirens* ($2n = 6x = 66$). The other known polyploid species or races of conifers are *Juniperus chinensis pfitzeriana* ($2n = 44$) and *J. squamata meyeri* (Khoshoo, 1959). To this short list, Hair (1968) has recently added *Fitzroya cupressoides* (Cupressaceae) with $2n = 4x = 44$.

In the pteridophytes, on the other hand, polyploidy is present almost everywhere and reaches the highest levels known in the plant kingdom. The Polypodiaceae are mostly high polyploids. Polyploidy exists in the two living genera of Psilotales, namely, Psilotum and Tmesipteris. The species of Equisetum are homoploid at a high chromosome number level, $2n = $ ca. 216, so as to indicate ancient polyploidy. In the Lycopodiales, high polyploidy occurs in Lycopodium and Isoetes, but low diploid numbers ($2n = 18$) persist in Selaginella (Manton, 1950).

In the lower members of the plant kingdom, polyploidy exists in some mosses and algae. A perusal of the *Index to Plant Chromosome*

Numbers reveals examples in Funaria, Chara, and other genera (Ornduff et al., 1959–1968).

In the animal kingdom and fungus kingdom, by contrast with the plant kingdom, polyploidy is exceedingly rare. For animals this conclusion is supported by the analyses of White (1954) and Bungenberg (1957); for fungi it is evident from the published lists of chromosome numbers (Ornduff et al., 1959–1968).

Some bonafide naturally-occurring polyploid races and species in the animal kingdom are found among the invertebrates (White, 1954). The best authenticated cases are in various groups of arthropods. They are: *Artemia salina* (Crustacea), *Trichoniscus elizabethae* (Isopoda), *Otiorrhynchus spp.* (Curculionidae, Coleoptera), *Solenobia triquetrella* (Psychidae, Lepidoptera) and *Ochthiphila polystigma* (Chamaemyiidae, Diptera) (Bungenberg, 1957; Stalker, 1956). Probable cases of polyploidy are also found in Lumbricus and other genera of earthworms (Muldal, 1952; Bungenberg, 1957) and in a few lower vertebrates (Schultz, 1969).

Polyploidy, then, can be regarded as one of the distinctive features of the plant kingdom.

Range in Chromosome Number

Chromosome numbers in the angiosperms range from $2n = 4$ to $2n =$ ca. 500. Both extremes occur in the dicotyledons. The low number is found in *Haplopappus gracilis* ($2n = 4$) and the high one in a species of Kalanchoe ($2n =$ ca. 500) (Baldwin, 1938; Jackson, 1957). Another high polyploid in the dicotyledons is *Buddleia colvilei* with $2n =$ ca. 300 (see Darlington and Wylie, 1955). Chromosome numbers in the monocotyledons range from $2n = 6$ to $2n = 226$.

The frequency distribution of angiosperm species with different gametic chromosome numbers is given in Table 2. The table is based on a sample of 17,138 species listed in the *Chromosome Atlas* (Darlington and Janaki, 1945; Darlington and Wylie, 1955). This sample was analyzed and tabulated as shown here by Dr. Howard Latimer and myself in 1957. It would be possible today to expand the tabulation on the basis of the many new chromosome counts reported since the middle

1950s and compiled in the *Index* of Ornduff et al. (1959–1968). However, the sample incorporated in Table 2 is a large and representative
one, and the table therefore portrays the main statistical features of
chromosome numbers in the flowering plants.

We see in Table 2 that there are thousands of angiosperm species
with 14 or 15 or more pairs of chromosomes. The higher numbers
range in an almost continuous aneuploid series from $n = 14$ to $n = 85$;
the series extends on with many gaps to $n = 154$; and then jumps to
$n = 250$. The higher polyploid numbers are, as might be expected,
represented in fewer species (Table 2).

The highest chromosome numbers in the plant kingdom are found in
the ferns and fern allies (Manton, 1950; Löve and Kapoor, 1966, 1967).
Tmesipteris tannensis has over 400 chromosomes in its somatic complement (Manton, 1950). Still higher numbers occur in Ophioglossum.
The series in this genus starts with $2n = 240$ in 7 species. *Ophioglossum
vulgatum* and *O. thermale* have $2n = 480$; *O. azoricum* has $2n = 720$
(Figure 33); *O. nipponicum* has $2n = 960$; and *O. reticulatum* reaches
the highest known chromosome number with $2n = 1260$. (See Löve and
Kapoor, 1966, 1967, for several new counts and references to earlier
counts in Ophioglossum.)

Range in Ploidy Level

When we turn from range in chromosome number to range in ploidy
level we frequently encounter uncertainties. Chromosome numbers can
be counted. In many plant groups, ploidy levels can be determined
accurately by the comparative method. In many other plant groups,
however, ploidy level cannot be determined as such, but instead can
only be estimated from some basic number which is hypothetical
(Darlington, 1937a, p. 239; Tischler, 1954).

In a genus composed of species with $2n = 14, 28, 42$, and 56 chromosomes, there is no problem in identifying the tetraploids, hexaploids,
and octoploids. Uncertainties do arise in a genus containing only the
numbers $2n = 28$ and $2n = 56$. The obvious decision, to call the
carriers of these numbers diploids and tetraploids, respectively, may be
wrong; but the more likely suggestion, that they are tetraploids and

Table 2 · The number of taxonomic species of angiosperms with a given gametic chromosome number (Grant, 1963).

(Odd somatic numbers are expressed as the arithmetic half-value, i.e., $2n = 15$ as $n = 7.5$.)

Haploid number	Dicots	Mono-cots	Both	Haploid number	Dicots	Mono-cots	Both
2	1		1	20	326	356	682
3	7	8	15	20.5		2	2
4	99	35	134	21	238	222	460
5	159	66	225	21.5	1	6	7
5.5	1		1	22	271	69	340
6	291	117	408	22.5	12	12	24
6.5	1	1	2	23	89	28	117
7	775	598	1,373	23.5	1	3	4
7.5	11	5	16	24	307	147	454
8	1,117	181	1,298	24.5	2	6	8
8.5		5	5	25	31	33	64
9	1,151	235	1,386	25.5	14	1	15
9.5	3	2	5	26	123	51	174
10	533	283	816	26.5	1	1	2
10.5	58	14	72	27	82	96	178
11	1,081	174	1,255	27.5	5	1	6
12	810	328	1,138	28	156	110	266
12.5	2	4	6	28.5	9	1	10
13	708	112	820	29	14	24	38
13.5	20	7	27	29.5	1		1
14	756	408	1,164	30	87	147	234
14.5		6	6	30.5		3	3
15	215	154	369	31	7	24	31
15.5		1	1	31.5	8	7	15
16	491	162	653	32	78	40	118
16.5	6	19	25	32.5		1	1
17	353	43	396	33	45	25	70
17.5	65	11	76	33.5	1		1
18	502	276	778	34	93	24	117
18.5	1	2	3	34.5		4	4
19	194	68	262	35	21	45	66
19.5	5	11	16	36	97	57	154

TABLE 2 Cont.

Haploid number	Dicots	Mono-cots	Both	Haploid number	Dicots	Mono-cots	Both
37		11	11	59			
37.5		5	5	59.5		1	1
38	35	15	50	60	16	23	39
38.5	2	1	3	60.5		1	1
39	27	6	33	61		2	2
40	36	46	82	61.5		1	1
40.5	4	2	6	62	1	2	3
41	7	9	16	63	4	4	8
42	37	22	59	63.5	1		1
42.5	1	1	2	64	6	2	8
43	5	7	12	65	5	1	6
43.5	1	2	3	66	7		7
44	16	6	22	66.5	1		1
45	23	15	38	67			
45.5	1	3	4	67.5	1		1
46	10	2	12	68	7		7
47	1		1	69	3	2	5
47.5		2	2	70	2	4	6
48	28	13	41	71			
48.5		1	1	72	7	3	·10
49	2	1	3	73			
50	10	9	19	73.5		1	1
50.5	1	1	2	74	2		2
51	26	5	31	74.5		1	1
51.5		1	1	75	1		1
52	9	2	11	76	5		5
52.5	1	2	3	77	2		2
53				78	2		2
54	10	10	20	79			
54.5	1		1	80	5	2	7
55	2	3	5	81			
56	26	5	31	82	4		4
57	16	3	19	83			
58	3	2	5	84	1		1
58.5		1	1	85	4		4

TABLE 2 Cont.

Haploid number	Dicots	Mono-cots	Both	Haploid number	Dicots	Mono-cots	Both
85.5	1		1	97			
86				98			
87				99	1		1
88	2		2	100	6		6
89				112	1		1
90	3	4	7	113		1	1
91				119	1		1
92	1		1	132	1		1
93				150	1		1
94				154	1		1
95	1		1	250	1		1
95.5	1		1		11,987	5,151	17,138
96	1		1				

octoploids, must remain a matter of hypothesis. Yet the probably correct hypothesis is better than the probably wrong fact, in this situation, as a basis for estimating the level of ploidy or the frequency of polyploids in the plant group.

Manton (1950) found that 13 species of Equisetum belonging to two sections of the genus all have $2n = $ ca. 216. We would be misled by the simple facts of observation here if we were to conclude that polyploidy is unknown in Equisetum. Actually, all of the species of Equisetum seem to be stationed at the same high polyploid level. We cannot hope to know what the basic number is in this isolated and homoploid genus and hence cannot estimate the level of ploidy (Manton, 1950).

What levels does polyploidy reach in the angiosperms? Let us begin to deal with this question by citing some angiosperm genera containing high polyploids in which the basic number can be read directly from the cytotaxonomic record and the upper ploidy level can therefore be determined reliably.

The polyploid series in Potentilla ($x = 7$) runs from $2n = 14$ to $2n = 112$. *Potentilla haematochroa* with $2n = 112$ is 16-ploid. In Chrysanthemum ($x = 9$) the series contains every even multiple from

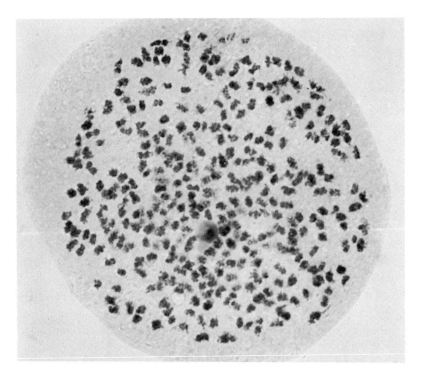

Figure 33 A spore mother cell of *Ophioglossum azoricum* with about 360 bivalents. (Löve and Kapoor, 1967; photograph by courtesy of Dr. Áskell Löve.)

$2x$ to $10x$. Then, after a break in the series, there are such high numbers as $2n = 160$ and 198. *Chrysanthemum lacustre* with $2n = 198$ would be 22-ploid. One subdivision of Senecio with a basic number of $x = 5$ has species at the lower and middle ploidy levels from $2x$ to $16x$. *Senecio roberti-friesii* in this group has $2n = $ ca. 180 and is thus approximately 36-ploid (Darlington and Wylie, 1955, for data and references).

Hair and Beuzenberg (1961) have recently reported a count of $2n = 263-265$ for *Poa litorosa* in New Zealand. The known primary basic number in Poa is $x = 7$. Consequently, *P. litorosa* is 38-ploid with a polyploid drop. This is the highest ploidy level known in the monocotyledons (Hair and Beuzenberg, 1961).

Buddleia (Loganiaceae) and Kalanchoe (Crassulaceae) are examples of angiosperm genera containing very high chromosome numbers in which the basic number and hence the ploidy level can only be inferred. Buddleia includes $2x$, $4x$, and $6x$ species on the base of $x = 19$. *Buddleia colvilei* with $2n = $ ca. 300 is therefore 16-ploid or nearly so. But $x = 19$ in Buddleia and related genera probably represents a secondary basic number of old tetraploid derivation. On this basis, *B. colvilei* is not 16-ploid but 32-ploid. In Kalanchoe (Crassulaceae) the predominant basic number is $x = 17$. The Kalanchoe species with $2n = $ ca. 500 is then approximately 30-ploid with aneuploid deviations. But if, as is likely, $x = 17$ is an old tetraploid number, we would have to double this estimate of ploidy level. Kalanchoe apparently reaches the higher level of approximately $60x$ (Darlington and Wylie, 1955, for data and references).

In the ferns and fern allies the actual or known basic numbers are generally high, as noted earlier. Manton (1950) gives the following basic numbers for four genera of Polypodiaceae: Asplenium ($x = 36$), Polypodium ($x = 37$), Dryopteris ($x = 41$), Polystichum ($x = 41$). Each of these genera has a polyploid series on its basic number. The *Polypodium vulgare* group ($x = 37$), for example, contains diploids, tetraploids, and hexaploids. The latter have $2n = 222$ chromosomes. But the actual basic numbers almost certainly represent an ancient polyploid condition. What the original or primary basic numbers were in these fern genera is unknown. It is safe to assume, however, that the existing polyploid ferns stand at a rather high level of ploidy (Manton, 1950).

Ophioglossum (Ophioglossaceae) presents a polyploid series starting at $2x = 2n = 240$ and ranging through the $4x$, $6x$, and $8x$ levels to over $10x$ in *O. reticulatum* with $2n = 1260$. It has been plausibly suggested that the original or primary basic number in Ophioglossum is $x = 15$. On this basis the existing "diploid" species are $16x$, and *O. reticulatum* is 84-ploid (Ninan, 1958; Löve and Kapoor, 1966).

Levan (1949) studied the viability of various artificially induced polyploid types of *Phleum pratense* ($x = 7$). He found that the timothy plants with $11x = 77$ chromosomes were viable, whereas those with $12x = 84$ had greatly reduced viability. A threshold for viability in this case lies between 77 and 84 chromosomes. Mitosis was disturbed in the $12x$ and $13x$ plants, suggesting that mitotic conditions may be an important factor determining this threshold (Levan, 1949).

Similar cytological and physiological difficulties must occur in high polyploids which arise in nature. Yet the levels of polyploidy found in nature far exceed the upper limits that can be reached in experimental cultures. No doubt the high polyploids existing in nature today are a selected sample of ancient lines which have successfully and gradually surmounted the various physiological difficulties.

Frequency

Our next task is to estimate the proportion of species of higher plants which are polyploid. Accurate estimates are difficult to make for frequency of polyploidy, just as they are for grade of polyploidy, and for the same reason. If we compute the percentage figures from members of cytotaxonomically observable polyploid series, in order to keep on solid ground of factual evidence, we inevitably lose track of many ancient polyploid species, and therefore seriously underestimate the frequency of polyploidy. The only way to guard against this bias is to make certain assumptions concerning original or primary basic numbers.

Among the earlier estimates for the angiosperms are those of Müntzing, Darlington, and Stebbins. Müntzing (1936) and Darlington (1937a, Ch. 6) stated in qualitative terms that about half the species of angiosperms are polyploid. Stebbins (1950, p. 300) gave a rough estimate that between 30 and 35% of angiosperm species are polyploid.

More recently I have approached the problem by combining the known frequency distribution of angiosperm species with diverse chromosome numbers, as shown in Table 2, with the plausible and widely accepted assumption that the primary basic numbers for this group are $x = 7$, 8, and/or 9 (Grant, 1963, p. 486). The gametic number $n = 14$ then marks a convenient dividing line for statistical purposes. Species with 14 or more chromosome pairs are classified as polyploids, and those with fewer than 14 pairs as diploids.

This method is not free from errors either, owing to the complicating effects of aneuploidy. There are some tetraploid species of angiosperms with $2n = 12$, 10 or even fewer pairs, which are derived from products of aneuploid reduction series. On the other hand, some species with $2n = 14$ or more pairs are diploid products of ascending aneuploidy. We do not know how many species in our sample belong in either category. However, it will be noted that the two sources of error are more or less compensating.

Computations made by the preceding method give us the following estimates. In the angiosperms as a whole, 47% of the species are polyploid. The estimated percentage frequency of polyploids in the dicotyledons is 43%, and that in the monocotyledons is 58% (Grant, 1963, p. 486).

The Gnetales have a high frequency of polyploidy. Khoshoo's data (1959) indicate that 38% of the Gnetales sensu lato are polyploid.

In the Coniferales, by contrast, only 1.5% of the species are polyploid, according to Khoshoo, and in the cycads there are no known polyploid species (Khoshoo, 1959).

For the pteridophytes we can use Manton's list of chromosome counts as a suitable sample (Manton, 1950, pp. 302–305). Taxonomic species in this list which contain two or more ploidy levels are counted as two or more species, for our present purposes, and hybrids are not counted. The sample as thus constituted contains 103 species representing all major groups of pteridophytes. Only five of these, among them three species of Selaginella with $2n = 18$, are primary diploids. The estimated frequency of polyploids in the pteridophytes is thus 95%.

Classification of Polyploids

The distinction between autopolyploids and allopolyploids (or amphiploids) is fundamental. It is also a qualitative distinction which oversimplifies the actual situation. Kihara and Ono (1926) originally expressed the distinction in broad terms. Autopolyploidy, in contradistinction to allopolyploidy, is the doubling of one and the same chromosome set.

In the first edition of *Recent Advances in Cytology*, Darlington (1932) restated the distinction as follows. An autopolyploid is a polyploid derived by doubling or adding the chromosome sets of a structural homozygote; hence it is an organism containing three or more sets of homologous chromosomes. By contrast, an allopolyploid is the product of doubling in a species hybrid; it is therefore a polyploid containing separate sets of nonhomologous chromosomes (Darlington, 1932, p. 169).

It was clear to the early students of the subject that a complete intergradation exists between homologous and nonhomologous chromosomes, between structural homozygotes and structural hetero-zygotes, and between homozygous individuals and interspecific hybrids. Autopolyploidy and allopolyploidy as defined above are thus the extreme members of a graded series. Further subdivision of these opposite classes to recognize various intermediate conditions may be convenient for many purposes.

Müntzing (1936, pp. 310 ff.), after discussing various definitions of autopolyploidy, arrived at the following concept. Autopolyploids have multiple chromosome sets which are identical structurally, but which may be different genically. Clausen, Keck, and Hiesey (1945, pp. 70 ff.) then went on to distinguish two main types of autopolyploids: those arising from fertile interracial hybrids and those arising from ordinary nonhybrid individuals.

Darlington (1932) recognized three nodal conditions of allotetra-ploidy on the basis of the degree of structural differentiation between the constituent genomes. The three conditions are exemplified by Raphanobrassica, *Primula kewensis*, and *Crepis rubra-foetida*. The main facts concerning Raphanobrassica and *P. kewensis* were presented earlier, in Chapter 13, to which the reader is referred for the details.

In the extreme case, as found in Raphanobrassica, the genomes are strongly differentiated structurally, so that the chromosomes do not pair in the diploid hybrid, but do form bivalents regularly in the tetraploid derivative. The intermediate condition, represented by *Primula kewensis*, is that where the chromosomes are somewhat less strongly differentiated structurally, so that they regularly pair in bivalents in both the diploid hybrid and the tetraploid derivative. Slight structural differentiation of the parental genomes, as between *Crepis rubra* and *C. foetida*, permits bivalent pairing in the diploid hybrid and extensive quadrivalent formation in the tetraploid derived from it. This third type of allopolyploid thus approaches an autopoly-ploid in its cytogenetic behavior (Darlington, 1932, p. 170).

Later authors have elaborated on this general theme. Clausen, Keck, and Hiesey (1945, pp. 70 ff.) classify amphiploids on two interrelated features, namely, the strength of the sterility barrier between the par-ental species, and the degree of chromosome pairing between the parental genomes. Raphanobrassica in their system is an interceno-specific amphiploid with nonpairing genomes; *Primula kewensis* is an interecospecific amphiploid with intergenomic pairing; and there are various intermediate combinations. Stebbins introduced the useful terms segmental allopolyploid and genomic allopolyploid to designate the *P. kewensis* type and the Raphanobrassica type of amphiploid respectively (Stebbins, 1947a; 1950, p. 315).

At the hexaploid, octoploid, or higher levels, it is possible for a plant to be simultaneously both an autoploid and an allopolyploid. This type of polyploid is aptly called an autoallopolyploid. It arises from a hybrid between two specifically and genomically distinct parents, one or both of which is an autoploid. Thus hybridization between an autotetraploid and a genomically different diploid will yield a triploid hybrid, which on doubling produces a hexaploid containing four sets of one genome and two sets of the other.

Let us summarize this discussion by listing the principal types of polyploids together with their genomic constitutions. Following the standard convention, a given genome is represented by a letter of the alphabet. Genomes A, B, etc., are strongly differentiated structurally so that their chromosomes usually fail to pair in the interspecific hybrids. Genomes or subgenomes A_s, A_t, etc., are differentiated structurally to an intermediate degree, so that partial or complete bivalent formation takes place in both the diploid hybrid and the amphiploid derivative. The diploid species which produce, singly or in combination, the various kinds of polyploids are AA, A_tA_t, BB, etc.

    I. Autoploids
        (1) Strict autoploid: $AAAA$
        (2) Interracial autoploid: $AAAA$
    II. Amphiploids
        (3) Segmental allopolyploid: $A_sA_sA_tA_t$
        (4) Genomic allopolyploid: $AABB$
        (5) Autoallopolyploid: $AAAABB$

It will be noted that types 2, 3, 4, and 5 are hybrid polyploids, whereas only type 1 is nonhybrid.

Criteria of Autoploidy and Amphiploidy

The principal criteria for distinguishing between autoploids and amphiploids are chromosome behavior, fertility, segregation ratios, and morphology. These criteria all break down in individual cases. Known amphiploids frequently exhibit the cytogenetic behavior of autoploids, and vice versa.

The most obvious aspect of chromosome behavior which is expected to differ as between autoploids and amphiploids is the presence or absence of multivalents at meiosis.

Multivalent formation, however, is limited by chiasma frequency, which in turn is affected by chromosome size and genetic conditions. Because of these limiting factors, autotetraploids nearly always have less than the maximum possible number of quadrivalents. Thus auto-tetraploid rye (*Secale cereale*, $x = 7$) shows a range of 0 to 6 quadrivalents per cell, where 7 IV would represent maximum pairing, and the average quadrivalent frequency varies between 2 and 4 IV in different strains (Müntzing and Prakken, 1941; Jain, 1960).

Gilles and Randolph (1951) compared the multivalent frequency in an autotetraploid line of maize (*Zea mays*, $x = 10$) at the beginning and the end of a 10-year period of selection for fertility. The newly induced autotetraploid plants had 8–10 IV in 89% of the pollen mother cells, whereas their tenth-generation descendants had 8–10 IV in only 52% of the cells. The frequency of bivalents increased as the quadrivalent frequency decreased. The artificial selection for fertility which brought about this change probably involved an unconscious selection for genic factors influencing the mode of chromosome pairing at meiosis. This case shows how autoploids in nature, and under natural selection, could come to have the type of bivalent pairing which is regarded as typical of amphiploids (Gilles and Randolph, 1951).

On the other hand, a segmental allopolyploid may and often does exhibit some multivalent formation. Tetraploid *Primula kewensis*, for example, usually forms bivalents, as we have seen earlier, but does have occasional quadrivalents. The occurrence of multivalents in this and other segmental allopolyploids is to be expected from the partial homology between chromosomes belonging to the different subgenomes.

If these homeologous chromosomes can pair in the diploid hybrid, they can also pair occasionally in the amphiploid derivative. Preferential pairing, in other words, may not be completely effective. This intergenomic (or so-called heterogenetic) pairing may simply produce bivalents. But, if the heterogenetic pairing is combined with normal homogenetic pairing in the same pairing partners, higher associations such as trivalents and quadrivalents will be formed. In a segmental allotetraploid of the genomic constitution $A_sA_sA_tA_t$, the four chromosomes belonging to any given chromosome type 1 may occasionally pair to form the quadrivalent $1A_s\text{-}1A_s\text{-}1A_t\text{-}1A_t$.

As regards fertility, a general rule was formulated by Darlington (1932, Ch. 7). There is an inverse correlation between the fertility of a diploid hybrid and that of its tetraploid derivative. Fertile hybrids produce sterile tetraploids; sterile hybrids produce fertile tetraploids; and hybrids of intermediate fertility give tetraploids of intermediate fertility. Autoploids can then be distinguished from amphiploids by the sterility of the former and the fertility of the latter.

This generalization is derived logically from the expected differences between autoploids and amphiploids in chromosome behavior outlined in the preceding paragraphs. The uneven segregation of homologous chromosomes from a multivalent in an autoploid leads to gametic sterility. Conversely, the regular segregation of homologs from bivalents in an amphiploid leads to gametic fertility.

But we have seen that the actual chromosome behavior in polyploids often deviates from expectation. Autoploids with genetically determined bivalent pairing may be highly fertile. Amphiploids, on the other hand, may be sterile for various reasons. An amphiploid which forms multivalents may undergo unequal chromosome segregation, like a typical autoploid, and thus produce unbalanced and inviable gametes. Or the amphiploid may show good bivalent pairing, at the cytological level of observation, but the pairing is heterogenetic. The pairing partners then segregate equally as to numbers but form genetically unbalanced recombination products. The amphiploid is again gametically sterile. Or, finally, genic sterility may be intermingled with chromosomal sterility in the diploid hybrid and be carried over into the amphiploid derivative.

The experimental evidence concerning the fertility of different types of polyploids is in line with the actual situation. The general rule of inverse correlation in fertility holds up in many cases but is also subject to numerous exceptions. Certainly many examples are known of autotetraploids with reduced fertility and allotetraploids with enhanced fertility. But many contrary examples are also known.

Sears (1941) produced an array of diploid hybrids and allotetraploid derivatives between ten diploid species of Triticum, Aegilops, and

Haynaldia ($n = 7$). In 18 hybrid combinations he could compare corresponding F_1s and allotetraploids with respect to chromosome pairing and fertility. The different diploid hybrids varied from high to low bivalent pairing. The allotetraploids showed a wide range in fertility from 100% to 0% fertility. But there was no constant relationship between the degree of pairing in the diploid F_1 and the degree of fertility in the tetraploid derivative. F_1 hybrids with low pairing may produce fertile amphiploids in some cases, or sterile amphiploids in others, depending on other contributing factors (Sears, 1941).

The expected difference between autoploids and amphiploids in breeding behavior is also deduced from the different extreme types of chromosome behavior described earlier. An autotetraploid should show tetrasomic inheritance in characters for which it is heterozygous, whereas an amphiploid should be constant and true-breeding.

Theoretical segregation ratios in polyploids are profoundly different from those of diploids. Let us consider a heterozygous autotetraploid with the genetic constitution *TTtt* for a gene *T* located close to the centromere on four chromosomes which are distributed at random to the gametes. This heterozygote is expected to give a zygotic ratio in F_2 of 1 *TTTT*: 34 heterozygotes: 1 *tttt*. If the gene *T* is 50 map units away from the centromere, the theoretical zygotic ratio in F_2 is 1 *TTTT*: 19.8 heterozygotes: 1 *tttt*. The ratios observed in experiments fall between these two extreme types of theoretical ratios (Lindstrom, 1936; Moens, 1964). In any case, the homozygous recessive class occurs in a much lower frequency in F_2 under tetrasomic inheritance (between 1/22 and 1/36) than under disomic inheritance (1/4).

In contrast to this situation in autoploids, an allotetraploid carrying *TT* on one chromosome pair in genome A, and *tt* on another chromosome pair in genome B, is not expected to segregate for *T* at all.

Tetrasomic inheritance, however, is not universally present in autotetraploids or necessarily absent in allotetraploids. Tetrasomic inheritance will be exhibited by a segmental allotetraploid if there is multivalent formation. On the other hand, an autotetraploid, particularly an old one, may have undergone gene divergence so as to convert the formerly homologous alleles into separate pairs of duplicate factors.

An amphiploid may show some segregation but not in tetrasomic ratios. Suppose that we have a segmental allotetraploid with the allele pair *TT* on a chromosome pair in one subgenome and the allele pair *tt*

Table 3 · Amphiploids that have been confused with one or both parents (Clausen, Keck, and Hiesey, 1945)

Amphiploid	Confused with
Spartina townsendii	S. alterniflora and S. stricta
Phleum pratense	P. nodosum
Poa annua	P. exilis and P. supina
Iris versicolor	I. virginica
Eriogonum fasciculatum ssp. foliolosum	E. f. typicum and E. f. polifolium
Rumex acetosella	R. angiocarpus and R. tenuifolius
Brassica napus	B. oleracea and B. campestris
Galeopsis tetrahit	G. speciosa and G. pubescens
Penstemon neotericus	P. laetus and P. azureus
Madia citrigracilis	M. gracilis
Artemisia douglasiana	A. suksdorfii and A. ludoviciana

in the other subgenome. This segmental allotetraploid will usually produce Tt gametes and $TTtt$ zygotes. But, occasionally, heterogenetic pairing (T/t and T/t), followed by independent assortment of the t alleles, will produce some tt gametes, and these can unite to give some $tttt$ zygotes. The segmental allotetraploid thus yields rare segregates toward the diploid parent carrying tt.

Gerstel and Phillips (1958) have carried out a series of ingenious experiments with synthetic amphiploids in Gossypium and Nicotiana which enable them to correlate segregation ratios with genomic affinities. Some amphiploids derived from distantly related species show no segregation. Other amphiploids formed from close relatives, like *Gossypium arboreum* and *G. herbaceum*, segregate in tetrasomic ratios. Still other amphiploids containing genomes which are well differentiated but not completely different yield rare segregates for particular marker genes (Gerstel and Phillips, 1958).

In gross morphology, finally, it is often supposed that an autoploid should resemble the parental diploid species, whereas an amphiploid should resemble the interspecific hybrid. But experimental autoploids do not always resemble the parental diploids. It is clear, moreover, that, if a segmental allopolyploid can segregate in the direction of one parental diploid for one gene, it can do likewise for other gene-determined characters as well, and eventually come to resemble a diploid species very closely in morphology and in ecology. Autoallopolyploids also tend to resemble particular diploid species in morphology.

Natural amphiploid species do, in fact, commonly vary in the direction of one or both parental diploid species. Clausen, Keck, and Hiesey (1945, p. 150) give a list of amphiploids which have been confused taxonomically with one or both parents. The examples are repeated here in Table 3. Many more examples have been added since 1945, and this pattern of variation can be said to be a common one (see Grant, 1964a). It shows that close morphological resemblance between a given polyploid and a given diploid is not a valid criterion of autoploidy.

Relative Frequency of Autoploids and Amphiploids in Nature

The identification of naturally occurring polyploids as autoploids or amphiploids is by no means a simple task, owing to the unreliability of the various criteria in borderline cases. Segmental allopolyploids and autoallopolyploids are likely to resemble autoploids in morphological characters and cytogenetic behavior. Even considerable experimental and cytogenetic work may fail to resolve the question conclusively. The problem is not impossible, however. By taking all criteria into consideration in the case of any given natural polyploid, and by resynthesizing that polyploid experimentally from its putative ancestors, it is possible to identify its nature and origin in a satisfactory manner.

The early literature on natural polyploidy can be said now, in the light of hindsight, to have oversimplified the problem. Both Darlington (1932, 1937a) and Müntzing (1936) concluded that autoploids are common in nature and important in plant evolution. This conclusion was based primarily on the widespread occurrence, in numerous taxonomic species of plants, of morphologically similar diploid and polyploid forms (Darlington, 1932, p. 209; 1937a, p. 226; Müntzing, 1936). It was supported by the evidence of multivalent formation in many of the same polyploid forms (Müntzing, 1936). But these two lines of evidence are now recognized to be insufficient to justify this conclusion.

Clausen, Keck, and Hiesey (1945) have critically surveyed the evidence in some twenty-eight thoroughly investigated plant groups containing natural polyploids. For our purposes here we will pool and simplify their conclusions (from their Chapters 6, 7, 8). Only three of the twenty-eight cases of natural polyploids are considered definitely to be autoploids, and three other cases are concluded to be probable autoploids. The clear-cut autoploids in the list are *Galax aphylla*, *Biscutella laevigata*, and *Zea perennis;* the probable autoploids are *Vaccinium uliginosum*, *Eragrostis pallescens*, and *Galium mollugo* and *G. verum*. The overwhelming preponderance of natural polyploids in the sample of well-investigated and critically reanalyzed cases are thus amphiploids (Clausen, Keck, and Hiesey, 1945).

Stebbins (1947a) then went over part of the same ground and reduced some of the probable cases to doubtful (i.e., *Eragrostis pallescens*) and some of the supposedly good cases to merely possible (i.e., *Zea perennis*). The only clear-cut autoploid remaining on the list in Stebbins' review is *Galex aphylla* (Stebbins, 1947a).

In the ferns and fern allies, Manton (1950) found evidence pointing to the widespread occurrence of hybridization and amphiploidy. But there is little evidence in the pteridophytes for natural autoploidy except perhaps in Psilotum (Manton, 1950).

It is only fair to add that some more recent authors still maintain that autoploidy is or could be more important in nature than is indicated by the above analyses (Gilles and Randolph, 1951; Darlington, 1956b, 1963). In recent years the list of known amphiploids has increased greatly and the list of probable natural autoploids has increased slightly.

The case of tetraploid *Vaccinium uliginosum* ($2n = 48$) has been reinvestigated by Rousi (1967), who concludes that it is probably an autoploid derived from *V. u. microphyllum* ($2n = 24$). Autoploidy is claimed for tetraploid members of the *Epilobium angustifolium* group ($x = 18$) on cytogenetic evidence which is extensive and suggestive but inconclusive (Mosquin, 1967). Many cases of alleged autoploidy continue to be reported in the taxonomic literature on the basis of the old and now discredited criteria.

One of the best analyzed and most interesting cases is the *Solanum tuberosum* group in the Andean region of South America. The tuber-bearing Solanums comprise a polyploid series of $2x$, $3x$, $4x$, $5x$, and $6x$ species on the base of $x = 12$ (Swaminathan and Howard, 1953). The diploid sexual forms are self-incompatible and are cross-pollinated by Bombus and other bees; they also reproduce vegetatively by tubers (Dodds and Paxman, 1962). The polyploids are mostly sterile as to pollen and seeds and are propagated vegetatively (Swaminathan and Howard, 1953; Swaminathan, 1954).

The cultivated diploid potatoes were formerly treated taxonomically as a series of different species—*Solanum phureja*, *S. stenotomum*, etc.— but recently, after restudy, they have been reduced to cultivars of *S. tuberosum*. *Solanum tuberosum* in this broader sense also includes the triploid variety *chaucha*, the tetraploid variety *andigena*, and variety *tuberosum* proper (4x) in Europe and North America (Dodds and Paxman, 1962).

The diploid cultivars grade into one another in morphological characters. There are no crossability barriers between them. The intervarietal hybrids have regular bivalent pairing. They are fertile in some combinations and have considerable pollen sterility in others. In short, these diploid forms are closely related inter se (Swaminathan and Magoon, 1961; Dodds and Paxman, 1962).

The autotetraploid nature of the common potato, *Solanum tuberosum tuberosum* ($2n = 48$), is indicated by several lines of evidence. The older evidence is suggestive, whereas more recent evidence almost clinches the case. It has long been known that the common potato has a high frequency of quadrivalents at meiosis. The maximum number is 9 IV per cell in the subvariety Chippewa (Swaminathan, 1954). The average frequency of quadrivalents is 4.4 IV in some subvarieties (Howard, 1961). In line with this cytological behavior is the occurrence of tetrasomic ratios for some genes in cultivated potatoes (Cadman, 1942).

Further evidence is furnished by the triploid hybrid between tetraploid *tuberosum* and diploid *stenotomum*. *Solanum tuberosum stenotomum* is the closest diploid relative of the cultivated potato and is believed to be ancestral to the latter (Swaminathan and Magoon, 1961; Howard, 1961). The triploid hybrid of *tuberosum* X *stenotomum* has an average of 8.2 III per cell. This high frequency of trivalents strongly suggests that the cultivated potato is an autotetraploid with *stenotomum* chromosomes in it (Howard, 1961).

But the strongest evidence for the autoploid nature of *Solanum tuberosum tuberosum* comes from the cytogenetic behavior of its reduced or polyhaploid form with $2n = 24$ chromosomes. Some polyhaploids show 12 II or full bivalent pairing in a majority of the cells, and have high pollen fertility. These polyhaploids also form fertile hybrids with some of the diploid cultivars (Swaminathan, 1954; Howard, 1961).

The next question is the type of autoploidy in the common potato. Is it a strict or an interracial autotetraploid? It is believed to be an interracial autoploid derived from *stenotomum* and some other closely related diploid cultivar (Swaminathan and Magoon, 1961). Dodds and Paxman (1962) suggest that *stenotomum* (2x) and *phureja* (2x) produced the interracial autoploid *andigena* (4x) in South America, and that *tuberosum* proper is a north temperate selection product from *andigena*.

Most of the polyploid species in the *Solanum tuberosum* group considered as a whole, however, seem to be hybrid derivatives. Thus *Solanum juzepczukii* (3x) is derived from *S. acaule* (4x) X *S. tuberosum stenotomum* (2x); and *S. curtilobum* (5x) from another interspecific hybrid (Swaminathan and Howard, 1953; Howard, 1961; Dodds and Paxman, 1962).

The best conclusion we can draw from the available evidence in the vascular plants is that amphiploidy is far more common and widespread than autoploidy. Among the few bona fide cases of natural autoploidy, moreover, at least some are interracial autoploids as, for example, *Biscutella laevigata* and *Solanum tuberosum tuberosum*. Hybridity of one sort or another is thus usually present in polyploids in plants.

The situation in the animal kingdom is entirely different. Polyploidy is a very rare condition in naturally occurring animals, as we saw earlier. And most if not all of the natural polyploid forms of animals appear to be autoploids (White, 1954).

CHAPTER *16*

Factors Promoting Polyploidy

Introduction · Intrinsic Advantages of the Amphiploid Condition · Modes of Formation · Polyploidy and Life Form · Primary Factors · Polyploidy and the Breeding System · Geographical and Ecological Factors · The Influence of Chromosome Size · The Influence of the Genotype · Conclusions

Introduction

In this book we are primarily interested in the problem of polyploidy from the standpoint of evolution studies. Why has polyploidy played such an important role in plant speciation? This happens to be a very complex question. It is also a question which has been extensively discussed by many different authors in many different ways.

Correlations have been traced between polyploidy and climate, latitude, elevation, type of habitat, life form, breeding system, hybridity, cell size, chromosome size, chromosome structure, sex chromosome mechanism, genotype, and other factors. Some of these factors are discussed in this chapter.

There is apparently no one single reason for the evolutionary success of polyploidy. On the other hand, the various polyploidy-promoting factors are not all equally important.

In attempting to analyze the evolutionary problem of polyploidy here, I have found it useful to distinguish three grades of controlling factors. They are: the primary, or most fundamental, conditions; the secondary factors, which are of considerable importance; and the tertiary, or relatively minor, factors. The conditions favoring polyploidy are presented in this order in the later sections of the chapter. By approaching the problem in this way, we can begin to put together a series of explanations.

Intrinsic Advantages of the Amphiploid Condition

It is well known that hybrids between well-differentiated races or species frequently exhibit superior vigor or viability or have enhanced physiological homeostasis. The genetic mechanisms underlying heterosis have been extensively studied and debated. We are concerned here, however, not with the genetic causes but with the phenotypic results of hybridity. Heterosis and physiological homeostasis are observable features of many hybrid combinations in the animal and fungus kingdoms as well as in the plant kingdom. These heterotic and homeostatic properties of many hybrids must often be adaptively valuable to their carriers.

Now hybrids do not breed true for their heterozygous condition by the normal sexual mechanism. If sterile, the hybrid does not reproduce sexually at all. If fertile, its valuable properties of heterosis and homeostatis are lost in the later sexual generations. But amphiploidy is a genetic system which provides a way out of the impasse (Darlington, 1932, Ch. 13, 16). The amphiploid recovers the lost fertility, and can reproduce sexually, yet breed true for a highly heterozygous genotype. It has the advantages of internal hybridity combined with those of true breeding.

When a given natural hybrid does possess heterotic or homeostatic properties, therefore, amphiploidy is a way of perpetuating those adaptively valuable properties throughout subsequent sexual generations. Moreover, amphiploidy does not prevent segregation and recombination entirely, except in the most extreme genomic allopolyploid, but permits the release of some new variations, thereby providing an opportunity for finer adaptive adjustments to arise and become established in the later polyploid generations.

The high frequency of polyploid species in higher plants, the vast majority of which are amphiploids and the remainder of which are largely interracial autoploids, bespeaks the successfulness of this genetic system in perpetuating adaptive hybrid genotypes within a framework of sexual reproduction.

A second characteristic of polyploids which may be advantageous in many situations is the greater buffering in their genotypes, as compared with diploids, owing to the presence of numerous duplications. The polyploid individual is better-buffered physiologically than the corresponding diploid. And it is better-buffered than the diploid in its breeding behavior, releasing its stored variability more slowly owing to tetrasomic inheritance. These buffering properties are true of autoploids as well as amphiploids.

An adaptive superiority of amphiploid species over their ancestral diploid species could be revealed by their comparative geographical distribution and abundance. Gustafsson (1946–1947, pp. 278 ff.) maintains that polyploid plants are not more widespread or abundant than their diploid relatives. Stebbins (1950, p. 347) maintains that polyploids are usually but not always more widespread than their diploid ancestors. Some statistical evidence is given to support the two dissimilar generalizations. Evidently more statistical evidence is needed.

We can also approach the question by comparing polyploids and diploids in a particular group which has been extensively studied cytotaxonomically. In the ferns, Manton (1950) found numerous cases in which the polyploids are abundant, but the diploids are rare, relictual, or extinct. Examples occur in Cystopteris, Polypodium, Dryopteris, and Asplenium (Manton, 1950).

The *Gilia incospicua* complex (Polemoniaceae) contains numerous diploid and tetraploid species in the deserts and mountains of Western North America. In this group the diploid species are generally restricted geographically and ecologically as compared with their amphiploid derivatives. Many of the diploid species are definitely relictual. The tetraploid species, on the other hand, are wide-ranging and abundant through large and diversified territories (Grant, 1964a). The tetraploid Gilias are also definitely more vigorous than their diploid relatives in cultivation in the experimental garden.

Modes of Formation

The two main modes of origin of the polyploid condition are somatic doubling in mitosis and nonreduction in meiosis (Heilborn, 1934).

In somatic doubling the chromosomes divide during mitosis, but do not separate to the poles, and are all included in one daughter nucleus. As a result, a $2n$ somatic cell gives rise to a $4n$ daughter cell. This somatic doubling may affect certain particular vegetative tissues which are not in the line of development to flowers, such as the root nodules in legumes, in which case the process does not concern us further here. Somatic doubling may also take place in the zygote or early embryo and change the whole plant. Or it may occur in a bud and consequently affect one branch of the plant. Tetraploid *Primula kewensis*, which we discussed briefly in Chapter 13, arose by somatic doubling of particular flowering branches on the diploid hybrid plant.

Nonreduction involves the failure of cell wall formation during meiosis. When this occurs in diploid pollen mother cells and embryo sac mother cells, diploid spores and gametes are formed. Tetraploid zygotes may then arise in one step after union of two unreduced gametes. This was the mode of origin of Raphanobrassica, as noted in Chapter 13.

Or the nonreduction may occur in one germ line at a time to give rise to either diploid pollen grains or diploid eggs. A new tetraploid individual can then develop in two steps. In the first step a $2n$ and $1n$ gamete give a $3n$ zygote, and in the next generation an unreduced $3n$ gamete fuses with a normal $1n$ gamete to produce the tetraploid. Artificial *Galeopsis tetrahit* arose in such a sequence of separate steps, as we saw in Chapter 13.

A less common mode of origin of the polyploid condition is polyspermaty, or the fertilization of an egg by two male nuclei. This process has been reported in *Listera ovata* and some other orchids (Hagerup, 1947).

Amphiploids usually arise from diploid hybrids, as in the examples of *Primula kewensis* and Raphanobrassica. It has also been suggested that, in certain groups, particularly in Zauschneria and Galium, amphiploids have originated as products of hybridization between specifically distinct autoploids (Clausen, Keck, and Hiesey, 1945). In such cases, if this is the correct interpretation, the doubling process occurred first and the hybridization took place later.

Polyploidy and Life Form

Müntzing (1936) surveyed the chromosome numbers in 48 herbaceous genera of angiosperms containing both perennial and annual species. He found a correlation between chromosome number and growth habit. As a general rule, the perennial species of a genus have a higher chromosome number than the annuals in the same genus. In many cases the higher chromosome numbers in the perennial species are clearly of polyploid origin. Thus we find polyploid perennials and diploid annuals in Helianthus, Euchlaena, Sorghum, and other genera (Müntzing, 1936).

Stebbins (1938) confirmed and extended this finding in a subsequent survey of 202 genera of dicotyledons. He tabulated chromosome numbers in each of three life-form classes, namely, woody plants, perennial herbs, and annual or biennial herbs. A high frequency of polyploid species is found in 52% of the herbaceous genera and in 36% of the woody genera. Hence the tendency for polyploidy is stronger in the herbaceous dicotyledons. And, when the sample of herbaceous genera is broken down into perennial vs. annual subsamples, the tendency for polyploidy is clearly stronger in the perennial fraction (Stebbins, 1938).

Gustafsson also found that the frequency of polyploidy is statistically higher in the perennial members than among the annual members of the Gramineae, Ranunculaceae, Cruciferae, Leguminosae, and Compositae in northern Europe (Gustafsson, 1946–1947, Ch. 15). It can be noted that the pteridophytes in which Manton (1950) found such a high frequency of polyploidy are predominantly herbaceous perennials.

Müntzing's (1936) way of explaining the observed correlation between polyploidy and perennial herbaceous habit was to suggest that the polyploid condition influences the life form. He pointed out that experimental autoploids generally have a slower growth rate than the parental diploids. This retardation of growth would, in many cases, convert an annual into a perennial on chromosome doubling (Müntzing, 1936; also Randolph, 1935).

Later authors have preferred to trace the connection in the opposite direction. The perennial growth habit, with the possibility of vegetative propagation, favors polyploidy (Stebbins, 1938; Gustafsson, 1946–1947, 1948; Darlington, 1956a, 1963). This interpretation of the facts is strongly supported by a correlation noted by Darlington. One of the very few natural polyploids in the conifers, *Sequoia sempervirens*, is also one of the exceptional conifers which regenerates from suckers (Darlington, 1956a).

In what way does vegetative propagation favor polyploidy? This question is answered in different ways by different authors. Vegetative propagation favors the survival of polyploid types which have already arisen, according to Darlington (1963, p. 60). Clonal reproduction reduces normal recombination but facilitates the functioning of unreduced gametes, according to Gustafsson (1946–1947, Ch. 15).

A simpler and probably more general explanation is that the long life span of a perennial herb, and especially one with means of vegetative propagation, gives a hybrid plant greatly enhanced chances of undergoing somatic doubling (Stebbins, 1938; Grant, 1956c). The perennial herbaceous hybrid *Primula kewensis* did not produce fertile tetraploid branches in its first year of existence, though it did so on three separate occasions in later years, and it is fair to assume that this hybrid might never have given rise to fertile amphiploid derivatives if it had had a short life span.

Primary Factors

Three important generalizations about the nature and distribution of polyploidy have now been established. First, polyploidy is a characteristic feature of the plant kingdom, and of this kingdom only, and here it occurs predominantly in long-lived perennial herbs. Secondly, polyploidy in plants usually takes the form of amphiploidy. And, thirdly, amphiploidy is a genetic system for perpetuating an adaptively valuable hybrid genotype by means of sexual reproduction. Indeed, it is probably the best possible genetic system for accomplishing this difficult task. With these generalizations before us to serve as a guide to our thinking, we can now define the primary polyploidy-promoting factors.

Polyploidy is promoted by a combination of three *primary* factors. These are: (1) long-lived organisms usually possessing means of vegetative propagation; (2) primary speciation accompanied by chromosome repatterning; and (3) the common occurrence of natural interspecific hybridization. These three factors have been discussed as phenomena in themselves, and without special reference to polyploidy, in previous chapters of this book; factor 1 in Chapter 1, factor 2 in Chapter 9, and factor 3 in Chapters 11 and 12.

It is important to recognize that polyploidy, as a common and widespread evolutionary development, depends on the joint action of all three factors. Any single factor, if lacking in any given group of organisms, will greatly retard or completely prevent the formation of polyploid species. Factors 2 and 3 are especially critical. Factor 1 is important but not indispensable and, where absent in annual herbaceous groups, can be compensated for by other secondary factors, as we shall see later.

The three primary factors occur characteristically and distinctively in the higher plants. And it is here also that polyploid species have evolved frequently and extensively. The same three factors are largely wanting or exceptional in the animal kingdom, where polyploidy is a very rare condition. The fundamental differences between higher plants and higher animals have thus led to fundamental differences in their respective patterns of evolution, as Gustafsson (1946–1947, Ch. 15) pointed out, and one of those differences is the occurrence of polyploidy.

Within the plant kingdom, polyploidy has an uneven distribution which can be correlated in many cases with the presence or absence of one particular primary factor.

Polyploidy is largely absent in plant groups conforming to the Ceanothus pattern of speciation. Examples in Quercus, Pinus, Ribes, Ceanothus, and other genera were described in Chapter 7. Many other nonpolyploid genera are found in the conifers, as Khoshoo (1959) has shown. These genera are all woody plants which have long individual lives and often reproduce vegetatively. Natural hybridization is frequent. But the related hybridizing species do not differ with respect to chromosomal rearrangements. Factors 1 and 3 are present, but factor 2 is absent.

Müntzing (1936) showed that polyploidy was definitely less frequent in annual herbs than in perennial herbs belonging to the same genus. Heiser (1950) found that polyploidy is relatively uncommon in the weed flora of Indiana and showed that this is correlated with the prevalence of an annual life cycle in successful weedy plants. Gustafsson (1946–1947, 1948) has also pointed out that annuals in general and weed floras composed chiefly of annuals in particular, have a low percentage of polyploid species.

Now annual herbaceous plants belonging to the Madia pattern of species relationships (see Chapter 7) are characterized by interspecific differences in chromosome structural arrangement. Natural hybridization is again common. But the individual organisms are short-lived and lack means of vegetative reproduction. Factors 2 and 3 are present, but not factor 1.

Polyploidy and the Breeding System

The main mode of origin of amphiploids in annual plants is nonreduction followed by union of the unreduced gametes (Clausen, Keck, and Hiesey, 1945). If the annual hybrid plant has an outcrossing breeding system, the chances of union of two unreduced gametes are very slight. This hybrid may initiate introgression but is unlikely to produce an amphiploid during its short life span. But if the annual hybrid plant is autogamous, the chances of two unreduced gametes uniting are much more favorable (Grant, 1956c).

Two annual sections of the genus Gilia, the sections Gilia and Arachnion ($x = 9$), contain a total of 39 species, all but one of which have been studied cytotaxonomically and genetically. As to ploidy in the 38 cytologically known species, there are 23 diploids, 14 tetraploids, and 1 octoploid (Grant, 1959, 1964a, 1965, for the basic data). The same species also range in breeding system from obligate outcrossers to autogamous inbreeders. It is significant that all of the 15 polyploid species are autogamous (Grant 1956c). There is one exceptional outcrossing race in one otherwise autogamous tetraploid species.

The ancestry of the amphiploid species of annual Gilia has been traced back in a number of instances to particular autogamous diploid species. Conversely, the autogamous diploid Gilias have produced proportionately more amphiploids than have the cross-fertilizing diploid Gilias.

The same correlation between autogamous breeding system and polyploid constitution is found in the annual sections of various other genera. A majority of the polyploid species in the annual members of Galeopsis, Madia, Microseris, Clarkia, Mentzelia, Amsinckia, and Escholtzia are autogamous. (See Grant, 1956c, for statement of the correlation and for additional literature references.) The observed correlation is in agreement with the idea that autogamy is a polyploidy-promoting factor of secondary importance which compensates to some extent for the short individual life span in annual plants.

Perennial herbs, in which polyploidy is common, are usually out-crossing and often self-incompatible. The correlation between the three conditions—growth habit, polyploidy, and breeding system—can be explained in different ways in this case.

Gustafsson (1946–1947), following D. Lewis (1943b), suggests that the style in a self-incompatible diploid acts in such a way as to favor the functioning of unreduced pollen grains produced by the same plant, thus leading to the formation of tetraploid or at least triploid zygotes. In my opinion a simpler and more direct explanation is quite sufficient. Hybrid plants with a perennial habit have a long time, and hence a fair chance, of undergoing polyploid doubling, usually by somatic doubling, but also sometimes by the union of unreduced gametes. In Gustafsson's view, then, the causal connection is between self-incompatibility and polyploidy, and perenniality is a correlated condition; whereas in my view the causal connection is between perenniality and polyploidy, and outcrossing is the correlated condition.

There is another angle to this question. Polyploidization converts a self-incompatible diploid into a self-compatible tetraploid in plants with the Nicotiana-type or gametophytic system of self-incompatibility (D. Lewis, 1943a, 1966). In Solanum and Nicotiana within the Solanaceae and in some members of the Scrophulariaceae and Leguminosae, the diploid species tend to be self-incompatible and the natural polyploid species self-compatible (D. Lewis, 1966).

H. J. Muller (1925) suggested that the establishment of a polyploid condition would be particularly difficult in dioecious organisms. One of the main reasons for this was supposed to be the breakdown of the sex-determining mechanism in tetraploids with the constitution XXYY for the sex chromosomes. Muller (1925) offered this suggestion in the hope that it would account for the rarity of polyploidy in animals, which are mostly dioecious, and its common occurrence in plants, which are mostly hermaphroditic.

Muller's hypothesis has been partly confirmed and partly not confirmed by the subsequent factual evidence. The few bona fide examples of polyploidy in animals are in fact found in groups which are either parthenogenetic (brine shrimp, sowbug, bagworm moth, Otiorrhynchus weevil, Ochthiphila fly) or hermaphroditic (earthworms) (Bungenberg, 1957; Stalker, 1956). On the other hand, the polyploid condition has succeeded in establishing itself in some dioecious groups of plants in both nature (Salix, Rumex, Galium) and experiment (Melandrium) (Westergaard, 1940; Löve, 1944; Stebbins, 1950, p. 367; Dempster and Ehrendorfer, 1965).

Geographical and Ecological Factors

It has long been known that the frequency of polyploidy in higher plants is correlated with latitude. The percentage of polyploid species within a regional flora increases regularly from lower to higher latitudes. This trend was first pointed out by Hagerup (1932) and Tischler (1935). The early data were subject to various errors which were eliminated in later studies without changing the general conclusion (Tischler, 1955).

The early surveys have also been greatly extended by more recent studies. The regional floras cited in the earlier papers have been restudied more completely, and new regional floras have been added to the sample. The individual studies are too numerous to list here; reviews with numerous references are given by Löve and Löve (1949, 1957) and Hanelt (1966). The newer evidence confirms the older generalization.

Table 4 gives the percentage of polyploid angiosperms in a series of regional floras ranging from 36° to 84° north latitude in Eurasia and the Arctic. The data are taken and condensed slightly from papers by Löve and Löve (1957) and Hanelt (1966), which can be consulted for further

Table 4 · Frequency of polyploid species of angiosperms in different latitudinal zones in Eurasia and the Arctic (Condensed from Löve and Löve, 1957; and Hanelt, 1966)

Area	Latitude, N°	Polyploids, %
Sicily	36–38	37.0
Rumania	44–47	46.8
Hungary	46–49	48.6
Pardubice, CSR	50	52.3
Central Europe	46–55	50.7
Schleswig-Holstein	54–55	54.5
Denmark	54–58	53.5
England	50–61	52.8
SW Greenland	60–62	74.0
Faroes	62	71.0
Iceland	63–66	71.2
Sweden	55–69	56.9
Finland	60–70	57.3
Norway	58–71	57.6
NW Alaska	68	59.3
Devon Island	75	76.0
Spitsbergen	77–81	74.0
Franz Joseph Land	80–82	75.0
Peary Land	82–84	85.9

details. The regional floras listed in the table have all been thoroughly studied cytotaxonomically. The floras are arranged in order of increasing latitude in the table. It will be seen that the frequency of polyploid species ranges from 37% in the Mediterranean region to 74% in Spitsbergen in a transect through Europe, and is generally high throughout arctic and subarctic zones (Löve and Löve, 1949, 1957; Löve, 1953; Tischler, 1955; Hanelt, 1966).

There is some evidence to indicate that the same trend toward increasing proportion of polyploids at higher latitudes occurs also in the southern hemisphere (Hair, 1966; Hanelt, 1966).

The frequency of polyploid species tends to be relatively high in high mountains the world over. Hanelt (1966) gives percentage figures ranging from 45 to 85% polyploids in various high montane floras of Eurasia, the Americas, and New Guinea. The polyploids sometimes have a higher frequency in high mountains than in the neighboring lowlands, but this is not always the case (Hanelt, 1966).

Löve and Löve (1967) have compared neighboring alpine and forest floras on Mt. Washington in the White Mountains of New Hampshire with respect to incidence of polyploidy. In the alpine zone above tree-line, 63.6% of the taxa of vascular plants are polyploid, whereas the surrounding forest vegetation has only 45% polyploids. The frequency of polyploidy is thus significantly higher in the alpine zone (Löve and Löve, 1967).

The facts have been interpreted in a variety of ways. Most of the interpretations are plausible. But it is more difficult to decide, and it is still undecided, which ones are true. Good discussions of the various explanations are given by Löve and Löve (1949, 1953), Johnson, Packer, and Reese (1965), and Hanelt (1966).

The first suggestion was that severe climates induce polyploid formation by means of temperature shocks (Hagerup, 1932). A plausible countersuggestion is that polyploids are better adapted than diploids to cold and other extreme environmental conditions (Löve and Löve, 1949, 1967; Löve, 1953). In the mountains of Iceland and in a valley in arctic Alaska, it has been found that plant communities living under more extreme environmental conditions have a higher rate of polyploidy than neighboring plant communities in more protected locations (Löve, 1953; Johnson and Packer, 1965). This correlation within a local area tends to support the Löves' hypothesis.

Still another countersuggestion starts by denying that polyploids are in fact better adapted to cold than are diploids. Gustafsson considers that the observed high frequency of polyploids in arctic and alpine floras is related primarily to the prevalence in such floras of a life-form class, namely, perennial herbs with vegetative reproduction, in which polyploidy is particularly common. The correlation between polyploidy and climate is therefore an indirect one on this view (Gustafsson, 1946–1947, 1948). But Löve and Löve (1949) show that the correlation between polyploidy and latitude holds up within the perennial life-form classes, which weakens Gustafsson's interpretation and strengthens their own.

Another plausible suggestion is that the causal connection is not between polyploidy and extreme climates per se, but between polyploidy and the type of habitat usually found in extreme climates. There are two versions as to what the causal connection is (Stebbins, 1950; Manton, 1950). Stebbins contends that polyploids are well adapted for colonizing newly exposed habitats. New open habitats have been created by the retreat of ice sheets in the far north and high mountains, but also by lava flows and mountain building, and have been colonized to a great extent by polyploids in all such disturbed areas, according to Stebbins (1950, Ch. 9). Manton suggests that the climatic or topographic disturbances of whatever sort promote interspecific hybridization and thus favor the formation as well as the establishment of new amphiploids (Manton, 1950).

Hanelt (1966) points out that earlier authors have tended to emphasize single factors in a problem which has multifarious aspects. It may well be that several ecological factors are acting together, and interacting with one another, to promote polyploidy at high latitudes and elevations and in other extreme or disturbed environments.

The Influence of Chromosome Size

The mechanics of chromosome and cell division probably acts as a limiting factor on polyploidy, as Darlington suggested long ago. Reciprocal relationships exist between cell size, chromosome size, and polyploid chromosome numbers (Darlington, 1932, pp. 204 ff.; 1937a, p. 84).

It is known that autoploids often have a slower growth rate than the related diploids, owing in part at least to retarded cell division (Müntzing, 1936). At some ploidy level, which differs from one plant group to another, the artificial polyploid crosses a viability threshold, which again is determined partly by mitotic conditions (see Chapter 15).

The cells forming nutritive tissues in embryo sacs and anthers often or usually contain endopolyploid nuclei. The cell is a product of repeated endomitotic divisions; it does not undergo cell division. Such endopolyploid cells attain ploidy levels far above those found in whole plants. Antipodal cells in the embryo sacs of certain members of the Ranunculaceae, Papaveraceae, and Compositae may be as high as 64-ploid (Müntzing, 1961, p. 320). The record is held by 4096-ploid nuclei in the embryo sac of *Phaseolus coccineus* (Darlington, 1965, p. 670.).

Individual cells which do not have to divide can clearly function at much higher ploidy levels than whole plants which must grow by cell division. This difference between terminal members of cell lines and growing plant bodies as regards tolerance of high ploidy levels is in line with Darlington's hypothesis.

The exigencies of cell mechanics may account for the absence or rarity of polyploidy in some Liliaceae and in some woody plants. Lilium, Fritillaria, and some other liliaceous genera with large chromosomes either have no polyploids or no high polyploids. It is possible that the chromosomes are too large for cell division with high polyploid numbers in such cases (Darlington, 1932, p. 477; 1937a, p. 84).

In woody dicotyledons the narrow cambium cells must restrict both chromosome number and chromosome size. The relatively low frequency of polyploidy in woody dicotyledons compared with perennial herbaceous dicotyledons, and the complete absence of polyploidy in some dicotyledonous genera and families, may be a result of the bottleneck of cambium cells (Darlington, 1937a, p. 84; Stebbins, 1938).

The Influence of the Genotype

Reduced chromosome pairing in a species hybrid sets the stage for nonreduction and amphiploid formation. Reduced pairing is a result of chromosomal rearrangements between the parental genomes. But chromosome behavior in meiosis is also under the control of specific genes, as numerous studies have shown (Darlington, 1937a; Rees, 1961; Lewis and John, 1963). It follows that the frequency of nonreduction and hence of polyploid production in an array of chromosomally sterile hybrids may be affected by genotypic differences with respect to particular meiotic genes (Swietlińska, 1960; Swietlińska and Zuk, 1965; Grant, 1965).

Genes controlling various aberrations of meiosis, such as asynapsis or abnormal spindle formation, are known in maize, Drosophila, and other genetically well-known organisms. Meiotic genes with the particular effect of interest to us here, namely the failure of the second meiotic division, so as to produce unreduced gametes, have been found in *Datura stramonium* and *Zea mays* (Satina and Blakeslee, 1935; Avery, Satina, and Rietsema, 1959; Rhoades and Dempsey, 1966).

A recessive mutant gene in *Datura stramonium* known as "dyad" causes dyads of unreduced spores to be formed at the end of meiosis in both the male and female lines. A diploid Datura plant homozygous for "dyad," when selfed, gave rise to some tetraploid progeny (Satina and Blakeslee, 1935; Avery, Satina, and Rietsema, 1959). Plants of *Zea mays* carrying the recessive allele of "elongate" in homozygous condition produce some unreduced eggs but normal haploid pollen. The first-generation progeny of "elongate" plants therefore consist mainly of triploids. These can go on to yield tetraploids and higher polyploids in subsequent generations (Rhoades and Dempsey, 1966).

The diploid F_1 hybrid of *Rumex thyrsiflorus* X *acetosa* shows individual variation in meiotic behavior. Some F_1 plants have normal meiosis and produce tetrads; others lack the second division and produce dyads of unreduced gametes. The segregation of dyad-forming types occurs also in the F_2 and B_1 generations. The F_2 progeny of hybrid individuals with normal meiosis are diploid ($2n = 14$). But the F_2 progeny of dyad-forming hybrid plants are tetraploids or near-tetraploids with $2n = $ ca. 25 (Swietlińska, 1960; Swietlińska and Zuk, 1965).

Eight species of Leafy-stemmed Gilias have been intercrossed artificially to produce 15 hybrid combinations which have reached the stage of flowering. Four of the 15 classes of interspecific hybrids have doubled spontaneously by the union of unreduced gametes to yield fertile or semifertile amphiploid progeny in F_2. The four amphiploid-producing hybrids all have *Gilia millefoliata* and/or *G. valdiviensis* as one or both parents. *Gilia millefoliata* and *G. valdiviensis* are two closely related diploid ($2n = 18$) species belonging to the same genome group. There is thus an unusual concentration of the *millefoliata* genome in the sample of artificial spontaneous amphiploids in the Leafy-stemmed Gilias (Grant, 1965).

One of the hybrids, *Gilia millefoliata* X *achilleaefolia*, doubled repeatedly in three replicate cultures. Replicate cultures of other hybrid combinations not containing *G. millefoliata* or *G. valdiviensis*, by contrast, have consistently failed to yield amphiploids (Grant, 1965).

It is significant in the light of these experimental results that two of the three known natural tetraploid species of Leafy-stemmed Gilia carry the *millefoliata* genome (Grant, 1965).

Levin (1968) finds a similar situation in the genus Phlox in eastern North America. There are five allotetraploid species of Phlox in this area. Four of the five tetraploid species contain the genome of the diploid species, *Phlox pilosa* (Levin, 1968).

It appears, therefore, that parental species differ genotypically in their ability to produce amphiploid progeny, in the experimental plot and in nature, where other factors are more or less equal. The genic contents of different genomes may be a tertiary factor affecting the systematic distribution of polyploids within a genus or section composed of diverse, sympatric, basic species.

Conclusions

Polyploidy is a common and widespread genetic system in higher plants. Within the subkingdom of vascular plants the polyploid condition is unevenly distributed with respect to systematic relationships, life form, climate, breeding system, chromosome structure and size, and other factors. The clue to the reasons for the evolutionary success of polyploidy in higher plants lies hidden in these various correlations and can be revealed by a proper analysis.

The vast majority of the naturally occurring polyploid species of plants that have been analyzed taxogenetically have turned out to be amphiploids. Now amphiploidy is a type of permanent hybridity. It is a genetic system which permits a species hybrid to bypass the sterility barrier and to breed true for its hybrid constitution. Amphiploidy is a genetic system enabling plants to exploit the intrinsic advantages of a highly heterozygous genotype, particularly heterosis and physiological homeostasis, in the numerous hybrid combinations where these advantageous properties are present.

The ability of a plant group to take advantage of the amphiploid genetic system in its evolutionary development depends on factors of various sorts and various degrees of importance. There are factors affecting the origin of a new polyploid and other factors affecting its subsequent establishment as a natural population. Internal and external factors are involved in both the origin and the establishment of the new polyploid.

Of primary and fundamental importance is a combination of three conditions. The first one is the existence of diploid species carrying different genomes or subgenomes. The second is natural hybridization between these species. The third is a long-lived perennial growth habit to increase the chances of somatic doubling; or, as a partial compensation in short-lived annuals, an autogamous breeding system to increase the chances of union of unreduced gametes. If any one of the three conditions is lacking in any given plant group, polyploidy is expected to be absent or rare. A survey of the distribution of polyploidy in relation to chromosome repatterning, hybridity, and perennial growth habit (or autogamy as a substitute in annuals) shows that polyploidy is, in fact, common only in plant groups possessing these characteristics.

Polyploidy is not uniformly present or uniformly common in all plant groups meeting the three primary conditions. The uneven distribution of polyploidy among plant groups which are comparable with respect to the primary factors points to the existence of other secondary and tertiary factors.

Severe climates, and high latitudes and high elevations as functions of severe climates, and stressful or disturbed habitats within a given climatic zone, are among the external polyploidy-promoting factors of secondary importance. Chromosomes which are large in relation to cell size, and cells which are small in relation to chromosome size and numbers, are among the internal secondary polyploidy-inhibiting factors. The presence or absence of genes causing failure of the second meiotic division and hence the formation of unreduced gametes is a tertiary factor which may explain the proportionately unequal contributions of different diploid species to the production of amphiploids within some genera.

CHAPTER *17*

Agmatoploidy

Introduction · Polycentric Chromosomes · Carex · Luzula · Cyperaceae and Juncaceae · Conclusions

Introduction

Chromosome number differences between individuals or populations take three forms: polyploidy, aneuploidy (in the strict sense), and agmatoploidy. Aneuploidy (sensu stricto) and polyploidy are numerical differences with respect to individual chromosomes or whole sets, respectively. The term agmatoploidy refers to differences in the number of independently assorting pairs of chromosome fragments in a group of organisms (Malheiros-Gardé and Gardé, 1951; Löve, Löve, and Raymond, 1957).

These types of chromosome number variation are correlated with the type of centromere. In most plants and animals the chromosome arms are attached to localized centromeres which perpetuate themselves and their chromosomes through a series of cell generations. The gain or loss of one or more chromosome pairs in the complement depends on the gain or loss of the corresponding centromeres. Conversely, chromosome arms or fragments cannot persist in the cell line or population unless they are attached to a centromere. Changes in chromosome number are aneuploid (sensu stricto) or polyploid.

A few groups of plants and animals have diffuse centromeres. The centromeric activity is not localized in a single region but is spread out through the length of the chromosome. Chromosomes containing diffuse centromeres have been termed polycentric (Lima-de-Faria, 1949; Bernardini and Lima-de-Faria, 1967). Fragments arising by transverse divisions of an original polycentric chromosome can perpetuate themselves through cell divisions, if their broken ends heal, and can thus become permanent components of the population. Diffuse centromeres open up the possibility of agmatoploidy (Lima-de-Faria, 1949; Löve, Löve, and Raymond, 1957).

Agmatoploid series cannot be distinguished from aneuploid series proper in a simple list of chromosome numbers. One may encounter $2n = 10, 12, 14 \ldots$ in either case.

Therefore the term aneuploidy in its general sense is still useful for descriptive purposes in dealing with nonpolyploid chromosome number variation. The phenomenon now known as agmatoploidy was formerly subsumed under aneuploidy, but was segregated out as its nature and special features became better understood (Malheiros-Gardé and Gardé, 1951). As a corollary of this terminological segregation, aneuploidy acquires a more narrowly defined usage as well as the original broad usage. It will be necessary in this chapter to use the term aneuploidy in both the strict and the general sense; the usage will be duly identified.

Polycentric Chromosomes

In the animal kingdom, chromosomes with diffuse centromeres are found in some Hemiptera or true bugs, i.e., in the Coccidae, Aphidae, and Heteroptera, and in the Acarina or mites (Lima-de-Faria, 1949). Among higher plants, diffuse centromeres occur in the Juncaceae

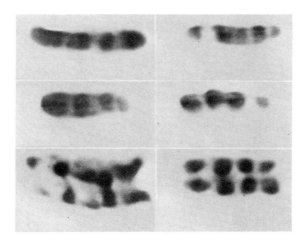

Figure 34 Mitosis in *Luzula campestris* ($2n = 12$). Lateral views showing chromosome orientation at metaphase and early anaphase. (Brown, 1954.)

(Luzula, Juncus) and Cyperaceae (Eleocharis, Scirpus, Carex) and have been suggested for the Zingiberaceae and Musaceae (Löve, Löve, and Raymond, 1957, for review with references).

Polycentric chromosomes containing diffuse centromeres differ in morphology and behavior from the ordinary type of chromosome with a localized centromere. It will be recalled that ordinary chromosomes exhibit a definite centromeric constriction at metaphase of mitosis, and that these centromeres become aligned on the metaphase plate, with the chromosome arms dangling in various positions. Polycentric chromosomes lack the constriction and, moreover, the whole chromosome becomes aligned on the metaphase plate. At anaphase ordinary chromosomes move toward the poles with the centromeres in the forward position and drawing the arms behind. The polycentric chromosome, by contrast, drifts toward the pole in a nearly single plane or even with its ends ahead of the midsection (Löve, Löve, and Raymond, 1957).

The above-mentioned features of polycentric chromosomes are exhibited by mitotic cells of *Luzula campestris* (Juncaceae) and are shown in Figure 34 (Brown, 1954).

There has existed a question as to whether the centromeric activity is continuously or discontinuously distributed throughout the length of a polycentric chromosome. In other words, is the diffuse centromere, fully or incompletely diffuse? Recent evidence in *Luzula purpurea* supports the latter possibility. The chromosomes of this species apparently contain numerous centromeric segments (Bernardini and Lima-de-Faria, 1967).

It is well known that acentric fragments of ordinary chromosomes tend to lag during cell division and sooner or later become lost from the viable cell lines. In organisms with diffuse centromeres, however, chromosome fragments could be expected to complete the normal metaphase and anaphase movements. Håkansson induced chromosome breakage by X-ray treatment in *Eleocharis palustris* (Cyperaceae). He found, as expected, that the chromosome fragments, even small ones, arranged themselves on the equatorial plate at metaphase and moved to the poles at anaphase (Håkansson, 1954, 1958).

Spontaneous chromosome breakage in organisms with polycentric chromosomes may therefore lead to the formation of ascending agmatoploid series. A species group may show much variation in the number of chromosome pairs. And the chromosome sets containing the larger numbers will include at least some smaller-sized members (Håkansson, 1954; Löve, Löve, and Raymond, 1957).

This is not to say that agmatoploidy is an inevitable result of fragmentation of polycentric chromosomes. Other factors, cytological as well as selective, are undoubtedly involved. White (1957a, 1957b) has called attention to one probably important cytological factor, namely, the role of the telomere or chromosome end. If the fragments do not acquire normal healed ends, they may not survive long (White, 1957a, 1957b).

In fact, aneuploidy (sensu lato) does not occur universally in all groups of organisms which possess polycentric chromosomes. Among Heteroptera with diffuse centromeres, some families such as the Pentatomidae and Corixidae show great constancy in chromosome number, while other families have a wider range of chromosome numbers (White, 1957a). Various species of Luzula, Eleocharis, and Carex have retained low basic numbers. There is relatively little aneuploidy (sensu lato) in Eleocharis.

Carex

The genus Carex is remarkable for its long and nearly continuous aneuploid series (sensu lato). The series ranges from $n = 6$ to $n = 56$

and includes every gametic number from $n = 12$ to $n = 43$ (Heilborn, 1932; Stebbins, 1950, p. 453; Davies, 1956). Short, consecutive aneuploid series (sensu lato) are found within several of the sections of Carex. Examples are Distantes ($n = 28$–37), Extensae ($n = 30$–35), Acutae ($n = 34$–42), and the *Carex caryophyllea* group ($n = 31$–34) (Davies, 1956).

The basic numbers $x = 6$, 7, 8, and 9 are known in Carex and an additional and perhaps original basic number, $x = 5$, is inferred (Heilborn, 1939; Wahl, 1940; Löve Löve, and Raymond, 1957).

Polyploidy occurs in Carex on the above and other secondary basic numbers. Thus the *C. siderosticta* group has forms with $n = 6$ and $n = 12$; the section Capillares ($x = 9$) contains diploid, tetraploid, and hexaploid species; and the *C. stenantha* group has forms with $n = 17$ and $n = 34$. There are polyploid series in Carex based on $x = 5, 6, 7, 8$, and 9 (Wahl, 1940; Löve, Löve, and Raymond, 1957).

Some of the observed aneuploid variation in Carex can therefore be attributed to various side effects of polyploidy (Wahl, 1940; Löve, Löve, and Raymond, 1957). Monobasic polyploidy on different base numbers will give similar but different derivative numbers, and dibasic polyploidy will fill in the gaps. Thus, different hybrid combinations of 7-paired and 8-paired diploids will produce a series of tetraploids with $n = 14, 15$, and 16. Polyploid drops and gains may extend the series further in both directions.

Natural hybridization is also known to occur in the genus Carex. Drury (1956) reports that nearly all of the species of section Vesicariae in Alaska are involved in hybridization. Natural hybrids occur between *Carex rostrata* on the one hand and *C. rotundata, membranacea*, and *physocarpa* on the other. Hybrid swarms between *C. rostrata* and *C. rotundata* in one part of Alaska are segregating out a new constant and fertile form which is increasing in numbers and developing into a new species of hybrid origin named *C. paludivagans* (Drury, 1956).

Hybridization and polyploidy can bring about true aneuploid variation. They have been shown to do so in the *Eleocharis palustris* group (Lewis and John, 1961). Aneuploidy in the strict sense is undoubtedly present also in Carex. But this does not appear to be the whole story.

Carex chromosomes do not have a visible localized centromere. On chromosome morphology they appear to have a diffuse centromere.

Moreover they often move end first to the poles at anaphase. These facts indicate that the chromosomes are polycentric, and suggest that fragmentation could contribute to the observed variation in chromosome number (Davies, 1956; Löve, Löve, and Raymond, 1957).

Heilborn (1924, 1932) compared *Carex pilulifera* and *C. ericetorum* with respect to total length as well as number of the somatic chromosomes. He found that the species with the higher number, *C. ericetorum* ($n = 15$), has shorter chromosomes than the lower-number species, *C. pilulifera* ($n = 9$). The number of long chromosomes in the complement decreases as the total number of chromosomes increases (Heilborn, 1924, 1932). Later workers have confirmed this observation in other species groups of Carex. The species with higher numbers usually have more short chromosomes than their low-number relatives (Davies, 1956; Löve, Löve, and Raymond, 1957).

The inverse relation between chromosome number and chromosome size could be due to several factors other than fragmentation. Consequently this line of evidence does not prove that fragmentation and agmatoploidy have contributed to the observed chromosome number variation in Carex. The size-number relation, however, when taken in conjunction with the occurrence of polycentric chromosomes and the unusually extensive development of aneuploidy (sensu lato), does suggest strongly that agmatoploidy has played a role in speciation in the genus Carex.

Luzula

The chromosomes of Luzula (Juncaceae) have been shown to possess diffuse centromeres (Malheiros, De Castro, and Câmara, 1947; Nordenskiöld, 1951, 1962; Brown, 1954). These and other authors have considered the possibility that chromosome fragmentation has contributed to chromosome number differences within this genus.

Chromosome numbers in Luzula range from $2n = 6$ in *L. purpurea* to $2n = 66$ and 72 in *L. pilosa*. The number $2n = 12$ is found in many species and is a basic number in the genus. There is an ordinary polyploid series on this base of $x = 6$ which reaches the octoploid level (Nordenskiöld, 1951).

Aneuploidy (sensu lato) is also present. Thus *Luzula spicata* has the contiguous numbers $n = 6$ and 7, and *L. orestra* has $n = 10$ and 11.

In each case the complement with the higher number includes two small chromosomes corresponding to one large chromosome in the lower-number complement. The aneuploid increase is therefore probably due to fragmentation and agmatoploidy (Nordenskiöld, 1951, 1956).

In addition, Nordenskiöld (1951) finds polyploid series of a special kind in which the total length of all chromosomes combined remains about constant. The individual chromosomes of a tetraploid are about half the size of those in the diploids; and the chromosomes of an octoploid are about half as large as those in the tetraploids. This situation is found in the *Luzula campestris* and the *L. spicata* groups. Nordenskiöld attributes the simultaneous changes in number and size to fragmentation of all the chromosomes in the ancestral complement. The numerical relations between species are polyploid, but the mode of doubling may be fragmentation rather than multiplication, and the series then constitutes a special type of agmatoploidy (Nordenskiöld, 1949, 1951, 1956, 1961).

Confirmation of this hypothesis of agmatoploidy is provided by the pairing behavior of large chromosomes with small ones in hybrids between species of different ploidy levels. *Luzula campestris* and *L. pallescens* have 6 pairs of large chromosomes; *L. sudetica* has 24 pairs of small chromosomes and is technically octoploid as to number. At meiosis in the interspecific hybrids, each of the six large chromosomes belonging to the diploid parental species pairs with several small chromosomes contributed by the high-number species, *L. sudetica* (Nordenskiöld, 1951, 1956, 1961).

Cyperaceae and Juncaceae

The family Cyperaceae possesses several unusual and distinctive cytological and embryological features. These features have been pointed out by Heilborn (1924, 1932) and many later students.

In the first place, polycentric chromosomes occur in some genera of Cyperaceae. Diffuse centromeres occur in Eleocharis and Carex, as we have already seen. Fimbristylis and some species of Cyperus, on the other hand, have localized centromeres. Both types of centromere apparently occur in Bulbostylis (Löve, Löve, and Raymond, 1957).

Agmatoploidy also occurs in the Cyperaceae as a correlated condition. Remarkably extensive and continuous aneuploid series (sensu lato) occur in Carex. Some aneuploidy is found in Eleocharis. But the species of Fimbristylis fall into an even or nearly even polyploid series.

In various members of the Cyperaceae, moreover, the two meiotic divisions occur in an order which is the inverse of that in other plants. The paired chromosomes do not disjoin in the first meiotic division, but do separate in the second division. In contrast to the normal sequence, division I is equational and division II reductional (Wahl, 1940; Davies, 1956).

Davies (1956) has pointed out that postreductional meiosis is closely correlated with diffuse centromeres in both the angiosperms and the insects. The explanation of this correlation is not clear.

A fourth peculiar characteristic of the Cyperaceae is the mode of pollen formation. In each tetrad of microspores, three degenerate and only one develops into a functional pollen grain (Heilborn, 1932; Wahl, 1940; Davies, 1956).

It is very interesting that most of these unusual cytological and embryological features appear again in the family Juncaceae.

Diffuse centromeres are found in Luzula, as mentioned earlier in this chapter, but not apparently in Juncus. Correspondingly, Luzula exhibits much aneuploidy (sensu lato), including some agmatoploidy, whereas Juncus has almost no aneuploidy. Postreductional meiosis occurs in the Juncaceae according to Nordenskiöld (1962). The four microspores all develop in the Juncaceae, but adhere together in a tetrad, which represents a condition reminiscent of that found in the Cyperaceae (Wulff, 1939; Maheshwari, 1949).

The foregoing cytological and cytogenetic similarities between the Juncaceae and Cyperaceae, taken together with various morphological and anatomical similarities, indicate that the two families are much more closely related to one another than was assumed in the older systems of classification, where the Juncaceae were allied to the lilies and the Cyperaceae to the grasses (Wulff, 1939; Maheshwari, 1949; Håkansson, 1954; Cronquist, 1968).

Conclusions

The term aneuploidy refers to differences between populations, individuals, or cells with respect to one or a few chromosomes. Aneuploidy thus includes diverse forms of chromosome number variation.

Aneuploidy includes some phenomena not discussed in this book. Among these are polysomic types and the presence or absence of supernumerary chromosomes and B chromosomes. It includes some other phenomena which have been discussed briefly in other chapters. Secondary aneuploidy is the loss or gain of one or more chromosome pairs in a polyploid, in other words, the polyploid drop or increase, as mentioned in Chapters 15 and 20. Primary aneuploidy is the loss or gain of one or more pairs of individual chromosomes in diploids by means of unequal reciprocal translocations followed by loss or gain of the corresponding centromeres. The results of primary aneuploidy, though not the mechanism, are referred to in Chapters 9 and 20.

The present chapter has discussed still another special type of aneuploidy, namely, agmatoploidy, which is the gain of one or more pairs of fragmented chromosomes in organisms with diffuse centromeres. It will be evident from the discussion in this chapter that our understanding of agmatoploidy is still in its early exploratory stages. The cytogenetic facts are not settled beyond dispute in many cases. The relation between agmatoploidy and plant speciation is unclear. We do not know how different agmatoploid populations are isolated reproductively. And we do not understand the adaptive value of agmatoploid changes.

CHAPTER 18

Agamospermy

Introduction · Embryological Pathways · Systematic Distribution · Breeding System · Hybridity and Agamospermy · Formation of Gametophytic Apomixis · Breeding Behavior · Segregation · Variation Pattern · Advantageous Features of Agamospermy

Introduction

Agamospermy is seed formation without fertilization. It is one of the main forms of apomixis, or reproduction without fertilization, in higher plants. Agamospermous reproduction in itself includes a wide variety of complex phenomena as will be described later. Agamospermy is of widespread occurrence in flowering plants, as indicated by the list in Table 5.

The subject of agamospermy was introduced briefly in one context in

Chapter 1 and in another context in Chapter 13. Here we shall review the embryological and cytogenetic phenomena as a basis for understanding the generation of variability in agamospermous plants. With this as a background, we can then go on to consider the evolutionary development of agamic complexes in Chapter 21.

For a more detailed treatment the reader is referred to Gustafsson's classic monograph, *Apomixis in Higher Plants* (1946–1947). Many additional facts of later vintage are given in the reviews of Stebbins (1950, Ch. 10), Nygren (1954), and Fryxell (1957).

Embryological Pathways

The various embryological pathways leading to agamospermous seed formation are shown in Figure 35. This diagram incorporates the classification and terminology of Gustafsson (1946–47), and reduces the complex phenomena to a form which makes broad comparisons possible.

The normal life cycle of a seed plant, which is ancestral to the various types of agamospermous life cycle, consists, as is well known, of an alternation of diploid sporophyte and haploid gametophyte generations; and these generations are separated by events of meiosis and fertilization which alternately reduce and restore the somatic chromosome number (Figure 35A). Agamospermous life cycles which perpetuate a more or less constant chromosome number over a series of seedling generations involve the bypassing of both meiosis and fertilization in the course of embryo formation.

This result is achieved in two ways, corresponding to the two main types of agamospermy known as gametophytic apomixis and adventitious embryony. In gametophytic apomixis a morphological gametophyte is present but is unreduced (Figure 35B). In adventitious embryony, by contrast, there is no gametophyte stage and the alternation of generations is accordingly eliminated (Figure 35C).

Under gametophytic apomixis there are two alternate pathways leading from the maternal sporophyte to the unreduced gametophyte, and two more pathways running from this gametophyte to the new embryo, as shown and labeled in Figure 35B. In diplospory the female gametophyte arises directly from an embryo-sac mother cell which does

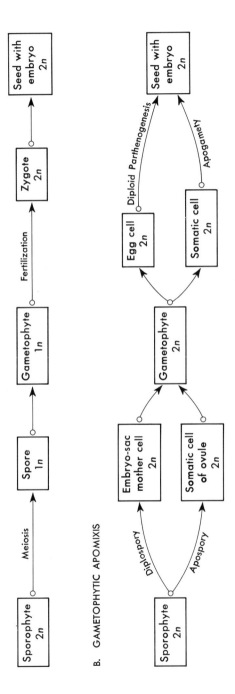

Figure 35 Modes of agamospermous reproduction. A normal sexual life cycle is shown for comparison.

A. NORMAL SEXUAL LIFE CYCLE

| Sporophyte 2n | →Meiosis→ | Spore 1n | → | Gametophyte 1n | →Fertilization→ | Zygote 2n | → | Seed with embryo 2n |

B. GAMETOPHYTIC APOMIXIS

Sporophyte 2n → Diplospory → Embryo-sac mother cell 2n → Gametophyte 2n
Sporophyte 2n → Apospory → Somatic cell of ovule 2n → Gametophyte 2n
Gametophyte 2n → Egg cell 2n → Diploid Parthenogenesis → Seed with embryo 2n
Gametophyte 2n → Somatic cell 2n → Apogamety → Seed with embryo 2n

C. ADVENTITIOUS EMBRYONY

| Sporophyte 2n | → | Somatic cell of ovule 2n | → | Seed with embryo 2n |

not undergo meiosis. Or the female gametophyte may arise from somatic cells in the nucellus or chalaza of the ovule (apospory). The female gametophyte in either case contains an unreduced egg and various accessory cells. In diploid parthenogenesis the unreduced egg cell produces the new embryo without fertilization. In apogamety some cell of the female gametophyte other than the egg gives rise to the embryo (Figure 35B).

Apogamous development of embryos has been described in a number of angiosperms. Gustafsson cites cases of synergid embryos in Taraxacum, Hieracium, Alchemilla, Alnus, and Poa; and cases of antipodal embryos in Hieracium, Elatostema, and Allium (Gustafsson, 1946–1947, p. 33). But, as Gustafsson notes, the phenomenon probably has little significance in nature in the angiosperms.

The remaining processes combine to give two complete lines of development which can be designated diplospory-parthenogenesis and apospory-parthenogenesis. Both sequences are well represented in their clear-cut form in the flowering plants (see Table 5). They also occur in mixed form in some angiosperm species where both diplosporous and aposporous embryo-sacs arise (Table 5).

In adventitious embryony, finally, the new embryo develops directly from the nucellus or integument of the ovule on the maternal plant, thus bypassing the embryo-sac entirely. Adventitious embryony is the type of agamospermy which most closely approaches the viviparous form of vegetative propagation. In both cases the new individual develops asexually in the inflorescence of the mother plant. But in adventitious embryony the daughter plant begins as an embryo within a seed, whereas in vivipary it develops from a bulbil arising within or in place of a flower.

In some agamospermous angiosperms, though not in others, pollination is necessary for apomictic seed development. The phenomenon wherein an apomictic flowering plant requires pollination as a prerequisite for seed formation is known as pseudogamy. In pseudogamy, pollination may provide a general stimulus to start embryo development, or, more commonly, pollination is followed by fusion of a sperm nucleus with the endosperm nuclei. This fertilization process then initiates the development of endosperm essential for proper embryo growth. The embryo itself, however, receives no genetic material from the pollen parent, and is entirely maternal in genotype.

An agamospermous plant may produce seeds exclusively by asexual processes, or it may occasionally produce sexual progeny. These alternatives mark the difference between obligate and facultative apomixis.

A single facultatively agamospermous plant may produce sexual seeds and asexual seeds in varying proportions in different years or even in the same year's batch. In other words, the sexual embryological pathway diagrammed in Figure 35A may be active in one line of seed development, while an alternative agamospermous pathway (Figure 35, B or C) is active in a different time or place on the same individual plant.

Hieracium aurantiacum in the Carpathian Mountains of Poland produces two types of viable ovules in the same flowering heads. Some ovules contain reduced embryo sacs resulting from normal meiosis, while others contain unreduced embryo-sacs of aposporous derivation. In the latter the embryo and endosperm develop autonomously, whereas in the former they require fertilization as a precursor to further development. Consequently a single flowering head produces a mixture of sexual and asexual achenes. The proportion of the two classes of achenes varies from plant to plant (Skalinska, 1967).

In the ferns, where agamospermy is of course absent by definition, an analogous role is played by apogamety, which is the prevailing mode of asexual embryo formation in this group. The sequence in apogamous ferns begins with the production of unreduced spores in the sporangia of the mother plant. These spores give rise to gametophytes carrying the unreduced chromosome number. A new sporophyte then develops from vegetative cells of the gametophyte without fertilization (Manton, 1950, Chs. 10–12).

Systematic Distribution

A list of agamospermous flowering plants is presented in Table 5. This list is rearranged from Gustafsson (1946–1947, App. II) and Nygren (1954), who give cross-references to the original papers. It should be borne in mind that many of the taxonomic species listed in Table 5 contain sexual as well as agamospermous forms. The species are grouped in the table first by mode of agamospermy and second by plant family.

Table 5 · Agamospermous flowering plants (Rearranged from Gustafsson, 1946–1947, Appendix II; and Nygren, 1954)

I. Diplospory-Parthenogenesis
  Dicotyledons
    Balanophoraceae
      Balanophora japonica ($2n = 94$–112); B. globosa (high number)
    Compositae
      Antennaria alpina ($2n = 84$); and 13 other species
      Chondrilla acantholepis ($2n = 15$); and 8 other species ($2n = 15, 20$)
      Erigeron annuus ($2n = 27$); and 2 other species ($2n = 27, 32$–36)
      Eupatorium glandulosum ($2n = 51$)
      Hieracium subgenus Euhieracium, numerous species ($2n = 18, 27, 36$)
      Ixeris dentata ($2n = 21$)
      Rudbeckia californica ($2n = 76$); and 5 other species ($2n = 76$)
      Taraxacum, numerous species ($2n = 24, 32, 40, 48$)
    Cruciferae
      Arabis holboellii ($2n = 14, 21$)
    Plumbaginaceae
      Statice olaeefolia confusa ($2n = 27$)
    Saururaceae
      Houttuynia cordata ($2n = 94$–104)
    Thymeliaeaceae
      Wikstroemia viridifolia ($2n = 27$)
    Urticaceae
      Elatostema eurhynchum ($2n = 52$); and 3 other species ($2n = 60, 80$)
  Monocotyledons
    Amaryllidaceae
      Cooperia pedunculata ($2n = 48$)
      Zephyranthes texana ($2n = 24$)
    Burmanniaceae
      Burmannia coelestis ($2n = 64$–72)
    Gramineae
      Calamagrostis canadensis ($2n = 42$–68); and 6 other species ($2n = 42$–112)
      Poa glauca ($2n = 42$–70); and 4 other species

TABLE 5 Cont.

II. Apospory-Parthenogenesis
    Dicotyledons
      Compositae
        Crepis acuminata ($2n = 33, 44$); and 8 other species ($2n = 22$–88)
        Hieracium subgenus Pilosella, numerous species ($2n = 27, 36, 42, 45$)
      Guttiferae
        Hypericum perforatum ($2n = 32$)
      Polygonaceae
        Atraphaxis frutescens ($2n = 88$?)
      Ranunculaceae
        Ranunculus auricomus ($2n = 16, 32, 40, 48$)
      Rosaceae
        Alchemilla arvensis ($2n = $ ca. 48); and 2 other species
        Malus hupehensis ($2n = 51$)
        Potentilla argentea ($2n = 14, 28, 35, 42, 56$); P. praecox ($2n = 42$)
        Rubus section Moriferus, numerous species ($2n = 21$–49)
        Sorbus chamaemespilus ($2n = 68$); S. hybrida ($2n = 68$)
      Urticaceae
        Elatostema acuminatum ($2n = 50$); E. latifolium ($2n = 60$)
    Monocotyledons
      Gramineae
        Paspalum dilatatum ($2n = 40$)
        Poa ampla ($2n = 63$); and 5 other species
III. Both Diplospory-Parthenogenesis and Apospory-Parthenogenesis
      Compositae
        Antennaria carpatica ($2n = 40$–42)
        Parthenium argentatum ($2n = 54$–144); P. incanum ($2n = 54$–90)
      Rosaceae
        Potentilla verna ($2n = 42, 49, 84$)
IV. Adventitious Embryony
    Dicotyledons
      Anacardiaceae
        Mangifera indica ($2n = 40$, ca. 52)
      Betulaceae
        Alnus rugosa ($2n = 28$)

TABLE 5 Cont.

Buxaceae
  Sarcococca pruniformis ($2n = 28$); and other species
Cactaceae
  Opuntia ficus-indica ($2n = 88$); and other species
Capparidaceae
  Capparis frondosa
Celastraceae
  Celastrus scandens ($2n = 46$)
  Euonymus latifolius; and other species
Euphorbiaceae
  Euphorbia dulcis ($2n = 12, 28$)
Guttiferae
  Garcinia mangostana ($2n = $ ca. 76)
Malphigiaceae
  Hiptage madablota
Myrtaceae
  Eugenia malaccensis ($2n = 22$); and other species
Ochnaceae
  Ochna serrulata ($2n = 35$)
Rutaceae
  Citrus, numerous species ($2n = 18, 27, 36$)
  Triphasia aurantiola
  Xanthoxylum bungei
Monocotyledons
  Araceae
    Spathiphyllum patinii ($2n = 18$)
  Liliaceae
    Allium odorum ($2n = 16, 32$)
    Hosta coerulea ($2n = 60$)
    Nothoscordum fragrans ($2n = 16$–$19, 24$)
    Smilacina racemosa
    Tulipa gesneriana ($2n = 24, 36$)
  Orchidaceae
    Nigritella nigra ($2n = $ ca. 64)
    Zeuxine sulcata ($2n = 44$)
    Zygopetalum mackayi ($2n = 40$)

The list is incomplete but fairly representative. We observe a wide range of plant families under each main type of agamospermy. Thus diplospory-parthenogenesis occurs in quite a few members of the Compositae and in numerous other families. Apospory-parthenogenesis is found in various Rosaceae and in other families. And adventitious embryony appears fairly frequently in the Liliaceae, Rutaceae, and other families.

Both diplospory-parthenogenesis and apospory-parthenogenesis are found together in the same species in *Parthenium argentatum* and some other plants (see Table 5). These two related modes of gametophytic apomixis occur in different species of Poa and in different subgenera of Hieracium.

Pseudogamy is a widespread condition in agamospermous angiosperms. The condition is found inter alia in species of Poa, Rubus, Potentilla, Ranunculus, Hypericum, Parthenium, Citrus, and Allium. Among the nonpseudogamous groups are Hieracium, Taraxacum, Antennaria, Crepis, and Calamagrostis (Gustafsson, 1946–1947, Ch. 6).

When the list of species in Table 5 is surveyed from the standpoint of geographical and climatic distribution, certain tendencies become apparent. Many of the examples of adventitious embryony are seen to occur in tropical and subtropical plants, whereas gametophytic apomixis is more common in plants of northern and colder regions.

With respect to chromosome number there is a very high proportion of polyploids in the sample of agamospermous flowering plants (Gustafsson, 1946–1947, Ch. 5). Many of these are high polyploids, such as *Antennaria alpina* with $2n = 84$. Others are odd polyploids, like the triploids in Crepis and Malus. Still others exhibit a range of aneuploid numbers varying around some polyploid mode. Thus polyploidy in one form or another is usually associated with agamospermy. Nevertheless, there are some agamospermous diploids in Citrus, Nothoscordum, Potentilla, and other genera.

Gustafsson (1946–1947) and Stebbins (1950) point out that asexual diploids occur more commonly in groups with adventitious embryony, like *Nothoscordum fragrans* and many species of Citrus, than in groups with gametophytic apomixis. The only known diploids with gametophytic apomixis seem to be some members of the *Potentilla argentea*, *Ranunculus auricomus*, *Hieracium umbellatum*, and *Arabis holboellii* complexes (Gustafsson, 1946–1947, Ch. 5; Stebbins, 1950, p. 391; Nygren, 1954).

The present author does not know of any recorded cases of agamospermy in the gymnosperms. No such cases are listed by Fryxell (1957) in his compendious review.

Manton (1950) lists the following examples of apogamous ferns (Polypodiaceae) with their levels of ploidy where known:

Dryopteris borreri ($2x$, $3x$, $4x$, $5x$), D. remota ($3x$), D. atrata ($3x$)
Phegopteris polypodioides
Cyrtomium falcatum ($3x$), C. fortunei ($3x$), C. caryotideum ($3x$)
Asplenium monanthes ($3x$)
Pellaea atropurpurea ($3x$)
Pteris cretica ($2x$, $3x$, $4x$).

These apogamous ferns all have high chromosome numbers ranging from $2n = 58$ in one form of *Pteris cretica* to $2n = 205$ in a form of *Dryopteris borreri*. It will be noted that most of them are odd polyploids, mainly triploids but occasionally pentaploids (Manton, 1950, pp. 195, 303–304). Many additional cases of apogamety in Pteris are reported by Walker (1962).

Breeding System

Gustafsson (1946–1947) surveyed 35 genera of flowering plants which contain both agamospermous species and sexual species. The purpose of this survey was to determine the type of breeding system in the sexual relatives of agamospermous plants.

The results can be grouped as follows. In ten of the partly agamospermous genera the sexual species are dioecious and/or monoecious. In 16 genera the sexual forms are self-incompatible. In 7 of the genera the floral mechanism promotes outcrossing. Self-fertilizing flowers are indicated in only 1 or possibly 2 of the 35 genera in the sample (Gustafsson, 1946–1947, Ch. 10).

Examples of agamospermous plants with dioecious or monoecious relatives are found in Elatostema, Balanophora, Wikstroemia, Antennaria, Euphorbia, and Alnus. Self-incompatible relatives of agamospermous plants occur inter alia in Calamagrostis, Zephyranthes,

Rubus, Parthenium, Taraxacum, and Hieracium. Burmannia is one of the exceptional genera containing self-fertilizing species related to agamospermous types (Gustafsson, 1946–1947, Ch. 10).

Self-incompatibility is seen to be the most common single breeding system in the sample of partly agamospermous genera. Dioecism and monoecism, which are relatively uncommon in the angiosperms as a whole, have a disproportionately high frequency of occurrence in partly agamospermous genera. The striking general conclusion emerges that outcrossing by means of one breeding system or another prevails in the sexual relatives of nearly all agamospermous flowering plants (Gustafsson, 1946–1947).

Hybridity and Agamospermy

A close association between agamospermy and hybridization has been noted by many students of this subject since the time of Ernst (1918). All agamospermous flowering plants which have been studied genetically have been found to be highly heterozygous. Most of them are species hybrids or hybrid derivatives. The relevant evidence has been reviewed previously by Gustafsson (1946–1947, Ch. 7) and Stebbins (1950, Ch. 10).

The first line of evidence for this conclusion, as pointed out by Ernst (1918) and later workers, is found in the variation pattern of most agamospermous species groups. Such groups often contain very large numbers of poorly demarcated microspecies. The different microspecies are so numerous and so interrelated morphologically as to make a satisfactory taxonomic treatment virtually impossible. The group as a whole exhibits the variation pattern of a greatly magnified hybrid swarm.

This characteristic type of variation pattern has long been known in the agamospermous species complexes in Hieracium, Taraxacum, Antennaria, Crepis, Rubus, Crataegus, Potentilla, Alchemilla, Poa, Calamagrostis, and Citrus.

In several species complexes containing both sexual and agamospermous populations, it has been observed that the agamospermous forms possess different combinations and recombinations of the morphological characters present in two or more sexual species. This evidence for hybrid origin of the agamospermous types has been obtained in Antennaria (Stebbins, 1932, 1935), the *Crepis occidentalis* group (Babcock and Stebbins, 1938), *Rubus* subgenus *Eubatus* (Gustafsson, 1943), Potentilla (see Gustafsson, 1946–1947, pp. 121–24), Parthenium (Rollins, 1944), and Calamagrostis (Nygren, 1946).

Direct evidence of natural hybridization is available in Antennaria (Stebbins, 1932, 1935), Parthenium (Rollins, 1944), and Hypericum (see Gustafsson, 1946–1947, pp. 124–25). Artificial hybrids match certain natural agamospermous types in Antennaria, Potentilla, and Rubus.

As noted earlier and as shown by Table 5, the great majority of agamospermous flowering plants are polyploids. We have learned elsewhere (in Chapter 15) that most naturally occurring polyploids in higher plants are amphiploids. Among agamospermous polyploids there is cytogenetic as well as morphological evidence for amphiploidy in Hieracium, Crepis, Rubus, Potentilla, Poa, and Calamagrostis (see Gustafsson, 1946–1947, Ch. 7).

Many agamospermous types are like species hybrids in being sterile or semisterile as to pollen and in having irregular meiosis or abortive anthers. Male-sterile agamospermous types occur, for example, in Taraxacum, Crepis, Hieracium, Ranunculus, Hypericum, Crataegus, Potentilla, Rubus, and Alnus (Gustafsson, 1946–1947, Chs. 6, 7).

The evidence of male sterility in nonpseudogamous apomicts, like Taraxacum, and Crepis, where the pollen does not function in agamospermous seed formation, can be construed in different ways. Stabilizing selection for male fertility is relaxed here, and consequently the absence of good pollen does not necessarily signify hybridity (Gustafsson, 1946–1947, Ch. 6; following Darlington, 1937a, Ch. 11). In pseudogamous apomicts like Potentilla, however, where some good pollen is necessary for seed formation, the production of much bad pollen after irregular meiosis is surely a strong indication of hybridity.

Facultative apomicts, on occasionally reproducing sexually, sometimes segregate a very wide range of types in the next or F_1 generation (see Figure 36). The extent of segregation in the F_1 progeny of the facultative apomict can be compared with that in the F_2 generation of a sexual species hybrid in the same plant group. Wide segregation of this sort as observed in Potentilla, Rubus, Sorbus, and Citrus reveals the hybrid constitution of the facultatively agamospermous mother plant. The breeding behavior of the agamospermous plant, in short, is like that of an interspecific hybrid.

Asexual embryo formation is associated with hybridity also in the ferns. Nearly all of the apogamous ferns studied by Manton are hybrids (Manton, 1950, Ch. 11).

Formation of Gametophytic Apomixis

The embryological pathways in gametophytic apomixis run close to the normal sites of meiosis and fertilization but involve the circumvention of both processes. Both meiosis and fertilization must be bypassed in any constant and persistent agamospermous line. Yet the two processes are independent. It is interesting, therefore, to find cases in which only one of the two successive phases of gametophytic apomixis occurs.

Solitary cases of apospory have been reported in *Leontodon hispidus, Picris hieracioides,* and some other sexual species of Compositae. Embryo-sacs with or without distinct egg cells arose from somatic cells of the ovule. However, these embryo-sacs do not give rise to embryos. Apospory is not followed by parthenogenetic development of the egg or apogamous development of its accessory cells (Bergman, 1935; Gustafsson, 1946–1947, pp. 42–43). Likewise, in a strain of *Parthenium argentatum,* unreduced eggs are produced by apospory or diplospory, but these eggs do not go on to develop parthenogenetically (Powers, 1945).

A complementary situation is haploid parthenogenesis, or the development of a reduced and unfertilized egg into an embryo which is haploid. This has been observed by Hagerup (1947) in several Danish species of orchids. This worker gives illustrations of parthenogenetic haploid embryos of *Listera ovata* (with $n = 17$), *Orchis strictifolius* (with $n = 20$), *Platanthera chlorantha* ($n = 21$), *Cephalanthera longifolia* ($n = 16$), and *C. damasonium* ($n = 16$) (Hagerup, 1947). The haploid individuals would be sterile and for this and other reasons would represent a transient condition in nature.

Haploid parthenogenesis is well known as a laboratory phenomenon in the animal kingdom, as in some plants; but here, unlike higher plants, it has also become established in the normal reproduction of several groups. In all Hymenoptera, for example, and in certain Coleoptera, Hemiptera, and Acarina, the male sex develops by haploid parthenogenesis (see Soumalainen, 1950; White, 1951).

The occurrence of one phase only in the two-stage process of gametophytic apomixis, as Gustafsson (1946–1947) points out, is not sufficient to put agamospermous reproduction on a continuing basis. The single and isolated phases do, however, represent tendencies in the direction of agamospermy (Gustafsson, 1946–1947, p. 42). An agamospermous reproductive system will arise if and when the two complementary tendencies of meiosis circumvention and diploid parthenogenesis become combined together in one organism.

At the genic level of determination it is plausible to suppose that the two tendencies, meiosis circumvention and diploid parthenogenesis, are controlled by separate genes or gene systems. Then gametophytic apomixis can be viewed as an expression of complementary gene interaction between these separate genes.

The apomixis-determining alleles of the separate genes or gene systems could be carried as unexpressed recessive factors in the ancestral sexual populations. Occasional crossing between carriers of opposite type would produce the recessive complementary genotype, which would behave as an agamosperm. The balanced combination of genes determining its agamospermous reproduction could easily be broken up again, however, in the outcross progeny of the new agamospermous plant (Powers, 1945; Gustafsson, 1946–1947, Ch. 8).

In adventitious embryony, where the new embryo develops from somatic cells in the ovule, a single change in embryological pathway suffices to bypass both meiosis and fertilization. The origin of adventitious embryony from an ancestral sexual cycle is therefore simpler embryologically and could be simpler genetically than the origin of gametophytic apomixis.

Breeding Behavior

The inheritance of gametophytic apomixis can be studied in the progeny of agamosperm X agamosperm crosses. Such crosses have been performed in a number of plant groups. Much of the recorded experimental evidence is preliminary and inconclusive. The results of some hybridization experiments in Potentilla, Poa, Calamagrostis, and Rubus, however, are of considerable interest and will be singled out for presentation in the following section.

In Potentilla it is possible to compare the results of intraspecific agamosperm X agamosperm crosses with those of interspecific crosses between agamosperms. The intraspecific crosses, *Potentilla verna* X *verna* and *P. tabernaemontana* X *tabernaemontana*, yield F_1 hybrids which are agamospermous like their two parents. In contrast, the interspecific crosses, *P. recta* X *adscharica* and *P. canescens* X *verna*, give F_1 generations composed predominantly or entirely of sexual hybrids. Here, therefore, the agamospermous condition is broken up by interspecific but not by intraspecific hybridization (cf. Gustafsson, 1946–1947, pp. 145–46).

Clausen and co-workers intercrossed six agamospermous species of Poa in eight hybrid combinations. The agamospermous parents were *Poa pratensis, ampla, compressa, canbyi, scabrella,* and *arida.* The offspring of the different F_1 individuals were grown as separate F_2 families. Some of the F_2 individuals were then selected as parents of separate F_3 families. The F_2 and F_3 families were observed for the presence or absence of segregation. Segregating families were classified as products of sexual reproduction in the parental individuals, whereas families which were uniformly like the maternal parent were classified as products of agamospermous reproduction (Clausen, Keck, Hiesey, and Grun, 1949).

The cross of *Poa ampla* X *pratensis* produced 35 F_1 plants which were individually progeny-tested to yield 35 F_2 families. Of these F_2 families, 20 were scored as agamospermous and 15 as sexual. A series of F_2 plants which were selected from both classes of families then yielded 13 agamospermous families and 9 sexual families in F_3. Other interspecific hybrid combinations of agamospermous Poas likewise segregated into agamospermous and sexual progenies in F_2 and F_3 (Clausen, Keck, Hiesey, and Grun, 1949).

In practice, many of the agamospermous families were not entirely uniform and matroclinous, but contained some sexual segregates. The proportion of sexuals in otherwise agamospermous F_2 families ranged from 10 to 90%. It follows that some agamospermous F_1 individuals are highly agamospermous and others are weakly so (Clausen, Keck, Hiesey, and Grun, 1949).

The main point, however, is that hybrids between agamospermous Poa species segregate in a complex fashion for the agamospermous condition. Reversions from agamospermy to sexuality occur in the F_2 and F_3 generations. Conversely, some sexual F_2 individuals give rise to agamospermous F_3s.

Several agamosperm X agamosperm crosses have been described in the European blackberries. The cross of *Rubus plicatus* X *caesius* produced an F_1 generation which was extremely variable. The F_1 hybrid of *Rubus polyanthemus* X *insularis* had good pollen and gave rise to an F_2 generation which segregated widely. Similar variation in F_1 or F_2 was observed in other agamospermous species crosses in Rubus. Here again agamospermy is broken up and sexuality is partially restored by hybridization (Gustafsson, 1943).

In Calamagrostis, Nygren (1951, 1962a) found that a cross between two sexual species, *C. epigeios* X *arundinacea*, gives F_1 and F_2 progeny which produce some diplosporous embryo-sacs. The genes for agamospermy are thus carried in sexual species and can be put together by hybridization. This is significant in showing how agamospermous species of Calamagrostis such as *C. purpurea*, which is a known hybrid derivative of *C. epigeios* and other sexual species, might originate in nature (Nygren, 1951, 1962a).

The experimental evidence shows that the inheritance of gametophytic apomixis is complex, as would be expected. Hybridization between sexual species can sometimes assemble a gene combination for agamospermy. Conversely, hybridization between two agamospermous species, or between an agamospermous and a sexual parent, can break up the same gene combination and restore sexuality in a part of the progeny. These sexual types can then go on to produce a new assortment of agamosperms in the next generation.

Segregation

The sexual progeny of facultative agamosperms generally exhibit a high degree of individual variation. This segregation is readily apparent in morphological characters. It extends also to chromosome numbers and physiological vigor.

Figure 36 shows the individual-to-individual variation in fruit characters in the sexual F_1 progeny of an Imperial grapefruit (*Citrus paradisi*). The pollen parent was lemon (*C. limon*).

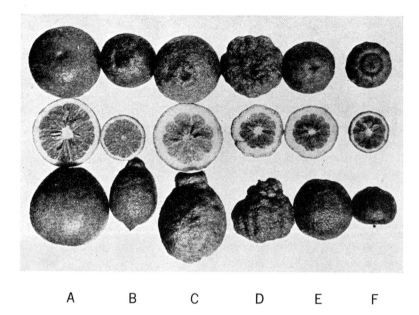

A	B	C	D	E	F

Figure 36 Segregation for fruit characters in the sexual progeny of a facultatively agamospermous Citrus plant. (A) Female parent, Imperial grapefruit. (B) Male parent, Lisbon lemon. (C–F) Four different F_1 individuals. (From Frost, after Gustafsson, 1946–1947.)

In Rubus section Moriferus, facultative agamosperms have been observed to yield very variable offspring following outcrosses in many hybrid combinations. Some of these, such as *Rubus plicatus* X *caesius*, which produced a segregating F_1 generation, were mentioned in the preceding section. Outcrosses between *R. plicatus* or other agamosperms, on the one hand, and the sexual diploid *R. tomentosus*, on the other, give F_1 generations which look like a segregating F_2 in morphology and in fertility (Gustafsson, 1943).

The *Rubus acutus* group is a complex assemblage of agamospermous microspecies in Scandinavia. Many of the microspecies are hybrid products—either F_1 hybrids or F_2 segregates—of crosses between *R. caesius* (4x) and various preexisting microspecies of the *R. acutus* group. The microspecies differ in characters of the leaves, stems, prickles, pubescence, etc.

Lidforss discovered a hybrid plant in Scania, Sweden, which was interpreted as an F_1 of *Rubus gothicus* X *caesius*. (*R. gothicus* is a microspecies in the *R. acutus* group.) The plant was fully fertile as to pollen. It was transplanted to the Lund Botanical Garden, where it produced a large F_2 generation and eventually some F_3 and F_4 offspring (Gustafsson, 1943).

The F_2 generation of *Rubus gothicus* X *caesius* was very interesting in that it contained a series of segregates resembling various named microspecies in the *R. acutus* group. Thus some individuals resembled *R. pruinosus* in having leaves with seven leaflets and straight prickles. Other F_2 plants were like *R. acutus, R. permixtus, R. wahlbergii, R. gothicus, R. polycarpus, R. ruedensis,* or *R. caesius;* and still others, as might be expected, were unlike any named naturally-occurring microspecies (Gustafsson, 1943).

Meiotic irregularities in a facultative agamosperm may lead to the formation of functioning gametes with different chromosome numbers. Zygotes then arise with numbers forming an aneuploid series. Such aneuploid variation has been observed in progenies in Poa and other grasses.

Grun (1954) determined the chromosome numbers of individual plants in the F_1 progeny of several interspecific crosses between facultatively agamospermous Poas. One cross between *Poa scabrella* X *P. pratensis*, where the parental individuals had $2n = 83$ and 68 chromosomes, respectively, yielded an F_1 generation with the following array of numbers: $2n = 65, 66, 67, 68, 68, 70, 70, 71, 71, 71, 74, 75, 75, 79$. Another cross between *P. ampla* ($2n = 64$) and *P. pratensis* ($2n = 68$) produced F_1 hybrids with numbers ranging from $2n = 60$ to $2n = 117$. Aneuploid variation of this sort was found in the F_1 generations of some 16 hybrid combinations in Poa (Grun, 1954).

Clausen and co-workers have observed the physiological vigor of progeny of interspecific crosses between agamospermous Poas. In general, the F_1 hybrids are vigorous, and so are most of their agamospermous F_2 descendants. But the sexual F_2 progenies of the same interspecific crosses, by contrast, contain numerous weak and inviable plants along with some vigorous individuals. This segregation and recombination for vigor in the sexual progeny of Poa hybrids represents normal hybrid breakdown. It is largely avoided in the agamospermous progeny of the same hybrids (Clausen, Keck, Hiesey, and Grun, 1949).

Variation Pattern

By far the chief source of new variations in a group of agamospermous plants is hybridization. The most common hybrid combinations in practice are sexual X facultative agamosperm and facultative agamosperm X facultative agamosperm. Even obligate agamosperms may engage in hybridization, however, as male parents if they produce any good pollen at all. The hybridizing parental plants may be genetically differentiated at any level from minor biotypes to distantly related species.

A second source of variation is autosegregation. Autosegregation refers to spontaneous changes in genotype in the egg cells that develop without fertilization. Thus, in Taraxacum, meiosis in an embryo-sac mother cell may proceed in an irregular manner so as to yield an egg containing more or fewer chromosomes than the mother plant. Or heterogenetic pairing and crossing-over in the embryo-sac mother cell may lead to a genotypically changed egg. Such eggs on developing parthenogenetically can then produce phenotypically aberrant plants. Autosegregation can occur in obligate as well as facultative agamosperms (Gustafsson, 1946–1947, Ch. 12).

A third source of variation is bud mutations leading to phenotypically altered branches. Such vegetative sports have been observed several times in the agamospermous blackberries (Gustafsson, 1943, p. 78; 1946–1947, p. 187).

The aberrant individuals, by whatever means they arise, can now go on to multiply asexually. The agamospermous system provides for the production of variability by various methods and, alternating with these bursts of variability, for the exact replication of any favorable new variant types by the asexual methods.

In the preceding section we saw that a single hybrid plant in the *Rubus acutus* group produced sexual F_2 progeny resembling various naturally-occurring agamospermous microspecies in the same species group. Different F_2 segregates could be classified as *R. acutus* sensu stricto, *R. pruinosus*, *R. polycarpus*, etc. The natural microspecies probably originated in a similar way, as sexual segregates from hybrids, and then reached the status of microspecies by agamospermous increase in numbers and area (Gustafsson, 1943).

In the preceding section we saw also that aneuploid variation in chromosome numbers occurs in sexual progenies of the *Poa pratensis* group. Here too a corresponding condition of aneuploidy is found in nature.

Löve (1952) counted the chromosomes of individual plants of a member of the *Poa pratensis* group, known as *P. irrigata*, growing in an area 10 meters in diameter in Iceland. In this small area he found plants with $2n = 82, 84, 87, 91, 95, 98, 105, 108, 111, 112, 113$, and 119 chromosomes (Löve, 1952). The *P. pratensis* taxonomic species as a whole over its extensive circumboreal area has an almost continuous aneuploid series from $2n = 22$ to $2n = 147$ (Löve, 1952; Grun, 1954; Clausen and Hiesey, 1958).

Advantageous Features of Agamospermy

Agamospermy is one of the genetic systems which is based on hybridity and which permits the reproduction of the hybrid genotype. It is appropriate to ask what are the particular adaptive advantages of the agamospermous genetic system.

In the 1930s, Darlington expressed (but did not develop) two viewpoints which have served as points of departure for discussions by subsequent authors. In the first edition of *Recent Advances in Cytology* (1932), he placed agamospermy under the heading of "the stabilisation of hybridity" (Darlington, 1932, p. 469). The stabilization-of-hybridity concept is not carried forward explicitly in later treatments of the subject (Darlington, 1937a, 1939, 1958). In its place, in *The Evolution of Genetic Systems*, agamospermy is presented as an "escape from sterility" (Darlington, 1939, pp. 108, 113; 1958).

It is necessary to note parenthetically that Darlington actually uses the term apomixis in the texts cited above. However, the term apomixis as used in the present book, following Gustafsson (1946–1947, Ch. 2), refers to a broader range of asexual processes than the apomixis of Darlington. Apomixis sensu Darlington is essentially the same as agamospermy in our usage.

The escape-from-sterility concept of agamospermy clearly has a wide application. Many agamospermous forms in different plant genera are sexually sterile or semisterile hybrids, as noted earlier in this chapter.

Odd polyploids, especially triploids, are usually sterile, and they are also frequent among agamospermous plants.

But there are agamospermous hybrids which are sexually fertile. This is the case in Citrus, where interspecific hybrids are meiotically normal and fertile (Webber and Batchelor, 1943; Reuther, Batchelor, and Webber, 1968). Yet many hybrid types in Citrus reproduce by adventitious embryony. Agamospermy is not an escape from sterility here.

The stabilization-of-hybridity concept thus had a wider and more general application to the facts of agamospermy than does the escape-from-sterility concept.

Suppose that natural hybridization produces some highly heterozygous genotype which possesses adaptively valuable heterotic properties. This genotype may be an F_1 hybrid or a later-generation derivative; it may be sexually sterile or fertile. In any case the particular heterozygous genotype of this hybrid would be broken up by normal sexual reproduction. But agamospermy enables the same hybrid to breed true to type by seeds. The agamospermous hybrid can produce exact copies of its adaptively valuable genotype in large numbers of seedling progeny.

The same copying process attributed above to agamospermy could also be carried out by vegetative propagation, and it sometimes is. But seeds generally possess certain adaptive features in respect to dormancy, resistance, and particularly dispersal, which are not shared to an equal degree by vegetative propagules in the same plant species. Agamospermy as a genetic system falls heir to these special advantages of seed reproduction.

Amphiploidy is another genetic system which can and often does produce copies of a hybrid genotype. Certain kinds of numerical hybrids arise in nature, however, which cannot be preserved by amphiploidy but can be maintained by agamospermy.

Odd polyploids—triploids, pentaploids, etc.—if adaptively valuable in their own right, can be kept alive through successive seedling generations by agamospermy, but not by amphiploidy. Natural hybrids between parental species with high polyploid numbers may be unable to double their chromosome number without exceeding the threshold limits for orderly cell division. Here again the agamospermous mechanism provides a way out of the impasse. The observed high frequency of both classes of polyploids, odd polyploids and high polyploids, among agamospermous plants may be construed as support for the idea that agamospermy can succeed in certain situations where amphiploidy would fail.

Agamospermy does not necessarily spell the end of sexuality entirely, as Gustafsson (1946–1947) and others have emphasized. Episodes of sexual reproduction may occur in the agamospermous lineage and give rise to new hybrid genotypes. The new hybrids thus formed can then reproduce themselves by another cycle of agamospermy. In this way new adaptive heterozygotes can arise from time to time within the agamospermous group and be mass-produced in their turn by the agamospermous mechanism.

Agamospermy, then, possesses certain advantages of its own in comparison with other competing genetic systems. Agamospermy can perpetuate a rare homoploid heterozygous segregate in a hybrid population of outcrossing plants, which sexual reproduction cannot do. It preserves odd polyploids and high secondary aneuploids, which amphiploidy cannot do. It perpetuates hybrids with high polyploid numbers where further doubling and amphiploidy would fail. It accomplishes asexual reproduction by means of seeds, which vegetative propagation cannot do.

PART V

Evolution of
Hybrid Complexes

CHAPTER *19*

The Hybrid Complex

Introduction · Classification · The Role of the Breeding System

---

Introduction

Many species groups belonging to different plant genera and different families exhibit a similar variation pattern. This common variation pattern possesses the following features. The plant group contains numerous and diverse forms which intergrade with one another in their morphological characters and ecological preferences. On the basis of intergradation, the diverse forms might be considered members of one polytypic species. But the total range of morphological and ecological variation within the group greatly exceeds that found in normal polytypic species in the same genus. Furthermore, some combinations of forms remain more or less distinct in some zones of sympatric contact. The intergradation is not like that between races.

Taxonomically, the group is "critical" and "difficult." No natural system of classification can be devised on field and museum methods alone; and no system devised by conventional methods works very satisfactorily for routine identification. When analyzed by these same conventional morphological and geographical methods, but from the viewpoint not of taxonomy but of population biology, the group is seen to be a complex of more or less poorly defined species, semispecies, and/or microspecies. The taxogenetic evidence when obtained frequently confirms the concept of an interconnected species group by revealing the existence of sterility barriers or other breeding barriers between some of the morphologically different and sympatric forms.

The plant groups exhibiting this characteristic variation pattern do not have any single genetic system in common. The component populations may be diploid or polyploid or a combination of both. Reproduction may be sexual, subsexual, or asexual. Different taxonomically critical plant groups are similar in the general features of their variation pattern but are not all alike in type of genetic system.

There is, however, another factor common to all plant groups which possess the above-described variation pattern. That factor is the widespread occurrence of natural hybridization within the group and the perpetuation of a considerable number of the hybrid derivatives. Accordingly, such groups can be referred to collectively as hybrid complexes. A hybrid complex is a species group in which natural hybridization has obscured the morphological discontinuities between the originally divergent ancestral forms (Grant, 1953).

Classification

Hybrid complexes can be classified according to the mode of stabilization of reproduction in the natural hybrids or hybrid progeny (Grant, 1953). In Chapter 13 we listed the known modes of hybrid speciation and we repeat the list here for ready reference.

The modes of stabilizing reproduction in hybrids are: (1) vegetative propagation; (2) agamospermy; (3) permanent translocation heterozygosity; (4) permanent odd polyploidy; (5) amphiploidy; (6) recombinational speciation; (7) hybrid speciation with external barriers.

These modes allow for true breeding of a highly heterozygous genotype (1–5) or homozygous segregate (6–7), under conditions of asexual (1–2) or sexual (3–7) reproduction.

If the process of hybrid speciation is repeated independently in different lineages, so as to yield two or more hybrid species from three or more original species, a hybrid complex develops. The type of hybrid complex corresponds to the mode of stabilization of hybrid reproduction.

1. Clonal Complex. The hybrids increase in numbers, asexually, by various means of vegetative propagation.

2. Agamic complex. The hybrids or hybrid derivatives increase in numbers wholly or partially by unfertilized seeds. The concept can be extended to include asexual ferns in which the hybrids reproduce by apogamety.

3. Heterogamic complex. The hybrids or hybrid derivatives are permanent translocation heterozygotes or permanent odd polyploids. The structural or numerical hybrids breed true for their heterozygous condition under sexual reproduction by regularly transmitting one set or subset of chromosomes through the male gametes and another through the female gametes. It will be noted that there are two types of heterogamic complex: those containing diploid translocation heterozygotes, and those composed of odd polyploids.

4. Polyploid complex. The derivatives of natural hybridization are sexually reproducing even polyploids.

5. Homogamic complex. The hybrid derivatives are sexual with a normal meiotic cycle. The daughter species of hybrid origin are diploid, or at least undoubled and homoploid with reference to the original parental species. The daughter species originate from the parental species by recombinational speciation or by hybrid speciation with external barriers. Consequently the homogamic complex may be composed of chromosomally intersterile species or interfertile species.

The foregoing classification of hybrid complexes is analytical. The hybrid complex which actually develops in a given plant group may contain a mixture of types. Many high polyploids and some low ones are agamospermous. Therefore agamic complexes are apt to be superimposed on and combined with polyploid complexes, as in the *Crepis occidentalis* group (Babcock and Stebbins, 1938). Similarly, recombinational speciation and amphiploidy may both occur in different branches of the same plant group. Then there results a hybrid complex which is partly homogamic and partly polyploid.

The Role of the Breeding System

Natural hybrids are formed between plant species possessing every type of breeding system. The outcome of the natural hybridization, however, is strongly influenced by the breeding system of the plants involved, and this influence may be carried over into the type of hybrid complex which eventually develops.

In annual herbs, where the time available for reproduction by a hybrid is short, the course of hybridization runs toward introgression under conditions of outcrossing and toward amphiploidy under conditions of self-fertilization (Grant, 1956c). The expected results of the breeding system are clearly evident in the annual Gilias, where the cross-fertilizing forms are strongly affected by introgression, whereas the autogamous members have produced the amphiploids. In hybridizing groups of annual plants, therefore, an autogamous breeding system promotes the formation of a polyploid complex.

Perennial plants with means of vegetative propagation have a relatively high frequency of polyploidy, as we saw in Chapter 16, and such plants of course also produce daughter clones. Extensive hybridization in a group of vegetatively reproducing perennial plants, therefore, is expected to yield polyploid and clonal complexes.

Self-fertilization in annuals and perennials alike is a factor promoting recombinational speciation, as pointed out in Chapter 14. Homogamic complexes composed of chromosomally intersterile species and microspecies are thus expected to develop in hybridizing groups of predominantly autogamous plants under favorable habitat conditions.

The sexual relatives of most agamospermous plants are dioecious, monoecious, or self-incompatible (Gustafsson, 1946–1947, Chs. 10, 15; 1948). Gustafsson notes that plants possessing one or the other of these breeding systems, and particularly hybrid plants, will have only a slight chance of perpetuating themselves sexually in isolated stands; whereas agamospermy and vegetative propagation are relatively easy ways out of the impasse of sterility. Obligate outcrossing in hybrids is thus one factor favoring the development of agamic and clonal complexes.

Autogamy prevails in the permanent translocation heterozygotes of the *Oenothera biennis* group and apparently in the permanent odd polyploids of the *Rosa canina* group. Each group is derived from meiotically normal and open pollinated ancestors. In each group, self-fertilization is a component of the genetic system which maintains the permanent hybridity. Ecologically, in the Oenotheras particularly, the hybrid derivatives have been successful in weedy or cultivated habitats.

It may be that the ancestral populations possessed indispensable heterotic properties and maintained these properties by continual cross-pollination. Then, on invading new habitats lacking the proper pollinators, they developed the genetic system of permanent hybridity based on self-fertilization of unlike gametes, as a compensation for the loss of cross-pollinators (Darlington, 1956b, 1963, pp. 51, 92). On this hypothesis the primary stimulus provoking the evolution of at least some heterogamic complexes has been a continuing dependence of the populations on heterosis throughout an impoverishment of the climate of pollinators; and the concomitant change in breeding system from outcrossing to selfing was a part of the readjustment of the whole genetic system to the new conditions.

CHAPTER *20*

The Polyploid Complex

Introduction · Methods of Analysis · Cytotaxonomic Examples · Biosystematic Examples · Aegilops-Triticum · The Gilia inconspicua Complex · Vestiges of Ancient Polyploid Complexes · Secondary Basic Numbers in the Angiosperms · Irreversibility · Suppression of Variability

---

Introduction

The same factors which promote the formation of one new tetraploid species from its diploid ancestors often operate to promote the origin of other new tetraploids from other diploids in the same species group. These factors can then lead the tetraploids to produce hexaploids and octoploids. The result of continued polyploidization in different anastomosing lineages is a polyploid complex.

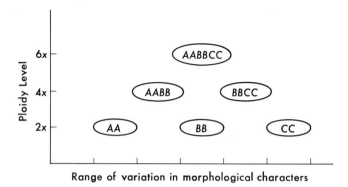

Figure 37 Structure of a simple polyploid complex.

A polyploid complex in one of its simplest possible forms could have the structure shown in Figure 37. The complex diagrammed here consists of three diploid species, their two possible allotetraploid derivatives, and one allohexaploid. Altogether there are six sibling species, isolated by chromosomal sterility barriers and perhaps by other associated breeding barriers, but sharing much genetic material in common, and therefore overlapping in their range of variation for morphological characters and ecological preferences.

Other possible versions of this simple polyploid complex might include autotetraploids as well as allotetraploids, autoallohexaploids as well as genomic allohexaploids, or allooctoploids as well as hexaploids. Given a larger array of diploid species as foundation stocks, evolution by means of polyploidy can build up a very extensive polyploid complex, containing numerous tetraploid and hexaploid species and reaching the higher ploidy levels in some derivative species.

Methods of Analysis

A polyploid complex can be identified as such by cytotaxonomic methods of study. The phylogenetic relationships between the constituent sibling species can be worked out by cytogenetic and taxogenetic methods. The phylogenetic analysis of the polyploid complex may be carried out to any degree of completeness. It will be useful for our present purposes to recognize four main stages of analysis.

The first stage is the identification of a polyploid complex as such. We are confronted with a large and taxonomically difficult group of plant species. The evidence of morphology and geographical distribution suggests the existence of sibling species; but their taxonomic boundaries, geographical areas, and genetic nature are unknown. Cytotaxonomic methods are called for. A preliminary series of chromosome counts may reveal a polyploid series within the plant group, giving us our first indication that the group is a polyploid complex.

Extension of the cytotaxonomic studies next enables us to fill in many details and round out the picture. This brings our understanding of the group to the second stage.

In the second stage of analysis we may learn that different plants, possessing similar morphological characters and occurring together in the same locality, have different even ploidy numbers. This information suggests close phylogenetic relationships between the sibling species standing at different ploidy levels. Or we may find a series of forms with different and nonintergrading morphological characters on the same basic ploidy level, and thus revealing the probable foundations of our polyploid complex. As chromosome counts from natural populations accumulate, we can begin to correlate a given condition of ploidy with particular morphological characters, biochemical constituents, or pollen grain or stomatal size. We can extrapolate to some extent from such correlations and predict the taxonomic and geographical boundaries of some of the sibling species. The main outlines of our polyploid complex can now be described on the basis of extensive cytotaxonomic work.

The third stage requires the introduction of experimental methods on a preliminary scale. Crosses are made between some of the sibling species, and the fertility of their hybrids is determined. This third of biosystematic stage of analysis is necessary in order to verify some of the relationships predicted previously on cytotaxonomic evidence. The biosystematic methods enable us to determine a few phylogenetic connections within the polyploid complex.

The fourth stage involves an extension of the experimental taxogenetic methods to all members of the complex. The ancestry of one tetraploid species from its diploid progenitors is worked out by the established methods of artificial hybridization, genome analysis, and resynthesis wherever possible. The same methods are employed to determine the ancestry of the other tetraploid members of the complex. They are used again for the higher polyploids. We can now trace the phylogenetic lines of descent from particular diploid species to particular polyploid species. The postulated phylogeny can be confirmed by cytogenetic analysis of artificial interspecific hybrids between the different polyploid species. The phylogenetic analysis of the polyploid complex is complete.

An analogy with chemical methods of analysis is helpful. We can identify a given unknown compound as a protein by finding that it contains certain constituent elements like nitrogen and sulfur. We can then determine its chemical formula by more refined analytical methods. Its structural formula, showing the bond connections between the elementary components, is revealed by still more detailed and laborious methods. Many kinds of proteins are known as to chemical formula but very few as to their complete structural formula.

So it is with the investigation of a polyploid complex. The identification of a plant group as a polyploid complex (stage 1) is relatively easy. A proper cytotaxonomic analysis (stage 2) is considerably more difficult and requires years of work. The biosystematic analysis (stage 3) is an even more difficult task which can scarcely be carried out by a single worker alone. Needless to say, the complete taxogenetic analysis of a polyploid complex (stage 4) is an enormously laborious and time-consuming project.

Certain common errors of methodology are worth noting. Cyto-taxonomists sometimes overestimate the reliability of descriptive cytotaxonomic methods as applied to a polyploid complex and, by the same token, underestimate the importance of the taxogenetic approach. The difference between cytotaxonomic and taxogenetic methods, as regards the amount of time and labor involved, is very great; but the additional effort of the experimental taxogenetic approach is absolutely necessary if the polyploid complex is to be analyzed phylogenetically with any degree of reliability.

Many cytotaxonomists misinterpret their findings by relating them to a taxonomic species concept instead of to a biological or evolutionary species concept. The cytotaxonomist discovers one set of diploids and tetraploids with similar morphologies; he treats them as different "chromosome races" of the same species. A second set of diploids and tetraploids, which are similar inter se but morphologically different from the first set, is treated as another species composed of two "chromosome races." The result is a taxonomic system which often confuses the natural relationships and the fertility relationships. The morphologically different tetraploid types may well be opposite segregates of the same segmental allotetraploid ($A_sA_sA_tA_t$) and, as such, should be recognized as conspecific. The interspecific boundary lines in such cases should be drawn, not through the middle of $A_sA_sA_tA_t$, but between it, taken as a whole, and the diploid species A_sA_s and A_tA_t.

Many plant groups have been identified as polyploid complexes. Quite a few polyploid complexes have reached the stage of thorough cytotaxonomic analysis, and some the stage of preliminary biosystematic analysis. An extensive but incomplete taxogenetic analysis has been carried out in only a small number of polyploid complexes.

In the next two sections we review the main features of some polyploid complexes which are known cytotaxonomically and biosystematically. Later we consider polyploid complexes in Aegilops-Triticum and Gilia which have undergone extensive taxogenetic analysis.

Cytotaxonomic Examples

Polyploid complexes in several branches of the genus Galium (Rubiaceae) have been analyzed cytotaxonomically by Ehrendorfer and his co-workers. Such polyploid complexes have been described in the *Galium pumilum* group in Europe (Ehrendorfer, 1949), the *G. anisophyllum* group in Europe (Ehrendorfer, 1964), the *G. incanum* group in Turkey and neighboring areas (Ehrendorfer, 1951), and the *G. multiflorum* group in western North America (Ehrendorfer, 1961; Dempster and Ehrendorfer, 1965).

The *Galium pumilum* group ($x = 11$) in Europe comprises a polyploid series ranging from diploid to octoploid. The diploid member of the complex, *G. austriacum*, occurs as a glacial relict in disjunct areas in the Alps, whereas the polyploids have widespread and continuous distributions (Ehrendorfer, 1949). The polyploid complex in the *G. anisophyllum* group in Europe contains every even ploidy level from $2x$ to $10x$. Here again the diploids occur in a series of scattered glacial refuges in the European mountains. The postglacial octoploids, by contrast, have a widespread and continuous distribution; and the tetraploids are intermediate between the diploids and the octoploids in geographical distribution pattern, being widespread but disjunct (Ehrendorfer, 1964).

Sanicula crassicaulis and its relatives (Umbelliferae) comprise another polyploid complex on the basic number of $x = 8$ (Bell, 1954). The taxonomic species *S. crassicaulis* includes tetraploids, hexaploids, and octoploids, which range widely on the Pacific slope of North America and Chile. The diploid species, *S. laciniata* and *S. hoffmannii* in coastal California, are the probable ancestors of tetraploid *S. crassicaulis*. Another diploid species, *S. bipinnatifida*, is believed on morphological evidence to enter into the ancestry of certain octoploid members of the complex (Bell, 1954).

A group of Asplenium ferns (Polypodiaceae) in eastern North America form still another polyploid complex on the base of $x = 36$ (W. H. Wagner, 1954). The diploid species, which possess the extreme morphological characters in this complex, are *A. montanum, platyneuron,* and *rhizophyllum* ($2n = 72$). The morphological intermediates are sterile diploid or triploid hybrids in some cases. Other fertile intermediate forms are tetraploids ($2n = 144$). Thus the tetraploid *A. pinnatifidum* is morphologically intermediate between *A. montanum* and *A. rhizophyllum* and is probably an amphiploid derivative of these two diploid species (W. H. Wagner, 1954).

The relationships in this group of Asplenium, as worked out by W. H. Wagner (1954) by cytotaxonomic methods, were later checked by chemotaxonomic methods (Smith and Levin, 1963). Certain biochemical substances in the various species were assayed by paper chromatography. The natural hybrids and allotetraploids were found to have combinations of the substances present in their putative parental diploids (Smith and Levin, 1963). The chemotaxonomic evidence confirmed the phylogeny based on cytotaxonomic evidence.

Other cytotaxonomically studied polyploid complexes are the *Polypodium vulgare* group (Polypodiaceae) in Europe and North America and the *Trillium kamtschaticum* group (Liliaceae) in Japan (Manton, 1950, 1951, 1958; Kurabayashi, 1958). Among the numerous cytotaxonomically analyzed polyploid complexes in the Gramineae we may mention Oryzopsis-Stipa in the northern hemisphere, the circumpolar *Poa alpina* group, and the *Bouteloua curtipendula* group of the

North American plains (B. L. Johnson, 1945; Nygren, 1962b; Gould and Kapadia, 1964).

Biosystematic Examples

The *Phacelia heterophylla* group consists of outcrossing perennial and biennial herbs which range widely throughout western North America and southern South America. The group is taxonomically difficult because of the existence of many different but intergrading morphological forms. Heckard (1960) has shown that it is a polyploid complex ($x = 11$) consisting of six known diploid species, numerous tetraploids, and some hexaploids.

Artificial hybridizations indicate that crossability and sterility barriers are well-developed between the diploid species. By contrast, the different tetraploid forms mostly cross fairly easily with one another to produce semifertile or fertile hybrids. These differences in degree of internal isolation are correlated with differences in the variation pattern. On the diploid level the taxa are generally separated by morphological discontinuities, whereas those on the tetraploid level are often connected by morphological intergradation (Heckard, 1960).

The polyploid members of the *Phacelia heterophylla* group are probably amphiploids derived from different hybrid combinations of the diploid species; this would further explain their morphological intergradation. There is no evidence as yet regarding the ancestry of the various tetraploid and hexaploid forms.

Heckard (1960) has grouped morphologically similar diploid and tetraploid forms together in the same species. The named entities are therefore taxonomic species but not biological or evolutionary species in many cases. The genetic evidence obtained by Heckard is sometimes in disagreement with the morphological evidence, as might be expected. Thus diploid *nemoralis* failed to cross artificially with tetraploid *nemoralis;* but tetraploid *nemoralis* crossed with tetraploid *californica* to produce highly fertile hybrids. The next steps in the analysis of this interesting polyploid complex will be to block out the biological species at the tetraploid and hexaploid levels and to determine their ancestry.

The genus Clarkia (Onagraceae) in western North America and temperate South America contains a more extensive polyploid complex

(Lewis and Lewis, 1955). The diploid species have $2n = 10, 14, 16$, and 18 chromosomes, the ancestral basic number being $x = 7$. These diploid species fall into several distinct species groups which could be and have been recognized as separate genera, such as Godetia, Eucharidium, and Clarkia sensu stricto. However, the phylogenetic lines, which diverge as branches on the diploid level, fuse in various combinations on the tetraploid and hexaploid levels. These anastomoses provide the rationale for treating the whole assemblage of species as a single genus, Clarkia sensu lato (Lewis and Lewis, 1955).

The postulated reticulate phylogeny of Clarkia, as understood in 1955, is shown diagrammatically in Figure 38. Much additional evidence has been obtained since 1955 by Lewis and his co-workers. Figure 38 shows the probable ancestry of the various polyploid species of Clarkia as inferred from the morphological and cytological evidence (Lewis and Lewis, 1955).

The *Achillea millefolium* group (Compositae) is a polyploid complex ($x = 9$) which is widespread in the northern hemisphere (Clausen, Keck, and Hiesey, 1948; Ehrendorfer, 1959). *Achillea millefolium* proper in the Old World and *A. borealis* in the New World are both hexaploid. The Old World *A. collina* and *A. virescens* and the North American *A. lanulosa* are tetraploid. Two European diploid species are *A. asplenifolia* of moist meadows and *A. setacea* of dry steppes. (Clausen, Keck, and Hiesey, 1948; Ehrendorfer, 1959).

The diploid species *Achillea asplenifolia* and *A. setacea* overlap geographically and, despite strong internal isolating barriers, have undergone natural hybridization to a limited extent. Introgression has taken place on the diploid level from *A. setacea* into *A. asplenifolia* in the Alps. And the tetraploid *A. collina* is an amphiploid derivative of these two diploids. *Achillea collina* then hybridizes with other tetraploid species such as *A. virescens*. Natural hybridization also occurs between the tetraploid and hexaploid species in certain combinations—*A. collina* ($4x$) X *A. millefolium* ($6x$) and *A. lanulosa* ($4x$) X *A. borealis* ($6x$)—to yield aneuploid progeny which further complicate the variation pattern of the group (Ehrendorfer, 1959).

Zohary and Feldman (1962) find a similar situation in the polyploid complex involving Triticum and Aegilops (Gramineae). Here too, as in Achillea, different allotetraploid species with different genomic constitutions hybridize naturally, thus adding to the complexity of an

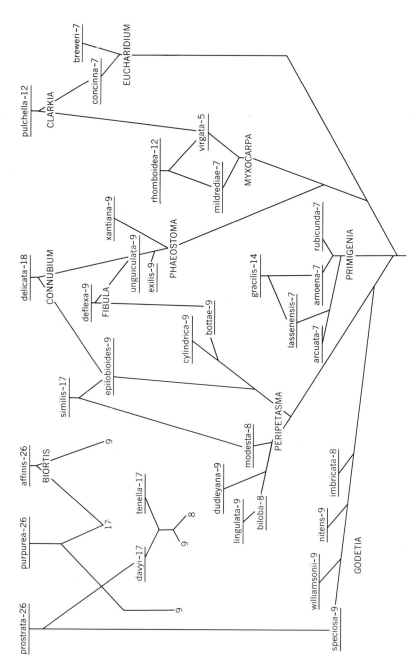

Figure 38 Phylogenetic relationships in the polyploid complex in the genus Clarkia. The sections of the genus are indicated by capital letters; the species are listed with their normal gametic chromosome numbers. (Lewis and Lewis, 1955.)

already complex variation pattern in the group (Zohary and Feldman, 1962).

Aegilops-Triticum

The grass supergenus Aegilops and Triticum has been extensively and thoroughly studied cytogenetically and taxogenetically because of the economic importance of wheat. The genomic relationships of the diploid species and the genomic constitutions of the polyploid species are summarized in reviews by Kihara (1954), Sears (1959), and Zohary and Feldman (1962).

The basic number in Aegilops-Triticum, as in the Hordeae generally, is $x = 7$. The supergenus contains numerous diploid and tetraploid species and several hexaploid species.

The diploid species are listed by genome group in Table 6. The polyploid species with their respective genomic constitutions are given in Table 7. The situation has been simplified for presentation here by ignoring subgenomic differences in various cases. The S genome in the tables is synonymous with genome B as cited in many papers.

The crisscross phylogenetic derivation of the tetraploid species can be visualized by reference to Table 7. Thus two diploid species carrying the DD and MM genomes have produced tetraploids with the constitution $DDMM$. One of these diploids (DD) and another (CC) have then produced another tetraploid $DDCC$, and so on. The different tetraploid species often have one common and one different diploid ancestor. The process of recombination of genomes is continued at the hexaploid level.

The phylogeny of the polyploids has been tracked down to particular ancestral diploid species in some instances. The best-known case is that of bread wheat, *Triticum aestivum* (or *T. vulgare*). Hexaploid wheat is known to contain a genome of *Aegilops squarrosa* (DD) and one close to that found in *Triticum monococcum* (AA); it probably derived its S (or B) genome from *Ae. speltoides* or *Ae. ligustica* (SS) (Sears, 1948; Sarkar and Stebbins, 1956). Similarly, *Aegilops triuncialis* is an allotetraploid derivative of *Ae. caudata* and *Ae. umbellata* in the Mediterranean region (Kihara, 1954).

The overlapping pattern of relationships between the polyploid

Table 6 · Genome groups found in different diploid species of Aegilops and Triticum (Kihara, 1954; Zohary and Feldman, 1962)

Genome group	Diploid species
Mt	Aegilops mutica
S	Aegilops bicornis, sharonensis, longissima, speltoides, ligustica
D	Aegilops squarrosa
C	Aegilops caudata
M	Aegilops comosa, uniaristata
Cu	Aegilops umbellata
A	Triticum boeoticum, monococcum

Table 7 · Genome constitution of polyploid species of Aegilops and Triticum. Different subgenomes of the same genome group not distinguished (Kihara, 1954; Zohary and Feldman, 1962)

Aegilops ventricosa, crassa 4*x*	*DDMM*
Aegilops crassa 6*x*	*DDDDMM*
Aegilops juvenalis	*DDCuCuMM*
Aegilops cylindrica	*DDCC*
Aegilops variabilis, kotschyi	*CuCuSS*
Aegilops triuncialis	*CuCuCC*
Aegilops columnaris, biuncialis, ovata, triaristata 4*x*	*CuCuMM*
Aegilops triaristata 6*x*	*CuCuMMMM*
Triticum dicoccoides, timopheevi, dicoccum, durum	*AASS*
Triticum spelta, aestivum	*AASSDD*

species of Aegilops-Triticum is due in the first instance to their independent and overlapping amphiploid derivations. This, however, is not the end of the story. There is evidence, both direct and indirect, for natural hybridization between different tetraploid species. This hybridization has altered the morphological characters and probably also the genomes of the species involved (Zohary and Feldman, 1962). Natural hybridization at the tetraploid level is thus superimposed on amphiploidy in the evolution of Aegilops-Triticum (Zohary and Feldman, 1962; Pazy and Zohary, 1965).

The Gilia inconspicua Complex

The *Gilia inconspicua* group ($x = 9$) is the segment of Gilia section Arachnion which consists of autogamous annuals. This group is widespread in western North America and recurs in southern South America. There are 23 known species of the complex in North America and 2 known species in South America; and there are good reasons for believing that additional unknown species exist in each area. Of the 23 North American species, 13 are diploid ($2n = 18$) and 10 are tetraploid ($2n = 36$). The tetraploids are amphiploids, and their ancestry has been worked out in most cases.

In Chapter 7 we introduced the *Gilia inconspicua* group as an example of a pattern of species relationships which is common in autogamous annual plants. One part of this group containing *G. transmontana* and its relatives was discussed and illustrated also in Chapter 2 from the standpoint of sibling species. Here we consider the group once again as a polyploid complex which has been extensively but incompletely analyzed taxogenetically.

The diploid species fall into five genome groups. Related species belonging to the same genome group usually carry differentiated subgenomes (V. and A. Grant, 1960).

F_1 hybrids between the tetraploid species have been analyzed cytogenetically in 17 hybrid combinations. Most of these hybrids display a medium degree of chromosome pairing at meiosis. Thus *Gilia malior* X *transmontana* has an average of 9 bivalents, and a range of 7 to 15 II, where 18 II would represent complete pairing. The degree of chromo-

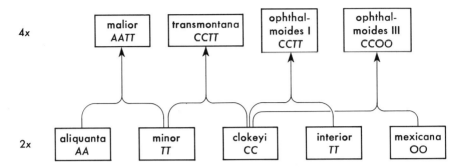

Figure 39   Phylogeny of a segment of the polyploid complex in the *Gilia inconspicua* group. The relationships shown are based on taxogenetic evidence, supported by morphological and ecological evidence, and confirmed in three of the four tetraploid species by resynthesis of the artificial amphiploid. (Drawn from data of V. and A. Grant, 1960; Grant, 1964a; Day, 1965.)

some pairing in this and similar hybrids suggests that the two tetraploid species involved share one basic genome in common and differ in their other genome (Grant, 1964c).

For some pairs of tetraploid species this suggestion has been confirmed and the common diploid genome has been identified by producing the appropriate triploid hybrids and finding the expected *Drosera* type of chromosome pairing. We would expect >9 II $+ <18$ I in triploid hybrids derived from crosses between tetraploid and diploid species belonging to an ancestor-descendant lineage.

Just this type of pairing was in fact observed in triploid hybrids of *Gilia sinuata* (4*x*) X *G. latiflora* (2*x*) and *G. modocensis* (4*x*) X *G. latiflora* (2*x*) (Grant, 1964c). Therefore the *latiflora* or T genome is present in both *G. sinuata* and *G. modocensis*. On morphological and ecological evidence it can be concluded that *G. latiflora* itself, and not some other member of the T genome group, was a diploid parent of both *G. sinuata* and *G. modocensis*. These two tetraploid species differ genomically and morphologically as regards their other diploid parent.

Day (1965) made a detailed taxogenetic study of two tetraploid species and their putative diploid parents in the *Gilia transmontana* subgroup. The morphological, ecological, cytological, and cytogenetic evidence pointed to a reticulate derivation of the two tetraploids, *G. transmontana* and *G. malior*, from three diploid species, as shown in Figure 39. Day then confirmed this phylogeny by artificially resynthesizing *G. transmontana* and *G. malior* from their diploid parents and showing that each artificial amphiploid is interfertile with the corresponding natural tetraploid species (Day, 1965).

Similar evidence was obtained for the ancestry of the tetraploid species known as *Gilia ophthalmoides* III. This species was believed to be derived from *G. clokeyi* (2x) × *G. mexicana* (2x). Spontaneous tetraploid progeny obtained from this diploid interspecific hybrid turned out to be good *G. ophthalmoides* III, thus confirming the phylogenetic hypothesis (V. and A. Grant, 1960, and unpublished). Additional evidence points to a partly common, partly different ancestry of a related tetraploid species, *G. ophthalmoides* I.

The reticulate phylogenetic derivation of four tetraploid species from five diploid species is shown in Figure 39. This reticulate pattern of phylogeny could be extended horizontally to include *G. sinuata*, *G. modocensis*, and other polyploid members of the *G. inconspicua* complex.

Vestiges of Ancient Polyploid Complexes

In the course of time the diploid species of a polyploid complex may die out. Or the lower polyploids may become extinct too, leaving the higher polyploids behind as the only living representatives of the complex.

The *Bromus carinatus* group (Gramineae) illustrates a polyploid complex which has reached an advanced stage of development. This group reaches high ploidy levels, *B. arizonicus* being 12x, but has suffered much extinction at the diploid and tetraploid levels. The ancestry of the 12x and 8x species has been worked out by Stebbins and collaborators (Stebbins and Tobgy, 1944; Stebbins, Tobgy, and Harlan, 1944; Stebbins, 1947b, 1956).

The *Bromus carinatus* complex belongs to the section Ceratochloa of the genus but contains genomes from two related sections, Bromopsis and Neobromus. The plants are perennial grasses, often bunch grasses, of North and South America. The basic number is $x = 7$.

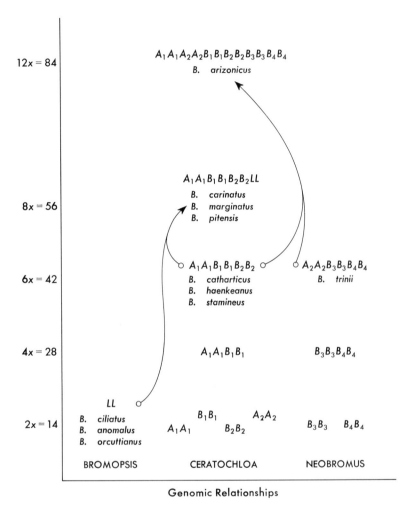

Figure 40 Phylogenetic relationships of high polyploid species of Bromus. Some of the basic genomes are unknown at the diploid or tetraploid level. (Redrawn from Stebbins, 1956.)

There are no known diploid or tetraploid species in Bromus section Ceratochloa. *Bromus catharticus* and its relatives in South America are allohexaploids carrying genomes derived from diploid species which are now extinct (Figure 40).

Bromus carinatus and *B. marginatus* in North America are allooctoploids. Three of their four genomes are homologous with those in the South American hexaploids. The fourth L genome is believed on cytological and morphological evidence to be derived from some North American diploid species of the section Bromopsis (Figure 40). The amphiploid origin of the octoploids probably took place in western North America in the late Tertiary, perhaps the Pliocene. It is necessary to suppose that some hexaploid species homologous with the present-day South American hexaploids occurred in North America in former times (Stebbins and Tobgy, 1944; Stebbins, 1947b).

Bromus arizonicus is an amphiploid at the $12x$ level. It, too, is homologous with the South American hexaploids in three of its six genomes (Figure 40). The other three genomes in *B. arizonicus* are probably derived from an allohexaploid in the related section Neo-bromus (Figure 40) (Stebbins, Tobgy, and Harlan, 1944).

The 12-ploid species, *Bromus arizonicus*, is a common and aggressive weed throughout much of its range in Arizona and California. Its diploid and tetraploid ancestors are extinct. Its hexaploid ancestor in the section Ceratochloa became extinct in North America, perhaps during adverse climatic changes in Pliocene-Pliestocene time, and survives today under more moderate climatic conditions in South America. Yet the 12-ploid derivative of these extinct diploids and lower polyploids is a successful and aggressive species today, showing how continued amphiploid doubling can give an old polyploid complex a new lease on life (Stebbins, 1947b, 1956).

We have seen earlier, in Chapter 15, that high polyploid numbers occur exclusively in nearly all of the main groups of ferns and fern allies. *Tmesipteris tannensis* has over 400 chromosomes in the somatic complement. Psilotum has $2n =$ ca. 100 and ca. 200 in different forms. The species of Equisetum all have $2n =$ ca. 216. A polyploid series in Ophioglossum starts at $2n = 240$ in several "diploid" species. The genera of Polypodiaceae have $2n = 60$ or more, usually many more, chromosomes in the sexual species. Lycopodium ranges from $2n =$ ca. 48 to 260. Selaginella and Isoetes are almost unique in the pterido-phytes in having some true diploid species with $2n = 9$ and 10, respectively (Manton, 1950).

It is clear that most of the main groups of recent pteridophytes—Tmesipterus, Psilotum, Equisetum, Dryopteris, etc.—are the remnants of geologically ancient polyploid complexes (Manton, 1950).

Secondary Basic Numbers in the Angiosperms

Many genera and some whole tribes and families of angiosperms have high basic numbers in the range $x = 16$ to 21. In such groups the obvious question arises whether the observed high numbers are secondary basic numbers of polyploid origin. This situation would be expected to occur in any case. The existence of ancient polyploid complexes which have lost their diploid ancestors in the course of time and extinction shows that it does in fact occur.

A good example of a secondary basic number of probable polyploid origin is provided by the Pomoideae. The four tribes of the Rosaceae have the following basic numbers: Rosoideae ($x = 7$), Prunoideae ($x = 8$), Spiraeoideae ($x = 9$), and Pomoideae ($x = 17$). The tribe Pomoideae thus stands out within the family in its cytological characteristics. Darlington therefore suggested that the tribe Pomoideae is a modified tetraploid derivative of more primitive 7-paired members of the Rosaceae (Darlington, 1932, p. 224). Stebbins later suggested, on the basis of morphological as well as cytological characteristics, that the Pomoideae ($x = 17$) could well be an allotetraploid product of the Prunoideae ($x = 8$) and the Spiraeoideae ($x = 9$) (Stebbins, 1950, p. 361).

Some examples of genera with high secondary basic numbers are: Gossypium (Malvaceae) with $x = 13$; Kalanchoe (Crassulaceae) with $x = 17$; Buddleia (Loganiaceae) with $x = 19$; and Fraxinus (Oleaceae) with $x = 23$. The existence of 7-paired and 6-paired species in the Malvaceae lends plausibility to the suggestion that the $x = 13$ of Gossypium is an old tetraploid number (Tischler, 1954). Likewise in the Loganiaceae there are species with $n = 6$, 7, or 8 whose genomes could be added in various ways to give the $x = 19$ of Buddleia (Darlington, 1956b).

Several authors have suggested that $x = 12$ in Nicotiana is an old tetraploid number derived by amphiploidy from hypothetical 6-paired ancestors (Darlington, 1937a, 1956b; Kostoff, 1941–1943; Goodspeed, 1954). Similar suggestions have been made for Solanum ($x = 12$) (Gottschalk, 1954; Darlington, 1956b). In the absence of 6-paired species in either Nicotiana or Solanum, these suggestions have to be considered purely speculative (Swaminathan and Howard, 1953; Rick and Butler, 1956). Certainly $x = 12$ is the obvious basic number throughout the family Solanaceae.

The same claims for old tetraploidy have been made by some workers and disputed by others in the case of Zea ($x = 10$) and the *Crepis occidentalis* group ($x = 11$).

The following families of woody dicotyledons have quite high basic numbers which have been plausibly considered by various authors to be of ancient polyploid derivation (Stebbins, 1950; Tischler, 1954; Darlington, 1956b):

Magnoliaceae	$x = 19$
Trochodendraceae	$x = 19$
Cercidiphyllaceae	$x = 19$
Salicaceae	$x = 19$
Hippocastanaceae	$x = 20$
Platanaceae	$x = 21$

Various families of woody dicotyledons have *moderately* high basic numbers in the range $x = 11$ to 16. Examples taken from the *Chromosome Atlas* of Darlington and Wylie (1955) are:

Garryaceae	$x = 11$
Fagaceae	$x = 12$
Aceraceae	$x = 13$
Betulaceae	$x = 14$
Ulmaceae	$x = 14$
Juglandaceae	$x = 16.$

Stebbins has suggested that these moderately high basic numbers ($x = 11$ to 16), as well as the still higher ones ($x = 19$, etc.) in many other woody dicotyledonous families, are secondary numbers established by events of amphiploidy in the early history of the angiosperms (Stebbins, 1938; 1947a; 1950, pp. 364–65).

This inevitably brings us to the moot question of the original basic number of the angiosperms. The original basic number which is suggested by most students of the question is $x = 7$ (Darlington and Mather, 1949; Darlington, 1956b; Raven and Kyhos, 1965; Hair, 1966; Ehrendorfer, Krendl, Habeler, and Sauer, 1968). Other suggestions place the original number in the range $x = 6$ to 8 (Stebbins, 1950, p. 365) or in the range $x = 7$ to 9 (Grant, 1963, p. 486; Ehrendorfer,

1964). I cannot see any compelling reason for picking $x = 7$ specifically. The diploid numbers $n = 7$, 8, and 9 occur together in many angiosperm families, including some primitive ones, and apparently represent a basic condition of the class. Perhaps the best conclusion which can be drawn with any degree of reliability from the available evidence is that the common denominator for the primitive angiosperms lies in the range $x = 6$ to 9.

With this as a premise, we can readily see that $x = 12$ in the Solanaceae, Fagaceae, etc., and other neighboring x numbers, *could* be of tetraploid origin. Some such chromosome numbers in this range probably are polyploid. In fact, *Haplopappus spinulosus australis* (Compositae) with $n = 8$ is known to be an allotetraploid based on $x = 4$ in Haplopappus section Blepharodon (Jackson, 1962).

But it is not safe to assume automatically that all moderately high basic numbers, i.e., $x = 11$ to 16, are polyploid in origin. This assumption ignores the existence of an alternative pathway to moderately high basic numbers, namely, ascending aneuploidy. Increasing aneuploid series have been reported in some woody families such as the Rutaceae and the Podocarpaceae (Smith-White, 1959a; Khoshoo, 1959).

We would expect ascending aneuploidy to occur in many groups of forest trees and shrubs for the following reason. A high diploid chromosome number is one of the means of maximizing the generation of recombinational variability in the plant population. And the ecological conditions in stable forest communities are such as to favor open recombination systems (Grant, 1958). Consequently, the dominant and subdominant species of forests are often exposed to selection for aneuploid increases. It is to be expected, therefore, that many high-number woody plants are true diploids.

Irreversibility

The evolutionary trend from diploid to tetraploid to higher polyploid is in the main an irreversible trend. This trend can be reversed partially by the polyploid drop mentioned in Chapter 15. We cited there the example of Hesperis, where $x = 7$ and tetraploid species have gametic numbers of $n = 14$, 13, and 12.

Polyploidy followed by polyploid drop has apparently occurred in several phyletic lines in the Proteaceae (Smith-White, 1959a; Johnson and Briggs, 1963). The probable basic number for the family is $x = 7$. The tribe Grevilleoideae starts with $n = 14$, which is assumed to be tetraploid, and runs down to $n = 11$ and 10 in Lomatia, Hakea, and some other advanced genera. Similar drops to $n = 13$, 12, and 11 occur in the tribes Proteeae and Persoonieae (Smith-White, 1959a; Johnson and Briggs, 1963).

Some cytotaxonomists have suggested the possibility of abrupt and complete reversals from the tetraploid to the diploid condition by means of polyhaploids (Raven and Thompson, 1964; De Wet, 1968). The haploid or polyhaploid progeny of a tetraploid parent is, of course, diploid. Such polyhaploids do arise spontaneously in polyploid plant populations. The polyhaploid derivative of an allotetraploid is, however, by definition and by observation, a chromosomally sterile diploid hybrid whose main chance of reproducing sexually is by doubling once again. Polyhaploids offer little hope for reversals in polyploid trends.

Suppression of Variability

The polyploid condition places certain restrictions on the generation of mutational and recombinational variability, as compared with the diploid condition, and these restrictions become tighter at the higher polyploid levels. The expression of new mutations is suppressed by gene duplication and polysomic inheritance. The formation of new recombination types is restricted by these factors and by preferential pairing.

Tischler points out that seedling mutants are common in Triticum monococcum ($2x$), are much rarer in T. durum ($4x$), and are quite rare in T. aestivum ($6x$) (Tischler, 1953).

The generation of variability is restricted but not blocked completely by the polyploid condition. Segmental allopolyploids can segregate to a limited extent for the interspecific gene differences. Polyploids can segregate mutant types rarely in polysomic ratios. Old polyploid species may undergo gene divergences in the course of time so that they behave genetically like diploids in the affected genes. Nevertheless, the variation-producing mechanism of the sexual process is inhibited in polyploids as compared with diploids.

The ability of polyploid species to respond to changing environmental conditions is therefore expected to be less efficient than that of diploid species. Certainly new polyploids are often very successful, more so than their diploid relatives, in various types of habitats, as we have seen in Chapter 16. They are successful by virtue of a genetic system which permits a particular hybrid genotype to breed true. But this genetic system also imposes restrictions on the release of new variations and thereby sets limitations on the evolutionary potential of the polyploid lineage.

As the mainly irreversible trend to higher ploidy levels continues in the phyletic line, these limitations of evolutionary potential must become greater. The polyploid system, as Stebbins states (1950, p. 366), is capable of producing numerous related species or even genera with generally similar characteristics, but has apparently played no role in the origin of new families or orders. Polyploid complexes have their periods of rise and of expansion within circumscribed limits followed in the course of time by their periods of decline and extinction.

CHAPTER *21*

The Agamic Complex

Introduction

The combined processes of natural hybridization and agamospermous reproduction lead, in the course of time, to the formation of an array of microspecies carrying different recombinations of the characters of the parental species. The resulting structure is an agamic complex in which the agamospermous hybrid derivatives are superimposed phylogenetically on the original sexual species.

The structure of a simple agamic complex is shown diagrammatically

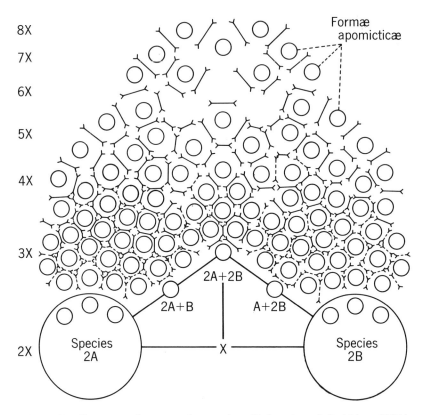

Figure 41 Structure of an agamic complex. (Babcock and Stebbins, 1938.)

in Figure 41. We see the numerous agamospermous derivatives bridging the gap and obscuring the morphological discontinuity between the diploid sexual species. The most intricate variation patterns known in plants are those found in agamic complexes. Thousands of Latin names have been published in specific or infraspecific rank in the agamic complexes of Rubus and Hieracium.

The species problem in an agamic complex is partly soluble. The diploid sexual species, if still extant, can be discovered, named, and mapped. Particular agamospermous derivatives which present a fairly uniform and distinctive character combination throughout a definite geographical area should also be recognized and named as taxonomic species.

But the species problem is also insoluble in some parts of the agamic complex. Between the diploid sexual species and the more successful and extensive agamospermous species lies a host of agamospermous variants of lesser rank. These variants range from minor biotypes to localized agamospermous microspecies. Which of them to recognize taxonomically, and in what taxonomic rank, are questions of judgment and convenience concerning which different specialists will often disagree.

The phylogenetic analysis of an agamic complex is also beset with many difficulties stemming from the great complexity of the problem. The stages of analysis outlined in Chapter 20 for polyploid complexes can be applied quite well to agamic complexes. Some agamic complexes, like the *Crepis occidentalis* group, have undergone a good cytotaxonomic analysis, and some, like Rubus section Moriferus, a good biosystematic analysis, but no large agamic complex has yet been thoroughly explored taxogenetically.

The Crepis occidentalis Complex

The *Crepis occidentalis* group ($x = 11$) is an assemblage of perennial herbs in western North America. The plants grow on arid mountain slopes with a center of distribution in the Columbia plateau and a range extending from the Pacific slope to the Rocky Mountains. The group contains sexual diploids and numerous agamospermous polyploids (Babcock and Stebbins, 1938).

Babcock and Stebbins (1938) have analyzed the *Crepis occidentalis* group on the basis of herbarium studies, field work, cytotaxonomic methods, and observations of live plants transplanted from nature. Experimental hybridizations were not possible, and therefore direct evidence concerning the genetic relationships between the forms is not available.

Babcock and Stebbins group the variations in the *Crepis occidentalis* species into nine taxonomic species. They are: *Crepis acuminata, bakeri, barbigera, exilis, intermedia, modocensis, monticola, occidentalis,* and *pleurocarpa*. Most of these taxonomic species contain a diploid sexual population and various agamospermous types which are more or less similar in morphology to the diploid. The agamospermous forms, where named at all, are treated in the trinomial category of forma apomictica. The named taxonomic species in the *C. occidentalis* group are thus composite and artificial entities.

There are seven known sexual diploid species in the group. They are treated as infraspecific units in the following taxonomic species: *Crepis acuminata, bakeri, exilis, modocensis, monticola, occidentalis,* and *pleurocarpa.* The sexual diploid populations comprise the basic biological species in the *C. occidentalis* group. The seven diploid species are well demarcated morphologically. Only one of them (*C. acuminata* 2*x*) is widespread geographically; the remaining six are restricted to relatively small areas. They overlap sympatrically in various parts of the Pacific Northwest (Babcock and Stebbins, 1938).

The agamospermous members of the *Crepis occidentalis* group are all polyploids on the base of $x = 11$. In ploidy level they range up to octoploids. They include the various possible odd polyploid types (3*x*, 5*x*, 7*x*). Morphologically, some of the agamospermous types closely resemble a single diploid species, but most of them combine the characteristics of two or more diploid species. It is in this superstructure of polyploid agamosperms that morphological intergradation between the various extreme character combinations is found (Babcock and Stebbins, 1938).

The agamospermous populations differ from one another in characters of the leaves, stems, heads, flowers, and achenes. Some are local and others widespread in geographical distribution. Those which have spread their distinctive character ensembles over any sizable geographical area have been accorded informal taxonomic recognition. They have been given trinomial Latin names in the informal category of forma apomictica (apm.), as mentioned earlier.

One example of a distinctive agamospermous microspecies is *Crepis pleurocarpa* apm. *grayi* in northeastern California. It is probably a pentaploid. On morphology, it is believed to be a hybrid derivative of the sexual diploid forms of *C. pleurocarpa* and *C. occidentalis.* Another agamospermous microspecies in the mountains of northeastern California is the probable triploid known as *C. acuminata* apm. *sierrae.* Altogether there are 113 of these named apomictic forms in the monograph of Babcock and Stebbins. This number represents only a fraction of the total array of distinct agamospermous populations of Crepis in western North America (Babcock and Stebbins, 1938).

The number of agamospermous microspecies is relatively high in areas near the center of distribution of the sexual populations. The sexual and facultatively apomictic populations are a continual source of production of new agamospermous microspecies in the central areas. On the periphery of the distribution area of the entire complex, by contrast, there are fewer microspecies, but each one has a more extensive range. Here certain particular agamospermous types have evidently been favored by natural selection, on the basis of their fitness for the regional environment, and have multiplied and colonized large areas (Babcock and Stebbins, 1938).

The phylogenetic history of the *Crepis occidentalis* complex can be reconstructed in its broad outlines from indirect evidence. The present geographical distribution of each of the sexual diploid species is correlated with the distribution of particular climatic and edaphic provinces which are known to have developed in the Pliocene and early Pleistocene. And therefore the diploid species probably established their respective ranges in the same periods, from early Pliocene to early Pleistocene, in response to drying of climate and outpouring of lava (Babcock and Stebbins, 1938).

The diploid species, after becoming established, hybridized in different combinations to produce allotetraploid derivatives. The process of hybridization continued, probably up to recent times, leading to the formation of agamospermous microspecies of secondary and tertiary hybrid origin, for some of these microspecies have high polyploid numbers and complex affinities. Later, during the Pleistocene, the polyploid agamosperms spread widely at the expense of the diploid sexuals, which became restricted to two centers of distribution in the Pacific northwest (Babcock and Stebbins, 1938).

The European Blackberries

The large genus Rubus ($x = 7$) is subdivided into four subgenera, of which Eubatus, comprising the blackberries and dewberries, is the most complex. The subgenus Eubatus has developed varied assemblages of species and microspecies in several temperate regions of the world: Europe, Asia, the Mediterranean region, eastern North America, the

Pacific coast of North America, and the Cordilleran region of Central and South America.

The Eurasian and eastern North American blackberries form one subgroup of Eubatus which is treated formally as the section Moriferus. The European blackberry flora is the best-known part of this section and is the central subject of discussion here. It is subdivided further into two natural branches known, respectively, as the Moriferi Veri and the Rubi Corylifolii. However, the European blackberry flora is not a distinct and natural unit in itself, apart from the eastern American and Asiatic members of the section Moriferus, and some species in these other blackberry floras are undoubtedly connected phylogenetically with their European relatives (Gustafsson, 1942, 1943, 1946–1947).

The blackberries in general (Eubatus) and the European blackberries in particular (Moriferus in part) have been studied by a succession of taxonomists and geneticists: Focke, Lidforss, Gustafsson, and others. The stimulus for such studies, both taxonomic and genetic, has been provided by "the almost inexhaustible variation of the blackberries" (Gustafsson, 1946–1947, p. 244). In an effort to cope with this variation taxonomically, some four or five thousand Latin names have been published over the years. These names refer to entities ranging from widely distributed species to highly localized variant forms.

In order to illustrate the amount of variation in the European blackberries, Gustafsson summarizes a system of classification by an earlier student, Sudre. Sudre's treatment of the Moriferi Veri contains 691 microspecies grouped around 109 taxonomic species. There are about 1930 named varieties in the Moriferi Veri in Europe alone (Gustafsson, 1946–1947, p. 245). The other main branch of European blackberries, the Rubi Corylifolii, contains one taxonomic species (*Rubus caesius*) but "innumerable" microspecies (Gustafsson, 1943).

The European blackberries form a polyploid series ($x = 7$). The ploidy levels present are $2x$, $3x$, $4x$, $5x$, $6x$, and perhaps $7x$. There are in addition some secondary aneuploids with numbers deviating from a polyploid condition. Most of the microspecies are tetraploids. A few triploid and pentaploid microspecies have attained wide distributional areas (Gustafsson, 1943, 1946–1947).

The eastern North American blackberries likewise form a polyploid series on $x = 7$, including triploids, pentaploids, and hexaploids as well as some diploid species (Gustafsson, 1943). The Pacific coast blackberries, by contrast, are all high polyploids ranging from the $6x$ to $12x$ level (Brown, 1943; Clausen, Keck, and Hiesey, 1945).

Most of the polyploids in the European blackberries are facultative agamosperms. Many of them are known to be hybrids or hybrid derivatives, as indicated in different cases by the evidence of morphological characters, pollen sterility, odd polyploid condition, and/or wide segregation on reverting to sexual reproduction. Agamospermy has also been shown to exist in the eastern American blackberries. In addition to agamospermous seed formation, the blackberries also spread, as is well known, by vegetative propagation (Gustafsson, 1943, 1946–1947).

Most of the agamospermous types in Europe form geographically localized microspecies. Some of these local microspecies are distinct units in themselves; others hybridize and intergrade with related microspecies. A few of the European agamospermous types have formed geographically widespread populations and hence have emerged as agamospermous species.

The diploid sexual species form the phylogenetic foundation of the agamic complex in the European blackberries. The five known diploid species in western Eurasia are *Rubus bollei, incanescens, moschus, tomentosus*, and *ulmifolius* ($2n = 14$). Some others such as *R. alnicola* are suspected of being diploids. Still other sexual diploids in eastern North America, like *R. allegheniensis*, may have played a role in the evolutionary development of the agamic complex in Europe (Gustafsson, 1943, 1946–1947).

The diploid sexual species of European blackberries are all members of the Moriferi Veri. They are distinct from the agamospermous microspecies and species and they are distinct as between themselves. Some are widespread and others restricted in distribution. *Rubus tomentosus* and *R. ulmifolius* have wide ranges from southern Europe to Asia Minor; *R. moschus* occurs in the Caucasus Mountains; *R. incanescens* in the western Mediterranean region; and *R. bollei* in the Canary Islands. The probable diploid *R. alnicola* occurs in the Pyrenees and Jura mountains. Where these species meet in nature they frequently hybridize to produce sterile hybrids. Many of the agamospermous microspecies of Moriferi Veri are derivatives, ancient or recent, of such hybrids (Gustafsson, 1943, 1946–1947).

The total range of variation in the agamospermous Moriferi Veri in Europe greatly exceeds the morphological extremes found in the

European diploids. But some European agamosperms are closely related morphologically to certain eastern American diploids like *Rubus allegheniensis*. It can be concluded from this that the diploid sexual ancestors of the agamic complex in Europe include species now restricted to eastern North America as well as to various ranges in western Eurasia (Gustafsson, 1943, 1946–1947).

There are no known diploids in the Rubi Corylifolii. This group does include, however, the common European dewberry, *Rubus caesius* ($2n = 28$), an allotetraploid derived from diploid species probably now extinct. *Rubus caesius* is a facultative agamosperm. It hybridizes freely with other species and microspecies of blackberries. Yet it is isolated from them by barriers of hybrid sterility and inviability. Like the diploids in the Moriferi Veri, it forms a distinct natural species (Gustafsson, 1942, 1943, 1946–1947).

Rubus caesius hybridizes with the diploids *R. ulmifolius* and *R. tomentosus* and with other diploid and polyploid species of blackberries. Numerous agamospermous microspecies of Rubi Corylifolii are segregation products of previous hybridizations between *Rubus caesius* and various members of the Moriferi Veri. Other agamospermous Rubi Corylifolii are secondary backcross products of *R. caesius* and other Rubi Corylifolii. Thus the natural hybrid of *R. caesius* X *R. gothicus* (Corylifolii) gives rise by segregation to types resembling other named microspecies of Rubi Corylifolii such as *R. acutus*, *R. wahlbergii*, and *R. pruinosus* (Gustafsson, 1942, 1943, 1946–1947).

The European raspberry, *Rubus idaeus* ($2n = 14$), belonging to another subgenus, Idaebatus, hybridizes with *R. caesius* ($2n = 28$). The triploid hybrids are sterile. However, fertile or semifertile tetraploid and hexaploid hybrids sometimes arise from unreduced gametes. These hybrids by backcrossing to *R. caesius* carry some raspberry genes into the *R. caesius* group. Some of the agamospermous forms of Rubi Corylifolii appear to be hybrid derivatives of *R. idaeus* and *R. caesius*. Thus a diploid species outside the subgenus Eubatus has also contributed to the enrichment of variation in the European blackberry complex (Gustafsson, 1946–1947).

The diploid sexual species of blackberry in Europe tend to have relictual and southern distributions in the Mediterranean region today. Some of them are isolated geographically. The American diploids are of course isolated today from the European diploids. It is probable that the diploid species had wider ranges and broader overlaps in preglacial or interglacial times or both. The first wave of allotetraploids and agamospermous derivatives may have arisen then (Gustafsson, 1942).

This interglacial blackberry flora would have had to retreat during the last glaciation. But, when climatic conditions ameliorated in post-glacial times, it probably spread again and produced new generations of agamospermous hybrid derivatives. Today the agamospermous blackberries, unlike their diploid sexual ancestors, range widely throughout northern and eastern Europe. The agamospermous micro-species of hybrid origin have successfully colonized the colder parts of this continent which were recently covered by ice (Gustafsson, 1942, 1943).

The Rubus Moriferus Pattern

The agamic complexes in Rubus and Crepis illustrate a pattern found in a number of other north temperate plant groups. It will be convenient to designate this fairly common pattern as the Rubus Moriferus pattern in recognition of the most thoroughly analyzed example.

The essential features of the Rubus Moriferus pattern may be summarized as follows. There are hundreds or thousands of interrelated agamospermous microspecies. These microspecies are of hybrid derivation. They produce seeds by gametophytic apomixis and fre-quently also, to some extent, by sexual means. The agamospermous forms are polyploids, even or odd, odd polyploids being common. These polyploid agamosperms are superimposed phylogenetically and taxonomically on a series of distinct sexual diploid species, which are extant. The sexual diploids, however, tend to have restricted or even relictual distributions, whereas the agamospermous derivatives are more aggressive and widespread.

Among the plant groups conforming to the Rubus Moriferus pattern is the genus Taraxacum. This widespread genus occurs in both the old and new worlds and both the northern and southern hemi-spheres. Its center of distribution lies in Eurasia, where it extends east to Japan and north beyond the arctic circle. Many variant forms exist which can be grouped into some 50 or 60 taxonomic species.

The basic number in Taraxacum is $x = 8$. The ploidy levels found are

$2x$, $3x$, $4x$, $5x$, and $6x$. There are also some secondary aneuploid deviants. The polyploids are agamospermous, facultative, or obligate. Most forms of the common dandelion (*T. officinale* or *T. vulgare*) are triploid ($2n = 24$). It is the agamospermous Taraxaca that range into the arctic region and that show weedy tendencies in other floras (Darlington and Janaki Ammal, 1945; Sørensen and Gudjonsson, 1946; Gustafsson, 1946–1947, pp. 259 ff.; Fürnkranz, 1960).

The agamospermous Taraxaca present the familiar pattern of numerous morphologically similar microspecies. Gustafsson estimates that there are 450 microspecies in Scandinavia alone (Gustafsson, 1946–1947, p. 261).

Quite a few sexual diploid species have been discovered in Taraxacum. *Taraxacum kok-saghyz* ($2n = 16$) from Turkestan is a well-known example. Some but not all of these diploids are rare or narrowly endemic. Thus *T. pieninicum* in the Pieniny Mountains of Poland is probably a glacial relict (Malecka, 1961). Gustafsson lists 13 diploid species, mostly from central and eastern Eurasia (Gustafsson, 1946–1947, p. 260). In recent years additional diploid sexual species have been found in other areas. Examples are *T. pumilum* in Greenland, *T. pieninicum* in Poland, and *T. officinale* ($2x$ form) and *T. laevigatum* ($2x$ form) in Austria (Fürnkranz, 1960; Malecka, 1961).

The diploid species of Taraxacum cross readily but produce highly sterile hybrids in the combinations tested. Some though not necessarily all of these diploid sexual species, and their first wave of hybrid products, form the phylogenetic foundations of the agamic complex in Taraxacum (Gustafsson, 1946–1947, pp. 128, 260).

Some other agamic complexes conforming to the Rubus Moriferus pattern may be listed here. See Gustafsson (1946–1947, pp. 230 ff.) for review of most of these cases; and see Sax (1954), Beaman (1957), and Skalinska (1967) for accounts of Cotoneaster, Townsendia, and Hieracium Pilosella.

Hieracium subgenus Pilosella ($x = 7$) in Eurasia
Hieracium subgenus Euhieracium ($x = 7$) in the
   northern hemisphere
Alchemilla subgenus Eualchemilla ($x = 8$) in the
   northern hemisphere
Alchemilla subgenus Aphanes ($x = 8$) in Eurasia,
   the new world, and Australia

Townsendia $(x = 9)$ in western North America
Hypericum $(x = 8)$ in Eurasia
Parthenium $(x = 9)$ in southern North America
Sorbus $(x = 17)$ in Eurasia
Cotoneaster $(x = 17)$ in Eurasia
Antennaria $(x = 7)$ in the northern hemisphere
Calamagrostis $(x = 7)$ in Eurasia

The details, of course, differ from case to case. Thus the diploid sexual species are well represented in Hieracium Euhieracium but are rare in Alchemilla. The sexual diploids are well represented again in Sorbus and Cotoneaster, but here they are old diploidized allotetraploids $(x_2 = 17)$. In Antennaria $(x = 7)$, Parthenium $(x = 9)$, and Calamagrostis $(x = 7)$ the ancestral sexual species seem to be tetraploids.

It is interesting that most of our examples of the Rubus Moriferus pattern belong to the families Compositae and Rosaceae which have been very successful in the northern hemisphere. The composite examples are Crepis, Taraxacum, Hieracium, Townsendia, Parthenium, and Antennaria; the rosaceous examples are Rubus, Alchemilla, Sorbus, and Cotoneaster. In addition, the Gramineae are represented by Calamagrostis, and the Hypericaceae by Hypericum.

The Bouteloua curtipendula Complex

The grama grasses comprising the *Bouteloua curtipendula* complex range widely throughout the grasslands of North America from Canada to Mexico, and occur in other disjunct areas of Central and South America. The recent revision of Gould and Kapadia (1964) recognizes 12 species in the group. One of them, the taxonomic species *B. curtipendula* proper, is very widespread, occurring in a broad belt through the central United States and Mexico (Figure 42). The other eleven species have more restricted distribution areas (Figure 43) (Gould and Kapadia, 1964).

Chromosome numbers range from $2n = 20$ to 103 (see Table 8). The basic number is $x = 10$. There are diploids, tetraploids, hexaploids, and higher polyploids up to the $10x$ level, as well as many aneuploids (Gould and Kapadia, 1962, 1964).

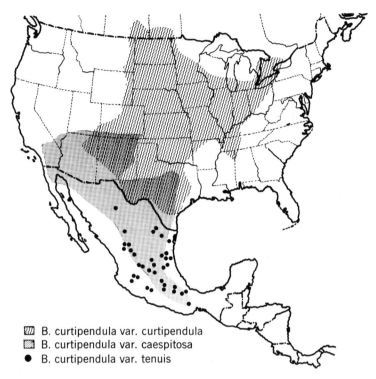

B. curtipendula var. curtipendula
B. curtipendula var. caespitosa
● B. curtipendula var. tenuis

Figure 42 Distribution of *Bouteloua curtipendula*. The chromosome numbers in the three named varieties are as follows. *B. c. curtipendula* (2n = 40, 41–64). *B. c. caespitosa* (2n = 58–103). *B. c. tenuis* (2n = 20, 40–42). (Gould and Kapadia, 1964.)

The higher polyploids and aneuploids in the *Bouteloua curtipendula* group are agamospermous, by means of apospory and pseudogamy, and they often have irregular meiosis with a low degree of pairing. There are no known sexual hexaploids or sexual aneuploids near the hexaploid level in the group. The diploids, tetraploids, and lower aneuploids are sexual, as far as known. Meiosis is regular in the diploids and some of the tetraploids, and moderately irregular in some other tetraploids and low-number aneuploids (Harlan, 1949; Freter and Brown, 1955; Gould, 1959; Mohamed and Gould, 1966).

In one sample of chromosome counts, Freter and Brown (1955) found that the dividing line between sexuality and agamospermy in this group

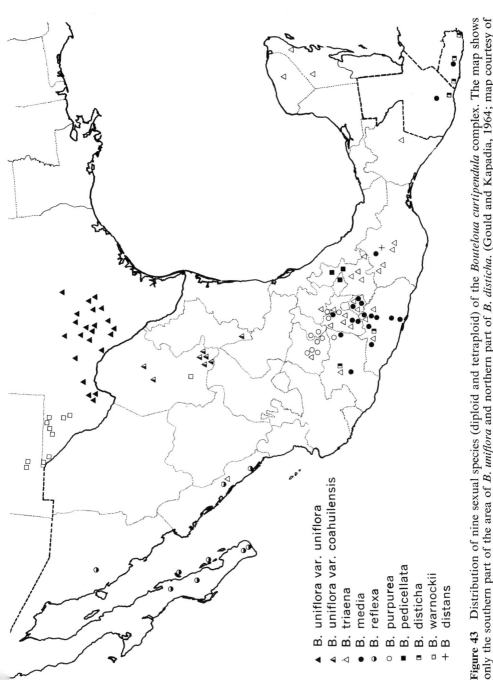

Figure 43 Distribution of nine sexual species (diploid and tetraploid) of the *Bouteloua curtipendula* complex. The map shows only the southern part of the area of *B. uniflora* and northern part of *B. disticha.* (Gould and Kapadia, 1964; map courtesy of Dr. Frank Gould.)

◄ B. uniflora var. uniflora
◄ B. uniflora var. coahuilensis
◁ B. triaena
● B. media
◑ B. reflexa
○ B. purpurea
■ B. pedicellata
◪ B. disticha
▢ B. warnockii
+ B distans

Table 8 · Chromosome numbers in the *Bouteloua curtipendula* complex ($x = 10$) (Gould and Kapadia, 1962, 1964; Gould, 1966)

Species	Distribution area	2n Number(s)
B. triaena	Southern Mexico	20
B. media	Central Mexico to Honduras	20
B. uniflora	Utah to north-central Mexico	20 (30 in a hybrid)
B. warnockii	Western Texas to New Mexico	21, 22, 23, 25, 28, 38, 40
B. reflexa	Baja California to Sonora	20
B. pedicillata	Central Mexico	Probably diploid on basis of pollen size
B. disticha	Tropical Mexico to Peru	As above
B. vaneedenii	Caribbean islands	As above
B. juncea	Caribbean islands	Unknown, possibly diploid
B. distans	Oaxaca	As above
B. purpurea	Central Mexico	40
B. curtipendula		
var. tenuis	Mexico	20, 40, 42
var. curtipendula	Southern Canada to Texas	40, 41, 42, 43, 44, 45, 46, 48, 50, 51, 52, 53, 54, 55, 56, 58, 59, 60, 62, 64
var. caespitosa	Southwestern U.S. to southern Mexico; also South America	58, 60, 62, 64, 69, 72, 77, 78, 80, 82, 84, 85, 86, 87, 88, 90, 91, 92, 94, 96, 98, 100, 102, 103

lies at $2n = 52$. All plants with more than 52 somatic chromosomes appear to be agamospermous. Plants with 52 or fewer chromosomes are sexual in the known cases. However, low-number agamosperms may yet be discovered (Freter and Brown, 1955; Gould, 1959).

The diploid sexual species such as *Bouteloua uniflora, B. triaena, B. media*, and others occur mainly in subtropical to tropical latitudes and have relatively restricted distribution areas. The same is true of the sexual tetraploids, *B. purpurea* and *B. warnockii* (in part) (Figure 43). The sexual diploid and tetraploid forms of *B. curtipendula tenuis* also occur in Mexico (Figure 42).

In contrast to the diploids, the members of the group with high chromosome numbers, namely, *Bouteloua curtipendula curtipendula* ($2n = 40$ to 64) and *B. c. caespitosa* ($2n = 58$ to 103), have distribution areas which are very extensive and more northerly, extending through the plains to southern Canada (Figure 42). As noted earlier, the plants with high numbers in the approximate range $5x$ to $10x$ are agamospermous. The map in Figure 44 shows that these agamospermous types of *B. curtipendula* are distributed over a wide area in the southwestern United States and northern Mexico (Gould, 1959).

An additional feature of the high-number Boutelouas is the existence of extensive aneuploid series. Table 8 shows an almost continuous aneuploid series with only a few short gaps in both *B. curtipendula curtipendula* ($2n = 40$ to 64) and *B. c. caespitosa* ($2n = 58$ to 103). The euploid numbers are the most frequently encountered ones in the samples of chromosome counts, and the various aneuploid numbers occur as deviants from these euploid nodes (Gould and Kapadia, 1962, 1964).

This extensive aneuploidy results from a combination of processes: natural hybridization, irregular meiosis, and agamospermous reproduction.

Natural hybridization between populations with different numbers leads to the segregation of aneuploid types in subsequent generations. This is the probable cause of the simple aneuploid series observed in the sexual species *Bouteloua warnockii* (see Table 8). *Bouteloua warnockii* ($2x, 4x$) hybridizes with the high-number *B. curtipendula caespitosa* in west Texas (Gould and Kapadia, 1964). The more extensive aneuploid series within *B. curtipendula* itself is attributed to more complex hybridizations within this taxonomic species and between it and various other species of the group (Gould and Kapadia, 1962; Kapadia and Gould, 1964).

Plants of *B. curtipendula* with high chromosome numbers often have irregular meiosis, as mentioned earlier. A meiotic feature of particular interest to us here is the occurrence of very unequal segregations at first anaphase, most of the chromosomes going to one pole and few to

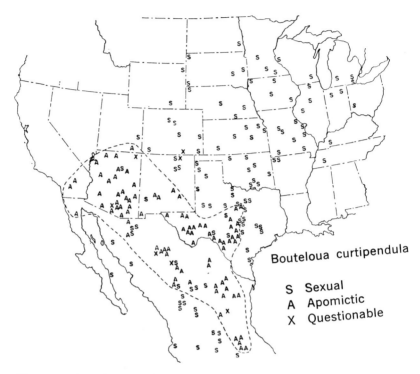

Figure 44 Distribution of sexual and agamospermous forms of *Bouteloua curtipendula*. (Gould, 1959.)

the other (see Figure 45). The meiotic products with the higher numbers apparently yield viable pollen grains. These in turn could produce aneuploid progeny. The various aneuploid types, once formed, can then be perpetuated by agamospermy (Gould and Kapadia, 1962, 1964).

In summary, the *Bouteloua curtipendula* group is a sexual polyploid complex at the tetraploid level, based on *B. uniflora*, *B. curtipendula* 2*x*, and various other diploid species. Superimposed on this sexual polyploid complex is an agamic complex ranging from the 5*x* to the 10*x* level. And superimposed on the euploid agamospermous forms is an extensive and nearly continuous series of secondary aneuploids.

The circumboreal *Poa pratensis* and *P. palustris* groups are agamic complexes which resemble the *Bouteloua curtipendula* group in their

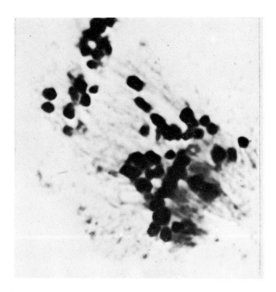

Figure 45 First anaphase in a dividing pollen mother-cell of a plant of *Bouteloua curtipendula caespitosa* with a high chromosome number. Note the very unequal segregation of chromosomes. (Gould and Kapadia, 1964; photograph courtesy of Dr. Frank Gould.)

general features. The sexual forms are low polyploids in Poa or, more rarely, diploids ($x = 7$). The higher polyploids are agamospermous or viviparous. And the euploid nodes are connected by an almost continuous series of secondary aneuploids. We saw earlier that the taxonomic species *P. pratensis* ranges from $2n = 22$ to $2n = 147$ (see Chapter 18).

Secondary aneuploidy is also prevalent in the agamospermous *Potentilla gracilis* complex ($x = 7$) in western North America. The chromosome numbers in this complex range from $2n =$ ca. 56 to ca. 109 (Clausen, Keck, and Hiesey, 1940, Ch. 3).

The pattern found in Bouteloua, Poa, and Potentilla can be classified as a variant form of the more common Rubus Moriferus pattern. It is a variant pattern in which secondary aneuploidy is developed on an extensive scale.

Citrus

The genus Citrus belongs to the subtribe Citrinae in the subfamily Aurantioideae of the Rutaceae. The subtribe Citrinae is a group of 13

Table 9 · Classification of the genus Citrus (Rearranged from Swingle, 1967)

I. Subgenus Papeda
    1. C. ichangensis. Ichang papeda. China
    2. C. latipes. Khasi papeda. India and Burma
    3. C. micrantha. Small-flowered papeda. Philippines
    4. C. celebica. Celebes papeda. Celebes and Philippines
    5. C. macroptera. Melanesian papeda. Southeast Asia to Polynesia
    6. C. hystrix. Mauritius papeda. Southeast Asia and Malaysia
II. Subgenus Citrus
    7. C. indica. Indian wild orange. Eastern Himalayan region
    8. C. tachibana. Tachibana orange. Japan
    9. C. medica. Citron. East Asia; now widely cultivated
    10. C. limon. Lemon. Southeast Asia; widely cultivated
    11. C. aurantifolia. Lime. Southeast Asia and Malaysia; widely cultivated
    12. C. aurantium. Sour orange. Southeast Asia; widely cultivated
    13. C. sinensis. Sweet orange. China and Indochina; widely cultivated
    14. C. reticulata. Mandarin orange. Southeast Asia and Philippines; widely cultivated
    15. C. grandis. Pummelo. Southeast Asia; widely cultivated
    16. C. paradisi. Grapefruit. West Indies; widely cultivated

---

genera of evergreen or rarely deciduous trees and shrubs bearing fruits with a juicy pulp. The majority of the Citrinae are native in southeast Asia, Malaysia, and Australia, and some members occur in Africa (Swingle, 1967).

Within the Citrinae, Citrus belongs to a group of six closely related genera with orangelike or lemonlike fruits. The genera in question are Fortunella, Eremocitrus, Poncirus, Clymenia, and Microcitrus, in addition to Citrus itself. These genera can be intercrossed. They occur from India and China through the Pacific island chains to Australia (Swingle, 1967).

The largest and most widespread genus in the group is Citrus with 16 recognized species in 2 subgenera. A synopsis of Swingle's (1967) classification of Citrus is given in Table 9.

The subgenus Papeda consists of wild species with small flowers and nonedible fruits. The pulp vescicles of the Papeda fruits contain acrid oil droplets. The subgenus Citrus, on the other hand, is characterized by large fragrant flowers and edible fruits, the pulp vescicles of which do not contain acrid oil droplets. *Citrus indica* and *C. tachibana* in this subgenus appear to be wild species. The remaining species of the subgenus Citrus (species 9 to 16 in Table 9) are commonly cultivated, and most of them, as for example *C. sinensis* and *C. paradisi*, are not known in the wild state. The probable original native areas of the cultivated species are given in Table 9.

It is important to realize that the taxonomy of Citrus, and particularly of the cultivated forms of the genus, is by no means as simple as Table 9 implies. The classification of the cultivated forms into definite species (species 9 to 16 in the table) is of necessity typological. As such, the classification system is useful as a preliminary reference frame, and it has been presented for this purpose, but it does not portray accurately the variation pattern of the group.

The variation pattern which has confronted and perplexed students of cultivated Citrus trees since the beginning of the 19th century has been one characterized by an absence of clear-cut species. One observes instead a series of different varieties, many of them hybrids of one combination or another, which connect the named species together into a network, a vast syngameon. The named species 9 to 16 in Table 9 are convenient nodal points in this network of forms.

The particular series of named species listed in Table 9 is not the only possible system which can be devised for Citrus. In fact, equally competent students of Citrus fruit trees have been divided for a century and a half between extreme lumpers and extreme splitters.

These opposite tendencies are well illustrated by the recent systems of Tanaka (1954, 1961) and Swingle (1967). Tanaka set out to name each and every morphological form and hybrid type as a species. As a result, for the genus as a whole, Tanaka recognizes 157 species, whereas Swingle has 16 (Tanaka, 1961; Swingle, 1967, pp. 364 ff.). Swingle, on the other hand, recognizes only the more prominent varieties and hybrids, and treats them as subordinate entities under the various taxonomic species (Swingle, 1967).

A closer comparison of the alternative systems of Swingle and Tanaka reveals that these systems are in pretty good agreement as regards the wild species in the subgenus Papeda. The divergence between them is greatest in the treatment of the cultivated Citrus trees.

It is in this portion of the genus, significantly, that the taxonomic confusion reaches its height.

The basic chromosome number in Citrinae, as in the Aurantioideae, generally is $x = 9$. The species of Citrus and related genera are nearly all diploid ($2n = 18$). Tetraploids and triploids occur as aberrant types in a number of cultivated forms of Citrus fruit trees (Darlington and Wylie, 1955; Reuther, Batchelor, and Webber, 1968, pp. 293–94).

The species of Citrus can be crossed artificially, and they hybridize spontaneously with one another in numerous combinations. Citrus also crosses with Poncirus, Fortunella, Microcitrus, and Eremocitrus. Many of the interspecific and even some intergeneric hybrids are sexually fertile. Other Citrus hybrids are sexually sterile but produce seeds asexually (Swingle, 1967; Reuther, Batchelor, and Webber, 1968). The Citrus group belongs to the Ceanothus pattern of fertility relationships.

Most cultivated varieties of Citrus are facultatively agamospermous and pseudogamous. The type of agamospermy is adventitious embryony from nucellar cells. Following pollination, nucellar embryos may develop alongside zygotic embryos. A seed then contains two or more embryos, a condition known as polyembryony, and these embryos may may be sexual and/or asexual in different proportions. Pollination may give segregating progeny (see Figure 36 in Chapter 18) or all matroclinous progeny, depending on the parentage (Webber and Batchelor, 1943; Gustafsson, 1946–1947; Reuther, Batchelor, and Webber, 1968).

In one series of tests the Eureka lemon, after pollination, yielded 223 seedlings, of which 33% were agamospermous and 67% sexual. In tests of other varieties the percentage of all seedlings that arose by nucellar embryony was found to be: Washington navel orange, 97%; Valencia orange 85%; Marsh grapefruit, 96%; Mexican lime, 78%; Dancy mandarin, 100% (Reuther, Batchelor, and Webber, 1968, pp. 304–305).

Poncirus trifoliata, which is also cultivated, is likewise facultatively agamospermous. Nucellar embryony has also been reported in the cultivated Fortunella. On the other hand, a strain of the wild species *Citrus ichangensis* and several cultivated varieties of *C. grandis* have shown no evidence of agamospermy in progeny tests reported to date (Reuther, Batchelor, and Webber, 1968, pp. 301–306).

It is significant for our story that some artificial interspecific hybrids

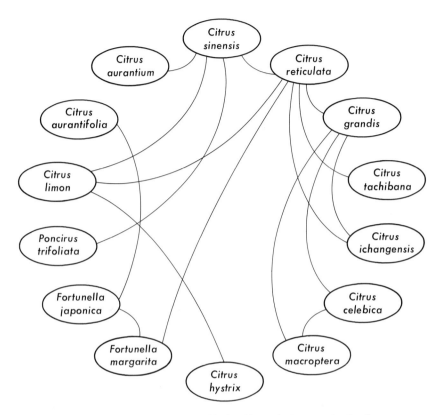

Figure 46 Spontaneous garden hybrids in Citrus fruit trees in the far east. Parentage inferred from morphological and field evidence. (Drawn from data of Swingle, 1967.)

in Citrus reproduce exclusively by nucellar embryony. This is the case in the F_1 hybrids of *Citrus paradisi* X *C. reticulata* (the tangelo) and *Citrus sinensis* X *Poncirus trifoliata* (the citrange). These hybrids breed true to type from seeds and are sometimes propagated by seeds for large-scale plantings (Swingle, 1967, p. 360).

Hybrid Citrus trees are common in home gardens and village groves throughout the far east. Numerous hybrid combinations are represented (see Figure 46). Some of the hybrid types are localized and sporadic, others have spread in cultivation. They have apparently arisen spontaneously in mixed plantings of older Citrus varieties and have then been preserved and sometimes propagated for their desirable characteristics (Swingle, 1967).

Figure 46, drawn from data of Swingle (1967), shows the large number of spontaneous hybrid combinations found in east Asia and the Pacific region. The parentage of these garden hybrids has been inferred from morphological and field evidence in most cases, and has been confirmed by controlled artificial crossings in a few cases. Some of the same hybrids, as well as some new ones like *Eremocitrus glauca* X *Citrus sinensis*, have been observed to arise spontaneously again in Citrus plantings in North America (Swingle, 1967).

The yuzu is a hybrid Citrus, probably derived from some cross of *C. reticulata* X *ichangensis*, which is widely cultivated in China and Japan. Chinese herbals indicate that the yuzu was grown in ancient China; it has been grown there since A.D. 1108 at least and probably since 237 B.C. (Swingle, 1967, pp. 389–90). The yuzu is strongly agamospermous, producing up to 95% nucellar seedlings (Reuther, Batchelor, and Webber, 1968, p. 305). Thus it can replicate its hybrid constitution asexually through successive seedling generations.

The grapefruit, *Citrus paradisi*, is not known in a natural or seminatural state in the far east, where the genus Citrus has its center of distribution, but apparently originated in the West Indies. Morphological characters suggest that the grapefruit may be a hybrid of *C. grandis* X *sinensis* (Swingle, 1967, p. 383). If this plausible but unproven view is correct, the grapefruit may have arisen as a spontaneous hybrid between pummelos and sweet oranges cultivated in the West Indies in the early period of European settlement. The grapefruit is strongly agamospermous, as noted earlier in this section, and would therefore perpetuate its hybrid genotype by seed reproduction.

The sources of variation in Citrus are manifold. Hybridization between preexisting species, semispecies, or varieties has already been mentioned. The hybrids can produce segregating progeny by sexual reproduction (see Figure 36 in Chapter 18). Bud sports are common and have led to new varieties of the Washington navel orange and other forms. Most cultivated varieties of Citrus give rise to a certain small proportion of autotetraploid seedlings (Reuther, Batchelor, and Webber, 1968).

The new variants, whatever their nature and origin, can then breed true from seeds by nucellar embryony, thus increasing the total number of agamospermous microspecies in Citrus. Unconscious human selection and, more recently, artificial selection have transformed some of the minor microspecies into widespread varieties and agamospermous species. By these processes the inedible wild species of Citrus have given rise to the varied array of cultivated Citrus fruit trees which now occur throughout the subtropical and warm temperate regions of the globe.

Citrus furnishes an example of a third pattern of agamic complex. In Citrus, by contrast with the Rubus Moriferus and Bouteloua patterns, the agamospermous types reproduce by adventitious embryony and remain predominantly on the diploid level.

Pteris

Apogamety is widespread in ferns, occurring in Adiantum, Asplenium, Dryopteris, Polystichum, Pteris, and other genera. This condition is particularly frequent in the large, worldwide genus Pteris, where 34% of the taxonomic species are apogamous (Walker, 1962).

Several different agamic complexes occur in Pteris. Some of these contain a mixture of apogamous and sexual forms; others consist predominantly or exclusively of obligately apogamous ferns. The presence or absence of hybridizing sexual forms is, as expected, correlated with the size and diversity of the agamic complex (Walker, 1962).

The *Pteris quadriaurita* group is a very large and complex one, containing both sexual and apogamous members. *Pteris quadriaurita* sensu stricto ($2n = 58$) and *P. multiaurita* ($2n = 58$), which are both sexual, hybridize in Ceylon; and these species or their hybrids also hybridize with a third member of the group, *P. confusa* ($2n = 58$), which is apogamous. The network of hybridizations extends beyond the central *P. quadriaurita* group to members of other species groups within the genus (Walker, 1958, 1962).

The *Pteris biaurita* group in east Asia consists entirely, as far as known, of obligate apogamous members. Most of them are "diploid" ($2n = 58$), and some are triploid ($2n = 87$). They form an array of true-breeding microspecies differing from one another in minor details of spore pattern and juvenile foliage. These morphological variations are grouped into a single distinct taxonomic species. The simple taxonomic structure of the agamic complex in *P. biaurita* is related to the absence or rarity of sexuality and natural hybridization in this group (Walker, 1962).

Pteris illustrates a fourth pattern of agamic complex which is fairly widespread in polypodiaceous ferns. In the fern pattern, asexual reproduction is by apogamety, usually obligate apogamety. The apogamous ferns may be diploid or polyploid but always have high chromosome numbers. The agamic complex frequently contains both sexual and apogamous members, or may consist exclusively of apogamous types.

Stages of Development

The relative stage of development of an agamic complex can be inferred from its degree of taxonomic complexity, extent of geographical distribution, and presence or absence of diploid sexual ancestors. On the basis of these criteria we can recognize four main stages of development: young, early mature, late mature, and old.

Parthenium presents an example of an agamic complex in a relatively young stage of development. The morphological variations can be grouped satisfactorily into just two taxonomic species, *P. argentatum* and *P. incanum*. The complex occurs in a single geographical province, ranging from Texas to Arizona and through adjacent parts of Mexico. The sexual ancestors of the complex, the tetraploid forms of *P. argentatum* and *P. incanum*, are both extant though somewhat restricted in distribution (Rollins, 1944, 1945, 1946).

An agamic complex containing hundreds or thousands of agamospermous microspecies distributed over a large geographical area is in what we can call a mature stage of development. Among mature agamic complexes a further distinction of degree can be drawn between those which have retained many or most of their ancestral diploid species and those which have lost most or all of their diploid sexual ancestors by extinction. The former are regarded as being in an early mature stage and the latter in a late mature stage.

In previous sections of this chapter we have considered some examples of early mature agamic complexes. A fairly good representation of diploid sexual species together with a very diverse array of agamo-spermous derivatives occurs in Crepis, Rubus, Taraxacum, Hieracium, and Citrus.

In a late mature agamic complex the diploids and the sexual poly-ploids are reduced and disappearing, but the agamospermous super-structure remains fully developed. Examples of this stage are found in Alchemilla, Potentilla, and Poa.

The agamic complex in Alchemilla subgenus Aphanes has one known diploid sexual species, *A. microcarpa* ($2n = 16$), and many agamospermous polyploids. In Alchemilla subgenus Eualchemilla, sexuality is very rare and polyploid agamospermy is richly developed (Gustafsson, 1946–1947, pp. 230, 254). The *Poa pratensis* and *P. palustris* groups have no known sexual diploids but consist chiefly of agamospermous and viviparous forms with polyploid and secondary aneuploid numbers (Gustafsson, 1946–1947, pp. 251–54). Similarly, the *Potentilla gracilis* group ($x = 7$) in western North America contains at least one sexual polyploid, *P. pectinisecta* ($6x$), and probably others, but no known diploids, and has numerous agamospermous polyploid types (Clausen, Keck, and Hiesey, 1940, Ch. 3).

An old agamic complex is depauperate, not only in sexual diploids and in sexual polyploids, but also in numbers of agamospermous microspecies. It appears to us in its taxonomic aspect as an isolated agamospermous species without known sexual relatives.

Houttuynia cordata (Saururaceae) in east Asia is the single species of its genus. It is high polyploid ($2n = 94$ to 98) and agamospermous, and has no known sexual relatives. It may well represent the remnants of an agamic complex which was formerly more abundant and varied and which has now become depauperate (Babcock and Stebbins, 1938).

Evolutionary Potential

Agamic complexes include some of the most successful plant groups in the world today. The complexes in Rubus, Hieracium, Poa, and similar genera have colonized large and diversified continental areas with unfavorable climates; and they have accomplished this task of coloni-zation by the generation of innumerable variations embodied in scores of species and hundreds or thousands of microspecies. It must also be recognized, however, that agamic complexes in their old stage have lost most of their former stock of species and microspecies, most of their capacity to generate new variations, and consequently most if not all of their ability to respond successfully to future environmental challenges.

From these opposite aspects of agamic complexes have flowed opposite judgments regarding their evolutionary potential. Darlington and Stebbins have been spokesmen for the idea that agamic complexes have a strongly restricted evolutionary potential, whereas Gustafsson has emphasized the high degree of evolutionary success which they can and do achieve in favorable instances.

Darlington states in *The Evolution of Genetic Systems* (1939, p. 113):

> Thus we see apomixis saves what can be saved when sexual fertility has been lost. Sexual reproduction provides in recombination the basis for the adaptation of all its posterity. Apomixis provides for its immediate progeny. It retains in some species a few relics of the sexual system which give it the means of postponing extinction and even developing a momentary efflorescence of new forms. But with the loss of sexual recombination the apomict, like the permanent hybrid, is cut off from ultimate survival. Apomixis is an escape from sterility, but it is an escape into a blind alley of evolution.

Agamospermy, then, is "a blind alley of evolution" (Darlington, 1939, p. 113); it is "evolutionary opportunism carried to its limit" (Stebbins, 1950, p. 414). Furthermore (Babcock and Stebbins, 1938, p. 62):

> The ultimate fate of an agamic complex of which the sexual ancestors have become restricted or extinct can be predicted; it will flourish as long as the conditions that existed during its formation prevail, but it will be unable to meet any new changes of environment, and will therefore in time become more and more restricted, and will finally die out.

But, says Gustafsson, the view that agamospermy puts an end to variation is only partly true (Gustafsson, 1946–1947, p. 183). As to the agamic complex being predestined to depletion of variation and eventual extinction, "is not this typical of all species and genera? Like separate individuals, species and genera are born, bloom and die" (Gustafsson, 1946–1947, p. 178).

Let us consider these divergent viewpoints in relation to the basic biological function of sexual reproduction. Sex is a genetic system for producing gene recombinations, which in turn comprise the chief source of hereditary variations in a population (see Chapter 1). The complete abandonment of the sexual process, by the change to obligate apomixis, cuts the population system off from its main source of new variations. And intermediate degrees of asexuality, as in facultative apomixis, restrict the generation of variability to an intermediate extent.

The sexual mechanism is, as a matter of fact, the chief source of new variations in an agamic complex as well as in a strictly sexual group. Now an agamic complex composed of facultative agamosperms and sexual diploids, like Rubus Moriferus, may display great variability and great evolutionary potential. But not all agamic complexes possess these features. An isolated complex of obligate apogamous ferns, like *Pteris biaurita*, presents a relatively stereotyped variation pattern (Walker, 1962). In the *Crepis occidentalis* complex, which contains obligate agamosperms in addition to facultative agamosperms and sexual diploids, the variability is great in and near the areas occupied by sexual populations, but is reduced to one or a few types in areas occupied exclusively by obligate agamosperms (Babcock and Stebbins, 1938).

Phytogeographic evidence indicates that certain particular agamospermous types can and do persist for long periods of time without changing much. *Taraxacum reichenbachii* and *T. dovrense* are two almost identical agamospermous microspecies which occur as glacial relicts in the Alps and Norway, respectively. They have diverged only slightly since before the last glacial period. Other ancient agamospermous microspecies are found in Hieracium and Alchemilla. Some agamospermous types must, on phytogeographic grounds, be at least 75,000 or 100,000 years old (Gustafsson, 1946–1947, Ch. 13).

Many agamospermous plants are polyploid, as noted previously. To the restrictions on recombination imposed by agamospermy per se, in such cases, must be added the restrictions imposed by the polyploid condition (see Chapter 20).

Retention of sexuality in a plant group is no guarantee of long-range evolutionary success. Sexual species groups decline and fall like apomictic ones, as Gustafsson appropriately notes, whereas partially asexual groups may be very successful over a long term. Moreover, the complete loss of sexuality in a group is not necessarily fatal for its immediate future prospects. But reversion to obligate apomixis does seem to spell extinction in the long run.

We quote, finally, the views of Stebbins (1950, p. 417) on this problem:

There is no evidence that apomicts have ever been able to evolve a new genus or even a subgenus. In this sense, all agamic complexes are closed systems and evolutionary "blind alleys" .... While sexual species may, during the course of their existence, give rise to entirely new types by means of progressive mutation and gene recombination, agamic complexes are destined to produce only new variations on an old theme.

CHAPTER *22*

The Clonal Complex

Introduction · Complexes with Reproduction by Stem Offshoots ·
Viviparous Complexes · Evolutionary Potential

Introduction

Taxonomic difficulties due to the asexual multiplication of diverse hybrid types are present in clonal complexes, as in agamic ones, but are usually present to a lesser extent in the clonal complexes. The latter, in which the hybrids reproduce by vegetative propagules rather than by agamospermous seeds, generally exhibit a less complicated variation pattern than agamic complexes.

In the first part of this chapter we describe a few examples of clonal complexes. It can be noted in passing that there is a need for more detailed studies of this type of hybrid complex. Finally we attempt to discuss the evolutionary potential of clonal complexes as compared with that of agamic complexes.

Complexes with Reproduction by Stem Offshoots

The *Opuntia phaeacantha* group ($x = 11$, Cactaceae) is a wide-ranging and variable complex in arid parts of the American southwest from California to Texas and Mexico. The plants are low, branched shrubs with large, colorful flowers. The latter are cross-pollinated by bees and beetles. In addition, as in the genus Opuntia generally, the plants multiply rapidly by means of their stem joints, which fall off the parent plant, are dispersed locally by sheet floods and animals, and develop into new plants.

Eight or more semispecies, differing from one another in flower color, type of spines, and branching habit, occur in the area between California and Texas. In California, *Opuntia occidentalis* occurs near the coast, *O. vaseyi* in the interior foothills, *O. megacarpa* in the western Mojave Desert, and *O. mojavensis* in the eastern Mojave Desert. In Arizona and other parts of the southwest, the group is represented by *O. phaeacantha*, *O. engelmannii*, *O. erinacea*, and other forms.

These semispecies overlap in range and hybridize in various combinations. For example, hybridization occurs commonly between *Opuntia phaeacantha* and *O. engelmannii* in Arizona, and between *O. occidentalis* and *O. vaseyi* in southern California (Benson, 1950; Munz, 1959). As a result of hybridization followed by the vegetative propagation of the hybrid progeny, derivative populations have grown up which blur the morphological distinctions between the parental populations.

This process can be seen on a microgeographical and microevolutionary scale in certain foothill localities in coastal southern California where *Opuntia occidentalis* and *O. vaseyi* meet and hybridize. *Opuntia occidentalis* has yellow flowers, yellow spines, and ascending branches, while *O. vaseyi* has orange-red flowers, brown spines, and prostrate branches. Some of the possible recombinations of these character differences are found in natural hybrid populations. But they are not intermixed randomly in the same population. Instead, one character combination has been proliferated as a clonal colony in one gulley, while a neighboring gulley is populated by a different clonal microspecies.

The *Stellaria longipes* group (Caryophyllaceae), containing several named species, ranges widely through the northern hemisphere.

Stellaria longipes proper ($2n = 52$) is fertile as to seeds. But *S. crassipes*, an arctic member of the group with a high polyploid number ($2n = $ ca. 104), is highly seed-sterile and reproduces by creeping subterranean runners. The high polyploid condition of *S. crassipes* suggests that it is derived from lower-numbered species in the group, perhaps as an amphiploid. And its present widespread arctic distribution suggests that it is a fairly ancient derived species which has spread by vegetative means (Hultén, 1943; Gustafsson, 1946–1947, pp. 212, 229).

A parallel situation is found in the *Potentilla anserina* group ($x = 7$). The tetraploid populations of this group reproduce sexually and fall into at least two intersterile species. But the hexaploid members are meiotically irregular and sterile as to pollen and seeds. They reproduce vigorously by runners. By this means of propagation the hexaploids have formed a series of clonal microspecies related to and derived from different tetraploid populations (Rousi, 1965).

Viviparous Complexes

Most species of Allium are sexual and seed-fertile. Vivipary, however, occurs fairly widely in this large genus. Various species reproduce by bulbils which arise in the sites of the flowers and replace seed formation. The degree of vivipary varies from partial to complete in different cases, and seed production is reduced correspondingly, being absent altogether in some viviparous species like *A. sativum* and *A. paradoxum*. Here we will briefly review the situation in two species groups, the *A. paniculatum* and *A. canadense* groups.

The *Allium paniculatum* group ($x = 8$) in Eurasia consists of *A. paniculatum*, *pulchellum*, *flavum*, *oleraceum*, and *carinatum*. The diploid sexual species in the group are *A. paniculatum*, *pulchellum*, and *flavum* ($2n = 16$). They occur in southern Europe (Levan, 1937).

The viviparous species, *Allium carinatum* and *A. oleraceum*, occur in northern Europe, and range in ploidy from $2x$ to $5x$. Viviparous *A. carinatum* is diploid in the Alps but triploid in the north of Europe. It is related morphologically to the sexual species *A. pulchellum* and *A. flavum*. Viviparous *A. oleraceum*, with tetraploid and pentaploid numbers ($2n = 32, 40$), is morphologically close to sexual *A. paniculatum* (Levan, 1937; Gustafsson, 1946–1947; Darlington and Wylie, 1955).

The *Allium canadense* group ($x = 7$) in eastern North America is another clonal complex containing several sexual forms ($2x$ and $4x$) and some viviparous derivatives ($3x$ and $4x$). In western North America the related *A. geyeri* complex ($x = 7$) also includes some sexual diploids and sexual tetraploids along with a series of viviparous polyploid derivatives ($4x$, $5x$, $6x$) (Ownbey and Aase, 1955).

The *Saxifraga stellaris* group is another large complex in the northern hemisphere. *Saxifraga stellaris* sensu stricto ($2n = 28$) and *S. ferruginea* are sexual species in the cold temperate zone. *Saxifraga foliolosa* ($2n = 56$, ca. 64) is viviparous. This asexual polyploid derivative is circumpolar in distribution, ranging farther north than its sexual relatives, and into the arctic zone (Gustafsson, 1946–1947, pp. 56, 106, 228).

Gustafsson (1946–1947, p. 233) cites several examples of viviparous species with high polyploid numbers and northern or arctic distributions which do not have any known close sexual relatives. Among these systematically isolated viviparous species are: *Polygonum viviparum* ($2n = 83$ to 110), *Cardamine bulbifera* ($2n = $ ca. 96), and *Saxifraga cernua* ($2n = $ ca. 66).

Vivipary is found in several genera of grasses, especially Agrostis, Deschampsia, Eleucine, Festuca, and Poa. It will be recalled from an earlier discussion (Chapter 18) that agamospermy is also well represented in the Gramineae. In Poa, agamospermy and vivipary are sometimes present together in the same species group.

The circumboreal *Poa alpina* group ($x = 7$) contains sexual, agamospermous, and viviparous forms. All three types of forms exist, in fact, within the taxonomic species *Poa alpina* itself, which includes diploid sexual populations ($2n = 14$), polyploids up to the hexaploid level, and various derived viviparous and agamospermous populations. *Poa bulbosa* in this group contains low-number sexual forms in the south of Europe and derived viviparous forms with tetraploid or hexaploid numbers in northern Europe (Gustafsson, 1946–1947, p. 225).

The *Deschampsia caespitosa* group ($x = 7$) is another large clonal complex ranging through the northern hemisphere and reaching the southern hemisphere in various places. Sexual tetraploid species or semispecies—*D. caespitosa, arctica, bottnica,* etc.—are widespread and are linked by hybridization. Vivipary crops up here and there throughout the range of the sexual populations. A related species, *D. alpina*, in the arctic zone exhibits several high polyploid and secondary aneuploid numbers ($2n = 39$, 41, 49, 56), and is almost entirely viviparous (Gustafsson, 1946–1947, pp. 224, 232).

Evolutionary Potential

Clonal complexes generally exhibit a simpler taxonomic structure than agamic complexes (see Webb, 1966, pp. 166–68). They usually contain fewer microspecies. The microspecies content of a hybrid complex in its mature stage may be taken as a measure of its evolutionary potential. Three main reasons may be suggested for the comparative differences between clonal and agamic complexes in microspecies composition and, hence inferentially, in evolutionary potential.

The most important single factor contributing to the relatively greater variability of agamic complexes, as compared with clonal ones, is probably the presence of some sexual processes in the former. The sexual processes may be immediate but intermittent, as in hybridization between different facultative agamosperms; or they may be delayed, as in reversions to sexuality in a normally agamospermous line, and in autosegregation. Because of their intermittent action and delayed effects the sexual processes in question are often and aptly termed subsexual. But in any case they greatly enhance the variability of the agamic complex.

Sexual processes play their role in the formation of the first generations of hybrids in a clonal complex. But thereafter, insofar as the hybrids reproduce predominantly or entirely by vegetative means, they are cut off from the segregation of new variant forms: The operation of subsexual processes is much reduced, and may be absent altogether, in the superstructure of a clonal complex.

A second factor to be considered is the relative efficacy of seeds and vegetative organs as dispersal units. The asexual derivatives in an agamic complex utilize the adaptations for dispersal of seeds and fruits. Their microspecies can consequently expand into medium or large areas in many cases.

The dispersal potential of the different kinds of vegetative propagules undoubtedly varies over a wide range from inflorescence bulbils, which may be able to migrate about as well as achenes, to stem runners, which are good only for short distances. We expect, therefore, that some clonal microspecies with efficient means of dispersal will attain widespread distributions. But by the same token there will be many other clonal complexes composed of plants without efficient means of long-range vegetative dispersal. In such complexes the variant forms which arise by hybridization may never spread beyond their local area of origin; they may not succeed in developing into microspecies of any importance in nature.

A third factor which may limit the development of many clonal complexes may be clonal senescence. Clonal senescence has been discussed from different points of view by Molisch (1938), Brink (1962), and Cameron and Frost (1968). The subject is still not well understood.

Certainly, many kinds of plants can reproduce vegetatively for hundreds or thousands of years without any apparent decline in vigor. A few examples were given in Chapter 1, and many more could be added (see Molisch, 1938; Gustafsson, 1946–1947). But this is not always the case (Molisch, 1938). Some plant groups such as Citrus, potatoes, and sugarcane exhibit symptoms of aging or degeneration after long-continued vegetative propagation (Mather, 1966; Cameron and Frost, 1968).

The observed clonal senescence is probably due mainly to the accumulation of virus diseases in the cytoplasm (Nygren, 1954; Mather, 1966; Cameron and Frost, 1968). Regular, directed gene changes in the somatic cells—some form of paramutation—may also be involved (Brink, 1962).

In Citrus the agamospermous seedling progeny are more vigorous than plants raised from cuttings. The asexual seeds are apparently free of the viruses which infect the stem tissues of the same plants (Nygren, 1954; Cameron and Frost, 1968). This observation is very interesting in relation to our present problem. It suggests that agamic complexes, by avoiding clonal senescence through seed reproduction, will often be able to develop further in time and space than clonal complexes.

In Chapter 21 we saw that the evolutionary potential of agamic complexes, while undoubtedly great, is restricted as compared with that of sexual groups. Here we see that the evolutionary potential of clonal complexes is even more limited.

CHAPTER 23

The Heterogamic Complex

Introduction

This chapter is concerned with two interesting and anomalous types of genetic system and with the hybrid complexes built up on these genetic systems. The genetic systems in question are permanent odd polyploidy, as exemplified by the *Rosa canina* group, and permanent translocation heterozygosity, as found in the *Oenothera biennis* group. The terms permanent odd polyploidy and permanent translocation heterozygosity are used in a special sense here and not in the purely literal sense of the words.

A pentaploid agamospermous Crepis plant or a triploid viviparous Allium plant is literally a "permanent odd polyploid," but it falls outside the scope of the concept as defined here. By permanent odd polyploid we mean a plant which breeds true for an odd euploid condition by sexual means.

A similar distinction should be made between permanent translocation heterozygotes as taken literally and as construed in the more restricted sense desired for our present purposes. A population may contain a high frequency of structural heterozygotes throughout successive generations as a result of chromosomal polymorphism combined with strong selection for the heterozygous genotypes. Well-known examples are the inversion heterozygotes of *Drosophila pseudoobscura* populations and the translocation heterozygotes of *Paeonia californica* populations. These examples approach, but do not quite attain, the condition which especially concerns us here. By permanent translocation heterozygote, in the more strict sense used in the present discussion, we mean a plant which breeds true by sexual means for one particular heterozygous condition, with little or no segregation of homozygous individuals.

The permanent odd polyploid and the permanent translocation heterozygote sensu stricto breed true, in most cases, by the regular transmission of one set of chromosomes through the male gametes and another through the female gametes. The opposite chromosome sets are reassembled by fertilization, often self-fertilization but sometimes cross-fertilization. In either case the numerical or structural hybridity is perpetuated from one sexual generation to the next.

The two genetic systems thus have some basic features in common, namely, continuing chromosomal heterozygosity, complementary genomes in the male and female gametes, and sexual fertilization, often self-fertilization; and these common features enable us to discuss them together in this chapter. The two genetic systems also give rise to similar types of hybrid complexes grouped together here under the common heading, heterogamic complex.

A heterogamic complex is a hybrid complex composed of heterogamic microspecies and their diploid and meiotically normal ancestors. Heterogamic microspecies are sexually reproducing, true-breeding numerical hybrids or structural hybrids. Hybridization between different ancestral species or semispecies, followed by establishment of

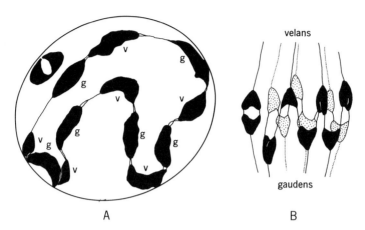

Figure 47 Chromosome pairing and disjunction in *Oenothera lamarckiana* (2*n* = 14). This microspecies carries the Renner complexes velans and gaudens. (A) Diakinesis, showing a ring of 12 chromosomes and 1 bivalent. (B) Metaphase I, with ring aligned on spindle for alternate disjunction at anaphase. (Emerson, 1935.)

permanent structural or numerical hybridity, produces an array of heterogamic microspecies which bridge the gaps between the ancestral forms. Subsequent cycles of hybridization between the heterogamic microspecies themselves can then further increase the taxonomic complexity of the group.

Permanent Translocation Heterozygosity

The numerous heterogamic microspecies belonging to the *Oenothera biennis* group in North America and Europe are diploid plants (2*n* = 14) with a short-lived herbaceous growth habit and a predominantly self-pollinating breeding system. They are permanent heterozygotes for successive whole-arm translocations (see Figure 15 in Chapter 5) and consequently form chromosome rings at meiosis (Figure 47). Most of the microspecies are complete translocation heterozygotes which have all 14 chromosomes joined into a single ring. Some microspecies, like *Oe. lamarckiana*, form rings of smaller size together with separate bivalents (Figure 47).

A ring-forming Oenothera plant, following self-fertilization, yields all ring-forming progeny. The expected classes of structural homozygotes are absent in the progeny of the translocation heterozygote. The latter breeds true by sexual means for its hybrid constitution.

The cytogenetics of Oenothera has been worked out by a succession of investigators. The anomalous breeding behavior of these plants was discovered by DeVries and explained by Renner in the classical period. The analysis was then greatly extended by Cleland and his students during a period of more than forty years.

Good reviews of the early work will be found in Cleland (1936) and Darlington (1937a, pp. 338 ff.). Much factual evidence is summarized in the later monograph of Cleland and co-workers (1950). A recent comprehensive review with numerous literature references is given by Cleland (1962); for a concise review see also Grant (1964b, pp. 185 ff. and 174 ff.).

It is convenient to describe the genetic system of permanent translocation heterozygosity here in terms of the classical subject, *Oenothera lamarckiana*. *Oenothera lamarckiana*, or *Oe. erythrosepala* as it has to be called in formal taxonomy, is widely distributed in North America and Europe, apparently as a garden escape (Figure 48) (Munz, 1949, 1965).

The chromosomes of *Oenothera lamarckiana* pair to form a ring of 12 and 1 bivalent at meiosis (Figure 47). This species is thus heterozygous for successive translocations on 12 of its 14 chromosomes.

The end arrangement of the *Oenothera lamarckiana* chromosomes can be expressed in terms of the arbitrary standard end arrangement found in the related structurally homozygous species, *Oe. hookeri*. The 14 arms of the 7 chromosomes in a standard race of *Oe. hookeri* are assigned consecutive numbers, as shown in Table 10. The two haploid sets are structurally homologous in most individuals of *Oe. hookeri*, which consequently exhibit regular bivalent formation at meiosis (Table 10).

The haploid sets in *Oenothera lamarckiana* differ from that in the standard or DeVries race of *Oe. hookeri* by several translocations (Table 10). Furthermore, the two sets in *Oe. lamarckiana* differ from one another with respect to successive translocations on all chromosomes except those with the 1-2 ends (Table 10). Pairing of homologous arms at meiosis thus accounts for the observed configuration of a ring of 12 and 1 bivalent (Figure 47).

Figure 48 *Oenothera lamarckiana* ($=$ *Oe. erythrosepala*). (Munz, 1949.)

Table 10 · Comparative genomic constitutions of *Oenothera lamarckiana* and *Oe. hookeri* (Cleland, 1950, pp. 222–25)

Species	End arrangement of chromosomes in each haploid set	Renner complex
Oe. hookeri		
DeVries race	1—2 3—4 5—6 7—8 9—10 11—12 13—14	standard hookeri
	1—2 3—4 5—6 7—8 9—10 11—12 13—14	standard hookeri
Widespread California race	1—2 3—4 5—6 7—10 9—8 11—12 13—14	hookeri Johansen
	1—2 3—4 5—6 7—10 9—8 11—12 13—14	hookeri Johansen
Oe. lamarckiana	1—2 3—4 5—8 7—6 9—10 11—12 13—14	velans
	1—2 3—12 5—6 7—11 9—4 8—14 13—10	gaudens

At metaphase I the ring-forming chromosomes align themselves on the spindle in a zigzag fashion for alternate disjunction (Figure 47B). At anaphase I, neighboring chromosomes in the ring separate to opposite poles and, conversely, the alternating chromosomes pass to the same poles. In this way the two differentiated genomes are re-assembled intact at the end of meiosis (Cleland, 1936, 1962).

The special types of genomes found in Oenothera are known as Renner complexes and are designated by different names. The two Renner complexes in *Oenothera lamarckiana* are velans and gaudens (Table 10). *Oenothera lamarckiana* thus has the heterozygous genomic constitution velans/gaudens. The result of alternate disjunction of the ring-forming chromosomes at meiosis is the segregation of the whole velans Renner complex to one daughter nucleus and the gaudens complex to the other (Figures 47B and 49). The velans/gaudens heterozygote, in other words, normally produces just two classes of gametes, velans and gaudens, as far as the structural arrangement is concerned (Figure 49).

Self-fertilization or sib crossing of *Oenothera lamarckiana* would be expected to yield three classes of progeny—velans/velans, velans/ gaudens, and gaudens/gaudens—in a Mendelian ratio of 1:2:1. Instead, only the heterozygous type is obtained, as noted earlier. The absence of the expected classes of structural homozygotes is due to the

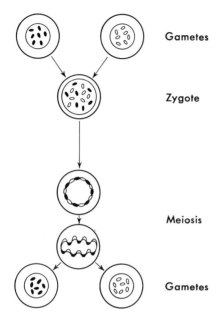

Gametes

Zygote

Figure 49 Absence of independent assortment of chromosomes belonging to different Renner complexes in Oenothera. (Cleland, 1964.)

Meiosis

Gametes

operation of a system of balanced lethals (Renner, 1925; Cleland, 1936, 1962).

The gene system consists of complementary lethal factors balanced in heterozygous condition. The double heterozygote $\left(\dfrac{+\ l_2}{l_1\ +}\right)$ is viable; but the homozygous recombinations $\left(\dfrac{l_1\ +}{l_1\ +}\right)$ and $\left(\dfrac{+\ l_2}{+\ l_2}\right)$ are not. Each set of complementary lethal factors is linked to one Renner complex. Taking the balanced lethals into consideration, the constitution of *Oenothera lamarckiana* can be written now as $\dfrac{\text{velans} + l_2}{\text{gaudens}\ l_1\ +}$. The velans/velans and gaudens/gaudens types are homozygous for lethal factors and die in the zygote stage.

Not only the lethal factors but also genes determining various phenotypic traits are linked to the diverse Renner complexes (see Cleland, 1962; Grant, 1964b). Furthermore, the genes borne on different chromosomes belonging to the same Renner complex are transmitted in a single linkage group. The translocation hybrid breeds

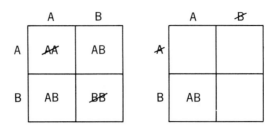

Figure 50   Two types of balanced lethal system in Oenothera. Left, system employing zygote lethals. Right, system with gametophyte lethals. (Cleland, 1964.)

true, therefore, not only for its particular structurally heterozygous constitution, but also and at the same time for a particular ensemble of phenotypic characters.

The several components in the genetic system of permanent translocation heterozygosity vary from species to species within the *Oenothera biennis* group. One variable is the size of the ring. The ring of 12 chromosomes found in *Oe. lamarckiana* is not typical of the group as a whole. A population of *Oe. irrigua* (= *Oe. hookeri hewettii*) in New Mexico regularly forms a ring of 8 + 3 bivalents. The most common configuration, however, in microspecies throughout the central and eastern United States, is a ring of 14 chromosomes (Cleland, 1950, pp. 218 ff.).

The balanced lethal system also shows variation. Lethals are absent or sporadic in western members of the group. Balanced lethals are characteristic, however, of the heterogamic microspecies in central and eastern North America and in Europe. Two types of balanced lethal system occur (Figure 50). In the first type, as exemplified by *Oenothera lamarckiana*, the lethals act in the zygote stage (Figure 50, left); in the other, which is widespread in North America, they operate in the gametophytes (Figure 50, right). Accordingly, the Oenothera plants with zygote lethals are semisterile as to seeds, and those with gametophyte lethals are semisterile as to pollen (Cleland, 1936, 1950, 1962).

A further consequence of a system of gametophyte lethals is that a given heterogamic microspecies regularly transmits one of its Renner complexes through the pollen and the opposite complex through the eggs.

Other examples of permanent and complete ring-forming hetero-zygotes are *Rhoeo discolor* (Commelinaceae), *Gayophytum heterozygum* (Onagraceae), and *Isotoma petraea* in part (Lobeliaceae) (Sax, 1931; Lewis and Szweykowski, 1964; James, 1965).

Sources of Variation

The causes of hereditary variation in the ring-forming Oenotheras are diverse. Occasional outcrossing between ordinarily inbreeding micro-species is one important source of new forms. Other processes operate to produce new variations within an inbreeding line itself. For, although *Oenothera lamarckiana* and its relatives generally yield uniform progeny after self-pollination, they do not do so exclusively, but give rise to a certain low proportion of new segregate forms, which comprise the mutations of DeVries.

In order to discuss the mutability of Oenothera it is necessary first to recognize that some genes are linked to a Renner complex in a ring-forming heterozygote, whereas other genes are not. The latter segregate in the normal way. In an incomplete translocation heterozygote like *Oe. lamarckiana*, the bivalent-forming chromosomes segregate inde-pendently of the ring-forming chromosomes, and the genes on the bivalent chromosomes are consequently not a part of the Renner complexes. Genes borne on the terminal pairing segments of the ring-forming chromosomes can cross over more or less freely from one Renner complex to the other in a complex heterozygote. Such genes are also independent of any particular Renner complex (see Cleland, 1962). These genes which recombine freely do not of course give rise to new "mutants" in Oenothera.

The genes determining the characteristic phenotypic traits transmitted in a block through any given Renner complex, and passed on uniformly in inheritance by any given complex heterozygote, must be located on the nonpairing differential segments of the translocation chromosomes (Darlington, 1937a, Ch. 9; Cleland, 1962). These are the proximal chromosome regions in the neighborhood of the translocation breaks; for an explanation of the chromosome mechanics involved see Grant (1964b, pp. 174 ff.). The balanced lethal factors must also be located in these differential segments. The translocation heterozygote then breeds true for the particular heterozygous gene combination in this part of the genome by means of the genetic system described in the preceding section.

Cytogenetic events occur at regular but rare intervals, however, to separate one or more genes on the differential segments from their respective Renner complexes and expose them to expression in new homozygous combinations. One such process is rare crossing-over and exchange of factors between Renner complexes. A gene may be crossed over occasionally into the opposite Renner complex. Self-pollination then yields some exceptional progeny which are homozygous for this gene and appear as new stable mutant types.

Another process is the occurrence of new reciprocal translocations within a given Renner complex. This can break up a large ring in a complex heterozygote into smaller rings and bivalents. Whole chromosomes and blocks of genes can then form homozygous segregates after self-pollination. The exceptional homozygous segregates arising by the above and other means constitute the diploid mutations in Oenothera (see Renner, 1941; Cleland, 1962).

The ring-forming chromosomes are subject to irregularities of separation at anaphase I. Nondisjunction occurs occasionally and gives some 8-chromosome gametes. They in turn yield some trisomic progeny with altered phenotypes. The majority of the mutants in *Oenothera lamarckiana* have turned out to be trisomics with $2n = 15$ (Cleland, 1962).

Hybridization between different complex heterozygotes is the second main source of new variations in Oenothera. Two heterogamic microspecies, each of which normally yields one class of offspring on selfing, often give rise to four classes of progeny in their first hybrid generation. Thus the cross of *Oe. lamarckiana* (velans/gaudens) X *Oe. strigosa* (deprimens/stringens) yields the four possible diploid combinations: velans/deprimens, velans/stringens, gaudens/deprimens, and gaudens/stringens (Darlington, 1937a, Ch. 9; Cleland, 1962).

Many of these new genome combinations, in the above and in other hybrid crosses, have more bivalents or small rings than either parent (Darlington, 1937a, Ch. 9; Cleland, 1962). Some gene pairs which are linked to Renner complexes in the parental microspecies, and are maintained there in a permanently heterozygous condition, will then become independently assorting in the intermicrospecies hybrids. Such genes can go on to segregate into new homozygous types in later generations. And, where the homozygotes are viable, these new segregates can become permanent additions to the variation pool of the hybrid complex.

Some first-generation hybrids between two heterogamic microspecies have complete or nearly complete rings associated with a new combination of balanced lethal factors. Such hybrids will breed true. They represent the potential beginnings of new hybrid microspecies originating as direct products of crossing between preexisting hybrid microspecies (Cleland, 1962).

The Oenothera biennis Complex

The North American species of Oenothera subgenus Oenothera are tall perennial to annual herbs, often weedy, with yellow vespertine flowers (Figure 48). They are all diploids with $2n = 14$. The group has been extensively studied from the cytogenetic standpoint by Cleland and his school, and from the taxonomic standpoint by Munz (Cleland, 1949, 1950, 1962, 1964; Munz, 1949, 1965). The *Oenothera biennis* complex comprises the derived heterogamic members of this species group.

The whole assemblage of North American Euoenotheras is taxonomically difficult owing to the large range of variations and the absence of clear-cut interspecific boundary lines. In his recent revision of this assemblage, Munz (1965) recognizes ten taxonomic species. Many of the species are polytypic with two or more subspecies. Cleland (1962, 1964) recognizes a similar array of taxonomic species, or "groups of races" as he terms them, on the basis of genome analysis. The systems of Munz and of Cleland, though not identical, are similar and well coordinated.

The North American Euoenotheras can be subdivided in the first instance into two series on their type of genetic system. A normal diploid sexual system occurs in the *Oenothera hookeri* group. The plants are open-pollinated, form bivalents at meiosis (with some exceptions), and lack balanced lethal factors. The *Oe. biennis* group possesses the derived genetic system of permanent translocation heterozygosity. These plants exhibit the large chromosome rings, associated with self-pollination and balanced lethals, as described in the preceding sections (Cleland, 1950, 1962, 1964).

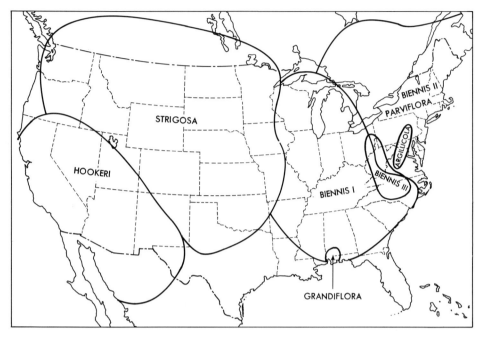

Figure 51 Geographical distribution of the taxonomic species of the *Oenothera hookeri-biennis* group in temperate North America. (Cleland, 1964.)

In the series of normal sexual diploids there are four species. They are *Oenothera elata* in central Mexico, *Oe. hookeri* in western North America, *Oe. grandiflora* in Alabama, and *Oe. argillicola* in the Appalachian region of the eastern United States (Figure 51). It should be noted that the bivalent-forming *Oe. grandiflora* referred to here is *Oe. grandiflora* L'Her. and not the *Oe. grandiflora* of DeVries; the latter is a ring-forming heterozygote (acuens/truncans) in the *Oe. biennis* group. It should be noted also that the bivalent-forming races of *Oe. hookeri* are mainly those of the Pacific region. Populations of *Oe. hookeri* farther east in New Mexico often have small rings and sometimes have lethals, and are thus transitional to the *Oe. biennis* group (Cleland, 1962; Munz, 1965).

The large ring-forming Oenotheras comprising the *Oe. biennis* group fall into three main taxonomic species in central and eastern North

America. They are *Oe. strigosa*, *Oe. biennis*, and *Oe. parviflora* (Figure 51). *Oenothera erythrosepala* (the *Oe. lamarckiana* of DeVries) occurs as a garden escape or weed in various parts of the United States as well as Europe (Figure 48). Other heterogamic microspecies such as *Oe. muricata* (a variant form of *Oe. biennis*) also occur in Europe. Two of the three main taxonomic species listed above are subdivided further according to their genome combinations and phenotypic characteristics into the population systems known informally as: *Oe. biennis* I, *Oe. biennis* II, *Oe. biennis* III, *Oe. parviflora* I, and *Oe. parviflora* II (see Figure 51). These species, however grouped and subdivided, are very variable (Cleland, 1962; Munz, 1965).

The phylogenetic roots of the *Oenothera biennis* group are clearly to be sought in the *Oe. hookeri* group. In ecological preferences and geographical distribution, *Oe. elata* in Mexico and *Oe. hookeri* in California are believed to be close to the ancestral stock. One of the end arrangements in *Oe. hookeri*, the so-called Johansen arrangement, which differs by one homozygous translocation from the arbitrary standard arrangement, is considered to be the ancestral genome (see Table 10). In eastern North America the same original genome occurs in *Oe. grandiflora* in Alabama, and a somewhat different genome in *Oe. argillicola* of the Appalachian region. These eastern bivalent-forming species are apparently relics of populations which diverged from *Oe. hookeri* (Steiner, 1952; Stinson, 1953; Cleland, 1962, 1964).

Two other ancestral populations are needed in order to account for the existence of the Renner complexes in certain members of the *Oenothera biennis* group. These unknown hypothetical populations are designated Population III and Population IV. They too must have diverged from the ancestral *Oe. hookeri* stock (Cleland, 1962, 1964).

The bivalent-forming ancestors of the *Oenothera biennis* group are thus a series of four population systems, labeled Populations I to IV, two of which (I, II) have left living remnants in *Oe. argillicola* and *Oe. grandiflora*, and the other two of which (III, IV) are now extinct. The diverse taxonomic species of the *Oe. biennis* group represent different permanent hybrid combinations of these four ancestral populations, as shown in Figure 52 (Cleland, 1962, 1964).

Thus *Oenothera parviflora* I is a hybrid derivative of *Oe. argillicola* X *Oe. grandiflora*, and *Oe. parviflora* II is the hybrid derivative of *Oe. argillicola* X Population III. Similarly, the hybrid of Population II X III

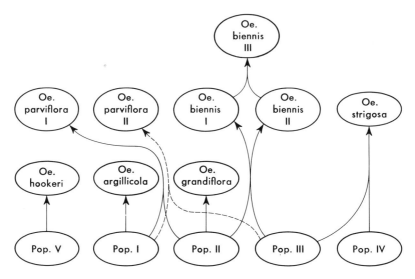

Figure 52 Phylogenetic relationships in the *Oenothera biennis* group. (Redrawn from Cleland, 1964.)

gave rise to both *Oe. biennis* I and *Oe. biennis* II; and these derivatives hybridized again to produce *Oe. biennis* III (Figure 52) (Cleland, 1962, 1964).

Several related but different Renner complexes are known on the male side in each of the taxonomic species of the *Oenothera biennis* group. The female-transmitted Renner complex also shows structural variation within each species. In *Oe. parviflora*, for example, there are eight known versions of the male-transmitted Renner complex and seven of the female-transmitted complex (Cleland, 1950, pp. 224–25).

The total number of Renner complexes in nature must be very great. These Renner complexes are assembled in diverse diploid combinations to make up the hundreds or thousands of heterogamic microspecies in the *Oenothera biennis* group.

Another heterogamic complex occurs in the Oenotheras of temperate South America (Hagen, 1950; Hecht, 1950).

Rhoeo discolor ($2n = 12$, Commelinaceae) from Central America is a complete permanent translocation heterozygote with a ring of 12 chromosomes (Sax, 1931). It is the only known species of its genus. It may well be the last remaining vestige of an ancient heterogamic complex (Darlington, 1939, p. 92; Stebbins, 1950, p. 433).

Permanent Odd Polyploidy

Most microspecies of the *Rosa canina* group in Europe are pentaploid ($x = 7$, $2n = 35$). At meiosis, 14 of the chromosomes pair to form 7 bivalents, and the remaining 21 chromosomes occur as univalents. The pairing behavior (7 II + 21 I) is the same in the pollen mother cells and embryo-sac mother cells. But chromosome distribution is quite different in the two cell lines, as shown semidiagrammatically in Figure 53 (Täckholm, 1922; Darlington, 1937a, pp. 460 ff.).

At first metaphase in the embryo-sac mother cells the 7 II + 21 I chromosomes are distributed in two groups with the 7 bivalents lined up on the equatorial plate and the 21 unpaired chromosomes assembled at the micropylar pole. The bivalent-forming chromosomes separate and pass to opposite poles of the spindle during first anaphase. Seven of the pairing chromosomes join the 21 unpaired chromosomes at telophase. The end of the first meiotic division thus usually finds 28 chromosomes at the micropylar pole and 7 at the other pole. The second meiotic division proceeds normally and yields a linear tetrad consisting of 2 large nuclei and 2 small ones. A large 28-chromosome nucleus at the micropylar end produces the functional embryo-sac (Figure 53) (Täckholm, 1922; Hurst, 1931; Darlington, 1937a).

In the pollen mother cells the univalents as well as the bivalents lie on the equatorial plate. In division I the bivalent-forming chromosomes separate and pass to the poles first, and then the univalent chromosomes divide and follow them to the poles. In the second division the pairing chromosomes divide and separate to the new poles. The nonpairing chromosomes also divide in division II, though belatedly, and the daughter chromosomes start to the poles, but most of them fail to reach the poles and remain behind (Täckholm, 1922; Darlington, 1937a).

Consequently meiosis in the male line yields daughter nuclei containing the basic complement of 7 pairing chromosomes plus varying numbers of nonpairing chromosomes. In one sample of 63 microspores determined for chromosome number, 9 cells had 7 chromosomes, and the remaining 54 cells had 8 to 22 chromosomes each. The microspores

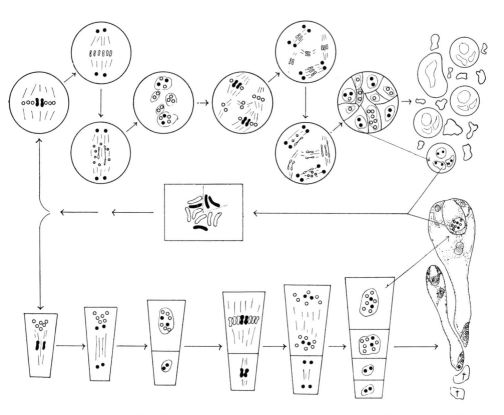

Figure 53 The course of meiosis in the male and female lines, and the fusion of complementary gamete types, in the *Rosa canina* group. Pairing chromosomes shown black, and univalent chromosomes white; only 2 bivalents and 6 univalents are shown. (Fagerlind, 1940.)

are irregular in size and shape, and most of them abort. The functional pollen grains are those with a set of 7 chromosomes from the pairing genome (Figure 53) (Täckholm, 1922; Darlington, 1937a; Fagerlind, 1940).

Most microspecies of the *Rosa canina* group have $2n = 5x = 35$ chromosomes, as noted previously (Täckholm, 1922; Darlington and Janaki Ammal, 1945; Darlington and Wylie, 1955). It was at first believed that the dog-roses maintain this constant pentaploid condition by agamospermy (Täckholm, 1922; Hurst, 1931), but they were later shown to be sexual (Fagerlind, 1940; Gustafsson, 1944). Several authorities state that they are self-fertilizing to a considerable extent (Darlington, 1937a, p. 463; 1963, p. 192; Stebbins, 1950, p. 338; Lewis and John, 1963, p. 338). Some outcrossing and hybridization also occur. More evidence on the quantitative frequency of inbreeding and outcrossing in these plants is desirable.

The self-fertilization of a pentaploid plant of *Rosa canina* is the union of a 28-chromosome female gamete with a 7-chromosome male gamete. This event of self-fertilization restores the pentaploid condition ($2n = 35$) and, furthermore, gives offspring like the parent plant in genomic constitution (7 II + 21 I). Cross-fertilization between sister pentaploid plants gives the same results. Thus the dog-roses breed true sexually for their odd polyploid condition (Figure 53).

Permanent odd polyploidy has also been discovered in *Leucopogon juniperinus* (Epacridaceae) in Australia (Smith-White, 1948). This species is a sexually reproducing triploid with $2n = 3x = 12$ chromosomes. At meiosis, two haploid sets form bivalents, whereas the third haploid set shows no tendency to pair. The plant thus behaves as an allotriploid with the genomic constitution AAB (Smith-White, 1948, 1955).

The pairing configuration at first metaphase is 4 II + 4 I in both the embryo-sac mother cells and the pollen mother cells. At first anaphase in both types of mother cells the univalents all pass to the same pole. This is the functioning pole in the embryo-sac mother cells and the nonfunctioning pole in the pollen mother cells. Consequently the embryo-sacs and egg nuclei carry 8 chromosomes, belonging to the A and the B genomes, while the functioning pollen grains carry 4 chromosomes of the A genome (Figure 54). Fertilization restores the somatic number $2n = 12$ and the genomic constitution AAB (Smith-White, 1948, 1955).

One of the conditions which has facilitated the origin of the balanced system of complementary gamete elimination in *Leucopogon juniperinus* is the peculiar mode of pollen development in these plants. In Leucopogon, as in other members of the tribe Styphelieae of the Epacridaceae, three microspores in each tetrad abort and only one develops into a mature pollen grain. The functioning pollen grain is one of the two 4-chromosome products of meiosis in *L. juniperinus* (Figure 54). This species can support a system of gamete elimination on the male side without any reduction in the total output of good pollen (Smith-White, 1955, 1959b).

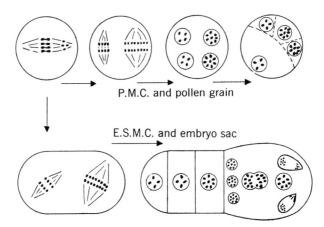

Figure 54 Formation of complementary types of gametes in the male and female lines in *Leucopogon juniperinus* (4 II + 4 I). The four univalents are included in the embryo-sac but not in the pollen. (Smith-White, 1959b.)

Leucopogon juniperinus is morphologically and cytologically uniform over a wide area in Australia. It is a single allotriploid species (Smith-White, 1955). The same genetic system of permanent odd polyploidy is elaborated on the much larger scale of a hybrid complex in the case of the *Rosa canina* group, as we shall see in the next section.

The Rosa canina Complex

The *Rosa canina* group comprises a distinct section (section Caninae) of the genus Rosa. The plants are thorny shrubs with small to medium or, rarely, large flowers (Figure 55). They occur as wild roses in low woods and fields throughout Europe and in neighboring parts of north Africa and western Asia.

The section Caninae contains numerous microspecies, some widespread and others localized, which grade into one another in a complex manner. Consequently the taxonomic treatment of the group differs from author to author, as is usual in treatments of hybrid complexes. Täckholm (1922, p. 289) recognizes 34 taxonomic species in 5 species groups. The treatment in Hegi's *Flora* reduces some of these taxonomic species to subspecies, but lists numerous varieties and formae under

Figure 55 Microspecies in the *Rosa canina* group. (A) *R. dumetorum* (5x). (B) *R. tomentella* (5x). (C) *R. agrestis* (5x). (D) *R. micrantha* (5x). (E) *R. rubrifolia* (4x). (Based on Hegi, 1909–1931, Vol. 4, Pt. 2.)

each specific or subspecific entity. Thus *Rosa canina* sensu stricto and *R. dumetorum* are treated as subspecies of *R. canina* sensu lato in Hegi's *Flora;* but *R. canina* sensu stricto has 7 varieties and 12 formae; while *R. dumetorum* has 5 varieties and 10 formae (Hegi, 1909–1931, Vol. 4, pp. 1031 ff.). It is generally agreed, therefore, that the group is taxonomically diverse and complex.

On morphological grounds the section Caninae is a close-knit unit well separated from other sections in the genus Rosa (Figure 55). For, on the one hand, the Caninae are quite distinct morphologically from other sections (Täckholm, 1922, Ch. 4) and, on the other hand, the microspecies intergrade in many combinations. Thus *R. canina* sensu stricto intergrades with *R. dumetorum, R. stylosa, R. pouzini,* and *R. chavini;* while *R. dumetorum* in turn also intergrades with *R. stylosa* and, in addition, with *R. coriifolia* and *R. tomentella* (Hegi, 1909–1931, Vol. 4, p. 991).

The basic chromosome number of the *Rosa canina* group is $x = 7$ as in other sections of the genus. However, there are no known diploids in the section Caninae. Most of the microspecies are pentaploids; a few are tetraploids or hexaploids (Täckholm, 1922).

In order to see the frequency distribution of ploidy levels in the *Rosa canina* group we will take Täckholm's (1922, p. 289) list of taxonomic species as a standard and annotate it for chromosome number according to the records published by Täckholm himself (1922, pp. 241–42) and by later workers (in Darlington and Janaki Ammal, 1945). Of the 34 taxonomic species: 15 are pentaploid, 3 are tetraploid, 3 are hexaploid, 4 contain both pentaploid and hexaploid forms, and 9 are uncounted.

Among the pentaploids ($2n = 35$) are *Rosa canina* sensu stricto, *R. dumetorum, R. micrantha, R. tomentosa,* and *R. stylosa* (in part). The tetraploids ($2n = 28$) are *R. rubrifolia, R. pomifera,* and *R. mollis.* Among the hexaploids ($2n = 42$) are *R. junzillii, R. pouzini, R. chavini,* and *R. stylosa* (in part).

The pairing behavior in the tetraploids is 7 II + 14 I and that in the hexaploids 7 II + 28 I. These types are thus like the pentaploids with their 7 II + 21 I in possessing both pairing and nonpairing genomes. And, like the pentaploids again, they transmit the univalent chromosomes through the egg cells (Täckholm, 1922).

Täckholm's view of the genomic constitution of a pentaploid microspecies, as based on the observed pairing relations of the chromosomes, can be expressed by the formula *AABCD*. Because of the meiotic behavior in some artificial hybrids between microspecies, he believed that the different microspecies of the *Rosa canina* group shared one genome in common, but differed in the others. Thus three pentaploid microspecies might be *AABCD*, *AACDE*, and *AADEF*, respectively. The pentaploid microspecies were regarded by Täckholm as direct products of hybridization between one ancestral diploid species and different octoploid species (Täckholm, 1922). In terms of genome formulas, the cross between *AA* and *AABBCCDD* would give the microspecies *AABCD*; the cross of *AA* X *AACCDDEE* would give microspecies *AACDE*; and so on.

Later cytogenetic work has modified these conclusions in some important respects. Gustafsson and Håkansson (1942) outcrossed three *Rosa canina* microspecies ($5x$ and $6x$) to the diploid species *R. rugosa* ($2n = 7$ II) in a different section. The artificial hybrids showed up to 14 bivalents at meiosis. This indicates homology between 3 rather than only 2 genomes within the polyploid dog-roses. They are autotriploid for one of their 5 or 6 chromosome sets (Gustafsson and Håkansson, 1942; Gustafsson, 1944). Fagerlind goes further in indicating some homology between all genomes in the polyploid dog-roses (Fagerlind, 1945).

The cytogenetic evidence of Gustafsson and Håkansson (1942) has led the first-named worker to suggest the following genome formulas. *Rosa canina* sensu stricto is $A_sA_sA_tCD$ and the related microspecies *R. rubiginosa* is *BBBCF* (Gustafsson, 1944). The *C* genome is found in *Rosa rugosa* in a different section and probably occurs more widely in the genus (Gustafsson and Håkansson, 1942). On the same and additional evidence Fagerlind would reduce some of the lettered genomes to subgenomes (Fagerlind, 1945; see also Blackhurst, 1948, and Wulff, 1954). In any case the *R. canina* microspecies seem to be autoallopolyploids with varying degrees of structural differentiation between their constituent genomes.

The genomic constitution of the polyploid dog-roses has an important bearing on the problem of the origin of new microspecies. The classical view of Täckholm, that the dog-roses are strict allopolyploids and are true-breeding on this account, led to the conclusion that the different microspecies are direct products of hybridization between different combinations of parental species (Täckholm, 1922). The potential array of microspecies is clearly limited by the number of hybridizing ancestral species on this view.

If, on the other hand, the dog-roses are autoallopolyploid in constitution, one highly heterozygous microspecies can spawn a series of new microspecies in its immediate progeny by the segregation of intergenomic gene differences. Thus in a pentaploid of the constitution $A_sA_sA_tCD$, the A_t chromosomes, which usually form univalents, will occasionally enter the pairing sets and will then proceed to transmit their genes through both the male and female gametes to the next generation (Gustafsson and Håkansson, 1942).

In short, one autoallopolyploid microspecies, which is normally true-breeding, can occasionally segregate one or more new true-breeding microspecies. And this process provides an explanation for the very large number of *Rosa canina* microspecies actually observed in nature (Gustafsson and Håkansson, 1942).

A plausible account of the phylogenetic history of the *Rosa canina* complex was given by Täckholm (1922). There are no known diploid species and no meiotically normal polyploids with the morphological characteristics of the section Caninae. The living forms all exhibit a derived genetic system. Hence the group must be fairly ancient. Furthermore, the living derived microspecies occupy areas in Europe which were formerly affected strongly by the ice age. The diploid ancestors of the *Rosa canina* group probably became extinct in the Pleistocene, and the existing hybrid complex grew up during and since that period (Täckholm, 1922, Ch. 5).

Evolutionary Potential

Darlington (1937a) called attention to some interesting cytogenetic similarities between the true-breeding odd polyploids of the *Rosa canina* group and the true-breeding translocation heterozygotes of the *Oenothera biennis* group. In each group there are some regularly pairing chromosomes or segments; they are the 7 bivalents in *Rosa canina* and the terminal pairing segments in *Oenothera biennis*. In each group there are also some nonpairing chromosomes or segments: the univalents in Rosa and the differential segments in Oenothera (Darlington, 1937a, p. 463; also Stebbins, 1950, p. 338).

A corresponding division of labor is found within the total chromosome complement in each group (Darlington, 1937a, p. 463). Some chromosomes or segments which undergo regular pairing and crossing-over are organs of variation. Other chromosomes or segments which do not pair carry the genes determining the constant features of the different microspecies.

Hybridization between bivalent-forming diploid species has produced in each instance an array of numerical or structural hybrids. The hybrids transmit one portion of their chromosome complement through the pollen and another through the eggs, and the two complementary types of gametes are often reunited by self-fertilization. In this way the hybrids increase in numbers and develop into heterogamic microspecies. The hybrid microspecies bridge the gaps between the ancestral species and give rise to a heterogamic complex in its primary stage. The microspecies next hybridize with one another to produce new secondary microspecies which further enrich the variability of the heterogamic complex in its later stages.

Considerable portions of the genotype are locked up in permanently heterozygous condition in the nonpairing chromosomal regions in *Rosa canina* and *Oenothera biennis*, and the plants normally breed true for these heterozygous gene combinations. But exceptional events of pairing and crossing-over in the same regions lead to the occasional segregation of new constant types. The exceptional segregates can be regarded as delayed products of hybridization in the usually true-breeding but highly heterozygous parental microspecies. These constant hybrid segregates, if adaptively valuable, can develop into new micro-species, thus increasing still further the total variability of the hetero-gamic complex.

A heterogamic complex in its mature stage contains thousands of microspecies which have originated directly or indirectly by hybridi-zation and have colonized areas of continental extent. Such heterogamic complexes, as exemplified by the *Rosa canina* and *Oenothera biennis* groups, are undoubtedly very successful in evolution like agamic complexes with which they can be compared.

Yet the fact remains that significant portions of the genotype are locked up in a heterogamic microspecies. Recombination is severely restricted by the same chromosomal mechanism that ensures the short-range evolutionary success. In the long run these restrictions can be expected to hamper critically the generation of new variations and leave the heterogamic complex stranded in an evolutionary blind alley (Darlington, 1939, p. 92; Stebbins, 1950, p. 441).

CHAPTER 24

The Homogamic Complex

Introduction · The Leafy-stemmed Gilias · Evolutionary Potential
· Reticulate Evolution · Implications for Angiosperm Phylogeny

Introduction

A homogamic complex is a hybrid complex in which the derived species
have originated by processes of hybrid speciation that do not involve
drastic changes in the genetic system. The known processes are recombinational speciation with chromosomal sterility barriers and hybrid
speciation with external barriers as discussed in Chapters 13 and 14.
A third possible process which should be looked for in future studies is
the formation by recombination of new gene-controlled sterility and
incompatibility barriers (see Chapter 13). Corresponding to these
various processes, the derived species are isolated from one another and
from the parental species by chromosomal sterility barriers, by external

barriers, by genic sterility and incompatibility, or by mixtures of internal and external barriers.

In any case the derived species remain on the same ploidy level as the ancestral species. Furthermore, the hybrid derivatives retain a normal sexual cycle associated with normal meiosis. If the parental species are sexual diploids, their daughter species are likewise sexual diploids.

In all other types of hybrid complexes it is possible to distinguish the derived species from the parental ones by the cytogenetic features or reproductive system of the former. This is not the case, however, in homogamic complexes, where the hybrid derivatives retain the original and ancestral genetic system.

Consequently, it is very difficult, though not impossible, to analyze a homogamic complex. Indeed, a homogamic complex may easily pass undetected as such, and many probably have done so. These inherent difficulties are reflected in the paucity of known examples to be reviewed in this chapter.

By the same token, however, the homogamic complex is fitted to play a special role in plant evolution. A consideration of this special role is our second main concern in this chapter.

The Leafy-stemmed Gilias

The Leafy-stemmed Gilias (Gilia section Eugilia) are a group of inter-sterile species of annual herbs which have been extensively studied taxonomically and cytogenetically. Most of the species are diploid ($2n = 18$); some are tetraploid ($2n = 36$). They are all sexual and range in breeding system from obligate outcrossers to predominant inbreeders (Grant, 1953, 1954b).

The relationships between *Gilia capitata* and *G. tricolor* in this group were described in Chapter 2 as an example of a pair of good biological species. In Chapter 7 we reviewed the fertility relationships within the section as a whole to illustrate a pattern which is common in annual plants, the Madia pattern. Here we describe the reticulate phylogeny of the group as an example of a homogamic complex.

We are concerned with the species of Leafy-stemmed Gilia in California, which is the center of distribution and probable center of origin of the section, and more particularly with the species on the

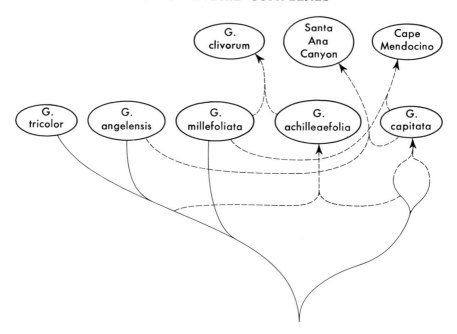

Figure 56 Phylogeny of the Leafy-stemmed Gilias in California. Dotted lines indicate species and populations of hybrid origin. All entities are diploid except *G. clivorum*, which is tetraploid. Further explanation in text. (Redrawn from Grant, 1953.)

California mainland, exclusive of the offshore islands. These species are shown in Figure 56. All of these species are diploid with the exception of *G. clivorum* ($2n = 4x = 36$).

The Leafy-stemmed Gilias fall into three species groups on morphological characters and ecological preferences. They are: the *G. tricolor* group including *G. tricolor* and *G. angelensis; G. millefoliata* and its relatives; and the *G. capitata* group. These three species groups reflect primary branchings in the phylogeny of the section (Figure 56). The extreme members are *G. tricolor* and *G. capitata* whose contrasting characteristics were described in Chapter 2 (see Figures 3, 4, and 56).

Gilia capitata itself is a complex polytypic species including an extraordinarily wide range of forms which fall into two extreme series and their hybrid intermediates. The extreme forms of *G. capitata* had diverged to the species level in past times and still preserve the species-like property of remaining distinct in certain zones of sympatric contact today. In other areas, however, the extreme forms are connected by bridging races of hybrid origin (Grant, 1950a, 1952a).

Gilia achilleaefolia is intermediate morphologically and ecologically between particular branches of the *G. capitata* and *G. tricolor* groups. Moreover it exhibits unusually great variability which is oriented in the direction of both putative parental groups. Some of these variant forms match known natural hybrids between the two species groups. All lines of evidence point to the hybrid origin of *G. achilleaefolia* shown in Figure 56 (Grant, 1954a).

Cytogenetic as well as morphological evidence indicates beyond any reasonable doubt that *Gilia clivorum* ($2n = 36$) is an amphiploid derivative of *G. achilleaefolia* and *G. millefoliata*, as shown in Figure 56 (Grant, 1952d, 1954c).

Populations morphologically intermediate between *G. capitata* and *G. angelensis* occur in Santa Ana Canyon in southern California, and populations morphologically intermediate between *G. capitata* and *G. millefoliata* on Cape Mendocino in northern California. These populations are of uncertain taxonomic status and are simply presented as such in Figure 56.

Figure 56 shows how hybridization can convert a branching phylogeny into a reticulate phylogeny without change of genetic system in the hybrid derivatives. And it shows successive strata of hybridizations. Species which are themselves of hybrid origin become the ancestors of a later generation of species and populations of hybrid origin.

Unanalyzed homogamic complexes undoubtedly exist in Pinus, Quercus, and many other plant groups which contain occasionally hybridizing diploid or homoploid species. The phylogenetic analysis of additional examples of homogamic complexes will be a fascinating task for future studies.

Evolutionary Potential

We have seen in previous chapters that most types of hybrid complex—polyploid, agamic, clonal, and heterogamic complexes—have a period of expansion followed by a period of decline. These successive stages of evolutionary success and evolutionary retrenchment are correlated with genetic systems which lock up genes in particular heterozygous combinations.

Amphiploidy, agamospermy, vegetative propagation, permanent translocation heterozygosity, and permanent odd polyploidy are methods of perpetuating particular favorable heterozygous gene combinations which have arisen by hybridization. They accomplish this end by imposing strong restrictions on gene recombination. The results are beneficial to the permanent hybrids during the period of expansion of the hybrid complex. But in the long run the permanent hybrids must pay the price for their opportunism. For, when the environment eventually changes profoundly, they may be unable to generate new gene combinations adapted to the new environmental conditions. The hybrid complex enters the period of decline.

Homogamic complexes differ from all other types of hybrid complex in that the hybrid derivatives are just as free from restrictions on recombination as are the ancestral species. The advantageous features of hybridization can be exploited by a homogamic complex without sacrificing the equally advantageous features of free gene recombination. It follows that a homogamic complex is not specially limited in its evolutionary potentialities, but may, on the contrary, continue to evolve progressively over a very long time span and give rise to new major branches (Grant, 1953).

When the ancestral diploid sexual species become extinct in an agamic, clonal, or heterogamic complex, and to a lesser extent in a polyploid complex, the hybrid derivatives are cut off from an important source of new variations. The loss is likely to prove crippling in the subsequent evolution of the hybrid complex. But, in a homogamic complex, by contrast, the original species may become extinct without jeopardizing the evolutionary potential of the derived species. Indeed, the ancestral species of one homogamic complex may well be the terminal products of a previous homogamic complex (Grant, 1953).

Figure 56 shows different strata of hybrid speciations in a homogamic complex. There is no reason to suppose that the known strata of hybridization are the only such strata in the phyletic history of this group. On the contrary, this and many other plant groups have in all probability evolved as a succession of homogamic complexes (Grant, 1953).

In summary, if agamic, clonal, and heterogamic complexes have a closed system of evolution, and polyploid complexes a restricted system, homogamic complexes can be said to possess an open-ended system of evolution.

Reticulate Evolution

The phylogeny of many or most groups of higher plants which have been studied carefully in the field and laboratory is seen to be reticulate rather than exclusively dichotomous and branching. The model of the phylogenetic tree cannot be carried over from zoology into botany without misrepresenting the relationships in many instances. In all major groups of higher plants we have to deal with phylogenetic networks rather than with phylogenetic trees.

The concept of reticulate evolution in the plant kingdom has been expressed and developed by several generations of botanists. The first authors to state that hybridization plays an important rôle in plant evolution, and to state this thesis as a generalization supported by scientific evidence, were Kerner (1894–1895, Vol. 2) and Lotsy (1916). Their early formulations suffered from the defects of pioneering attempts. Lotsy, for example, redefined species as pure lines and hence tended to equate hybridization chiefly, though not exclusively, with normal crossing. Kerner, of course, worked in the pre-Mendelian era, and moreover he refrained from committing himself fully or explicitly regarding evolutionary changes greater in magnitude than speciation. But Kerner and Lotsy exposed the problem clearly to view.

During the first part of this century a series of plant geneticists— Winge, Täckholm, Renner, Müntzing, Gustafsson, Manton, and others—approached the problem of hybridization and evolution experimentally and inductively, finding evidences of reticulate phylogeny in one case after another. Among the taxonomists of the same period who saw plant phylogeny as a network were Diels, Hayata, and DuRietz. Somewhat later the genetic and taxonomic approaches fused to form the modern biosystematic school. Members of this school who have contributed importantly to our understanding of reticulate evolution include Anderson, Stebbins, Heiser, H. Lewis, Ehrendorfer, and others.

Ehrendorfer has introduced the useful concept of differentiation-hybridization cycles to describe one aspect of the pattern of reticulate evolution. He emphasizes that the course of evolution in many plant groups is a cyclical alternation of two phases: divergence and hybridization (Ehrendorfer, 1958, 1959). We would go on to say that only the homogamic complex possesses a mode of reticulate evolution in which differentiation-hybridization cycles can continue indefinitely.

Angiosperm evolution has been macroevolution; it has also been reticulate evolution. The long-continued progressive evolutionary changes in the different main phyletic lines of the angiosperms can have occurred in combination with reticulate patterns of phylogenetic relationships if, and only if, these phyletic lines have evolved as successive homogamic complexes (Grant, 1953). For the homogamic complex is the only type of hybrid complex with an open system of evolution.

Implications for Angiosperm Phylogeny

This book has dealt with plant evolution at the level of species and hybrid complexes. Plant evolution at these levels is strongly affected, as Gustafsson pointed out (1946–1947), by the basic characteristics of the plants themselves: by the relatively simple organization of the plant body, the open system of growth, the ease of vegetative propagation, the facility for uniparental reproduction. These characteristics have left their stamp on plant microevolution, which, as we have seen, is generally reticulate.

In the concluding section of this book it may be worthwhile to attempt to extrapolate from plant microevolution to angiosperm phylogeny. The pattern of reticulate evolution revealed by genetic and taxonomic studies of species groups can shed some light on the problems surrounding the phylogenetic classification of the higher categories.

The phylogenetic classification of most orders and superorders of angiosperms remains confused after several centuries of taxonomic work. The difficulties and uncertainties stem from a reticulate pattern of morphological resemblances between different groups.

Consider four related groups designated A, B, C, D. Groups A and B share several morphological characters in common; another set of characters is common to B and C; and still another character combination unites C and D. Different systems of classification of the entire assemblage can and will be proposed by different students, depending on which characters are given first priority. Thus one system may contain the coordinate taxa A, B-C, and D, and another system the taxa A-B and C-D.

Difficulties of this general nature and only partially satisfactory solutions of this sort have characterized the search for a natural system of orders in one major segment of the angiosperms after another. In practice, the difficulties are greater than in our simple hypothetical model, since the reticulate pattern of morphological relationships usually extends to additional groups E, F, G, etc. The picture is well portrayed for a series of actual superorders and orders by Cronquist in *The Evolution and Classification of Flowering Plants* (1968).

It is significant that the tangled taxonomic structure described above, although common, is not universal in the angiosperms. Some large groups stand out as distinct phyletic units and have long been recognized as such. A good example is the very large family Compositae and its order Asterales. The basic dichotomy between the Dicotyledons and Monocotyledons has been generally recognized since the time of John Ray. The dichotomous phylogeny of higher animals affords another striking contrast to the reticulate pattern found in many groups of angiosperms.

We have then a reticulate pattern of morphological character resemblances in many broad segments of the angiosperms which is real and not an artifact of our methods of observation; for basic dichotomies can be discerned where they exist, even in large and diverse groups, by the same methods of observation.

Now the reticulate pattern of morphological relationships as seen at the ordinal level in angiosperms can be observed again at the intermediate level of genera and tribes in many angiosperm families. And the reticulate pattern of phenotypic resemblances at all levels is reminiscent of the reticulate pattern of genetic relationships found in species complexes of plants. This common pattern encourages us to look for common underlying causes inherent in plants themselves and in their modes of evolution.

A cause of reticulate patterns in both plant macroevolution and plant microevolution is convergence: the approach of one phyletic line toward another in respect to several or many characteristics.

Let us return to our hypothetical example. It will be recalled that groups A, B, C, D exhibit morphological similarities in the combinations A-B, B-C, and C-D. We often find an approximation to the following general situation. The related groups A and B differ in respect to characters in which A retains a more primitive and B a more advanced condition. Similarly, group C shows advancement in various characters over the related but more primitive group D. The resemblances between groups B and C, on the other hand, are in advanced or derived characters. These resemblances, in other words, are products of evolutionary convergence.

Evolutionary convergence can be brought about by selection without hybridization, or by the joint action of hybridization and selection. It is assumed in either case that separate lineages are exposed to selection for similar adaptive responses to the same or similar environmental conditions. Both modes of evolutionary convergence occur readily in plants.

The facility with which different plant groups can make a transition from one ecological niche to another, so as to yield convergences in whole character combinations, is demonstrated by hundreds of examples. Many separate genera of perennial herbs have given rise to annuals in Mediterranean climates; different mesophytic stocks yield convergent xerophytes in deserts; bee-pollinated genera have repeatedly produced hummingbird-flowered species in hummingbird territory, and so on.

The proneness of plant lineages to change over from one adaptive zone to another is correlated with the relative simplicity of the gene systems involved. Contrasting character combinations adapted to different ecological niches may be found at the racial level of divergence in some instances. Therefore change-overs or quantum shifts, which are major in ecological terms but relatively minor in genetical terms, can occur repeatedly in different phyletic lines under the guiding influence of parallel selection alone. Within the limits of crossability the production of convergences is stimulated still further by hybridization.

Bibliography

Addison, G. and R. Tavares. 1952. Hybridization and grafting in species of Theobroma which occur in Amazonia. *Evolution* 6: 380–86.

Allan, H. H. 1949. Wild species-hybrids in the phanerogams. *Bot. Rev.* 15: 77–105.

Allard, R. W., and S. K. Jain. 1962. Population studies in predominantly self-pollinated species. II. Analysis of quantitative genetic changes in a bulk-hybrid population of barley. *Evolution* 16: 90–101.

Allard, R. W., and P. L. Workman. 1963. Population studies in predominantly self-pollinated species. IV. Seasonal fluctuations in estimated values of genetic parameters in lima bean populations. *Evolution* 17: 470–80.

Alston, R. E., T. J. Mabry, and B. L. Turner. 1963. Perspectives in chemotaxonomy. *Science* 142: 545–52.

Alston, R. E., and B. L. Turner. 1962. New techniques in analysis of complex natural hybridization. *Proc. Nat. Acad. Sci.* 48: 130–37.

_____ and _____. 1963a. Biochemical Systematics. Prentice-Hall, Englewood Cliffs, N.J.

_____ and _____. 1963b. Natural hybridization among four species of Baptisia (Leguminosae). *Amer. Jour. Bot.* 50: 159–73.

Anderson, E. 1939. Recombination in species crosses. *Genetics* 24: 668–98.

_____. 1948. Hybridization of the habitat. *Evolution* 2: 1–9.

_____. 1949. Introgressive Hybridization. John Wiley, New York.

_____. 1953. Introgressive hybridization. *Biol. Reviews* 28: 280–307.

Anderson, E., and B. R. Anderson. 1954. Introgression of *Salvia apiana* and *Salvia mellifera*. *Ann. Missouri Bot. Gard.* 41: 329–38.

Anderson, E., and A. Gage. 1952. Introgressive hybridization in *Phlox bifida*. *Amer. Jour. Bot.* 39: 399–404.

Anderson, E., and L. Hubricht. 1938. Hybridization in Tradescantia. III. The evidence for introgressive hybridization. *Amer. Jour. Bot.* 25: 396–402.

Anderson, E., and B. Schafer. 1931. Species hybrids in Aquilegia. *Ann. Bot.* 45: 639–646.

Anfinsen, C. B. 1959. The Molecular Basis of Evolution. John Wiley, New York.

Avery, A. G., S. Satina, and J. Rietsema. 1959. Blakeslee: The Genus Datura. Ronald Press, New York.

Avery, P. 1938. Cytogenetic evidences of Nicotiana phylesis in the *alata*-group. *Univ. Calif. Publ. Bot.* 18: 153–94.

Babcock, E. B. 1947. The genus Crepis. *Univ. Calif. Publ. Bot.* Vols. 21, 22.

Babcock, E. B., and J. A. Jenkins. 1943. Chromosomes and phylogeny in Crepis. III. The relationships of one hundred and thirteen species. *Univ. Calif. Publ. Bot.* 18: 241–92.

Babcock, E. B., and G. L. Stebbins. 1938. The American species of Crepis. Their interrelationships and distribution as affected by polyploidy and apomixis. *Carnegie Inst. Washington Publ.* 504.

Babcock, E. B., G. L. Stebbins, and J. A. Jenkins. 1937. Chromosomes and phylogeny in some genera of the Crepidinae. *Cytologia, Fujii Jubilee Vol.* 188–210.

Baetcke, K. P., and R. E. Alston. 1968. The composition of a hybridizing population of *Baptisia sphaerocarpa* and *Baptisia leucophaea*. *Evolution* 22: 157–65.

Baldwin, J. T. 1938. Kalanchoe: the genus and its chromosomes. *Amer. Jour. Bot.* 25: 572–79.

Bateman, A. J. 1946. Genetical aspects of seed-growing. *Nature* 157: 752–55.

——. 1956. Cryptic self-incompatibility in the wallflower: *Cheiranthus cheiri* L. *Heredity* 10: 257–61.

Baur, E. 1914. Einführung in die experimentelle Vererbungslehre. 2nd ed. Borntraeger, Berlin.

——. 1932. Artumgrenzung und Artbildung in der Gattung Antirrhinum, Sektion Antirrhinastrum. *Zeitschr. ind. Abstammungs- u. Vererbungslehre* 63: 256–302.

Beadle, G. W. 1963. Genetics and Modern Biology. American Philosophical Society. Philadelphia, Pa.

Beaman, J. H. 1957. The systematics and evolution of Townsendia (Compositae). *Contrib. Gray Herbarium* (Harvard), No. 133.

Beaudry, J. R. 1960. The species concept: its evolution and present status. *Revue Canadienne de Biologie* 19: 219–40.

Beeks, R. M. 1962. Variation and hybridization in southern California populations of Diplacus (Scrophulariaceae). *Aliso* 5: 83–122.

Bell, C. R. 1954. The *Sanicula crassicaulis* complex (Umbelliferae); a study of variation and polyploidy. *Univ. Calif. Publ. Bot.* 27: 133–230.

Benson, L. 1950. The Cacti of Arizona. 2nd ed. University of Arizona Press, Tucson, Ariz.

——. 1962. Plant Taxonomy. Ronald Press, New York.

Benson, L., E. A. Phillips, and P. A. Wilder. 1967. Evolutionary sorting of characters in a hybrid swarm. I. Direction of slope. *Amer. Jour. Bot.* 54: 1017–26.

Bergman, B. 1935. Zytologische Studien über die Fortpflanzung bei den Gattungen Leontodon und Picris. *Svensk Bot. Tidskrift* 29: 155–301.

Bernardini, J. V., and A. Lima-de-Faria. 1967. Asynchrony of DNA replication in the chromosomes of Luzula. *Chromosoma* 22: 91–100.

Blackhurst, H. T. 1948. Cytogenetic studies on *Rosa rubiginosa* L. and its hybrids. *Proc. Amer. Soc. Hort. Sci.* 52: 510–16.

Blair, W. F. 1960. The Rusty Lizard: A Population Study. University of Texas Press, Austin, Tex.

Bradshaw, A. D. 1959. Population differentiation in *Agrostis tenuis* Sibth. I. Morphological differentiation. *New Phytol.* 58: 208–27.

Brainerd, E. 1924. Some natural violet hybrids of North America. *Vermont Agric. Exp. Sta. Bull.* No. 239.

Bridges, C. B. 1922. The origin of variations in sexual and sex-limited characters. *Amer. Nat.* 56: 51–63.

Brink, R. A. 1962. Phase change in higher plants and somatic cell heredity. *Quart. Rev. Biol.* 37: 1–22.

Brown, S. 1943. The origin and nature of variability in the Pacific coast blackberries (*Rubus ursinus* Cham. & Schlecht. and *R. lemurum* sp. nov.). *Amer. Jour. Bot.* 30: 686–97.

———. 1954. Mitosis and meiosis in *Luzula campestris* DC. *Univ. Calif. Publ. Bot.* 27: 231–78.

Brown, W. L., and E. O. Wilson. 1956. Character displacement. *Systematic Zool.* 5: 49–64.

Bungenberg de Jong, C. M. 1957. Polyploidy in animals. *Bibliographia Genetica* 17: 111–228.

Cadman, C. H. 1942. Autotetraploid inheritance in the potato; some new evidence. *Jour. Genetics* 44: 33–52.

Cain, A. J. 1954. Animal Species and Their Evolution. Hutchinson and Co., Ltd., London; and Harper and Row, New York.

Cameron, J. W., and H. B. Frost. 1968. Genetics, breeding, and nucellar embryony. In: The Citrus Industry. 2nd ed., Vol. 2. Ed. by W. Reuther, L. D. Batchelor, and H. J. Webber. University of California Press, Berkeley and Los Angeles, Calif.

Camp, W. H., and C. L. Gilly. 1943. The structure and origin of species. *Brittonia* 4: 323–85.

Carlquist, S. 1956. On the generic limits of Eriophyllum (Compositae) and related genera. *Madrono* 13: 226–39.

Carson, H. L. 1955. The genetic characteristics of marginal populations of Drosophila. *Cold Spring Harbor Symp. Quant. Biol.* 20: 276–87.

———. 1959. Genetic conditions which promote or retard the formation of species. *Cold Spring Harbor Symp. Quant. Biol.* 24: 87–105.

Carson, H. L., and W. B. Heed. 1964. Structural homozygosity in marginal populations of nearctic and neotropical species of Drosophila in Florida. *Proc. Nat. Acad. Sci.* 52: 427–30.

Caspari, E. 1948. Cytoplasmic inheritance. *Advances in Genetics* 2: 1–66.

Cave, M. S., and L. Constance. 1947. Chromosome numbers in the Hydrophyllaceae: III. *Univ. Calif. Publ. Bot.* 18: 449–65.

Chambers, K. L. 1955. A biosystematic study of the annual species of Microseris. *Contrib. Dudley Herbarium* (Stanford) 4: 207–312.

Clausen, J. 1926. Genetical and cytological investigations on *Viola tricolor* L. and *V. arvensis* Murr. *Hereditas* 8: 1–156.

——. 1951. Stages in the Evolution of Plant Species. Cornell University Press, Ithaca, N.Y.

——. 1965. Population studies of alpine and subalpine races of conifers and willows in the California high Sierra Nevada. *Evolution* 19: 56–68.

Clausen, J., and W. M. Hiesey. 1958. Experimental studies on the nature of species. IV. Genetic structure of ecological races. *Carnegie Inst. Washington Publ.* 615.

—— and ——. 1960. The balance between coherence and variation in evolution. *Proc. Nat. Acad. Sci.* 46: 494–506.

Clausen, J., D. D. Keck, and W. M. Hiesey. 1940. Experimental studies on the nature of species. I. Effect of varied environments on western American plants. *Carnegie Inst. Washington Publ.* 520.

——, ——, and ——. 1945. Experimental studies on the nature of species. II. Plant evolution through amphiploidy and autoploidy, with examples from the Madiinae. *Carnegie Inst. Washington Publ.* 564.

——, ——, and ——. 1948. Experimental studies on the nature of species. III. Environmental responses of climatic races of Achillea. *Carnegie Inst. Washington Publ.* 581.

Clausen, J., D. D. Keck, W. M. Hiesey, and P. Grun. 1949. Experimental taxonomy. *Carnegie Inst. Washington Yearbook* 48: 95–106.

Clausen, R. E., and T. H. Goodspeed. 1925. Interspecific hybridization in Nicotiana. II. A tetraploid *glutinosa-tabacum* hybrid, an experimental verification of Winge's hypothesis. *Genetics* 10: 279–84.

Clausen, R. T. 1959. Sedum of the Trans-Mexican Volcanic Belt. Cornell University Press, Ithaca, N.Y.

Clayton, E. E. 1950. Male-sterile tobacco. *Jour. Heredity* 41: 171–75.

Cleland, R. E. 1936. Some aspects of the cyto-genetics of Oenothera. *Bot. Rev.* 2: 316–48.

——. 1949. Phylogenetic relationships in Oenothera. *Hereditas, Suppl. Vol.*, 173–88.

——. 1950. Studies in Oenothera cytogenetics and phylogeny. *Indiana Univ. Publ., Science Series*, No. 16.

——. 1962. The cytogenetics of Oenothera. *Advances in Genetics* 11: 147–237.

——. 1964. The evolutionary history of the North American evening primroses of the "*biennis* group." *Proc. Amer. Phil. Soc.* 108: 88–98.

Clifford, H. T. 1954. Analysis of suspected hybrid swarms in the genus Eucalyptus. *Heredity* 8: 259–69.

Cockrum, E. L. 1952. A check-list and bibliography of hybrid birds in North America north of Mexico. *Wilson Bull.* 64: 140–59.

Colwell, R. N. 1951. The use of radioactive isotopes in determining spore distribution patterns. *Amer. Jour. Bot.* 38: 511–23.

Cook, S. A. 1962. Genetic system, variation and adaptation in *Eschscholzia californica*. *Evolution* 16: 278–99.

Cooperrider, M. 1957. Introgressive hybridization in *Quercus marilandica*. *Amer. Jour. Bot.* 44: 804–10.

Cottam, W. P. 1954. Prevernal leafing of aspen in Utah mountains. *Jour. Arnold Arboretum* 35: 239–48.

Crane, M. B., and K. Mather. 1943. The natural cross-pollination of crop plants with particular reference to the radish. *Ann. Appl. Biol.* 30: 301–08.

Cronquist, A. 1968. The Evolution and Classification of Flowering Plants. Houghton Mifflin, Boston, Mass.

Cuénot, L. 1951. L'Évolution biologique. Masson & Cie., Paris.

Dansereau, P. M. 1941. Études sur les hybrides de Cistes. VI. Introgression dans la section Ladanum. *Canad. Jour. Res., C.* 19: 59–67.

Darlington, C. D. 1932. Recent Advances in Cytology. 1st ed. Churchill, London.

——. 1937a. Recent Advances in Cytology. 2nd ed. Churchill, London.

——. 1937b. The early hybridisers and the origins of genetics. *Herbertia* 4: 63–69.

——. 1939. The Evolution of Genetic Systems. 1st ed. Cambridge University Press, Cambridge.

——. 1940. Taxonomic species and genetic systems. In: The New Systematics. Ed. by J. Huxley. Clarendon Press, Oxford.

——. 1956a. Natural populations and the breakdown of classical genetics. *Proc. Roy. Soc., B.* 145: 350–64.

——. 1956b. Chromosome Botany. 1st ed. George Allen and Unwin, London.

——. 1958. The Evolution of Genetic Systems. 2nd ed. Basic Books, New York.

——. 1963. Chromosome Botany and the Origins of Cultivated Plants. 2nd ed. Hafner, New York.

——. 1965. Cytology. Churchill, London.

Darlington, C. D., and E. K. Janaki Ammal. 1945. Chromosome Atlas of Cultivated Plants. 1st ed. George Allen and Unwin, London.

Darlington, C. D., and K. Mather. 1949. The Elements of Genetics. George Allen and Unwin, London.

Darlington, C. D., and A. P. Wylie. 1955. Chromosome Atlas of Flowering Plants. 2nd ed. George Allen and Unwin, London.

Darwin, C. 1859. On the Origin of Species. 1st ed. John Murray, London. Reprinted by Watts and Co., London, 1950; and by Harvard University Press, Cambridge, Mass., 1964.

——. 1868. The Variation of Animals and Plants under Domestication. 2 vols. John Murray, London.

Davies, E. 1956. Cytology, evolution and origin of the aneuploid series in the genus Carex. *Hereditas* 42: 349–65.

Day, A. 1965. The evolution of a pair of sibling allotetraploid species of Cobwebby Gilias (Polemoniaceae). *Aliso* 6: 25–75.

De Wet, J. M. J. 1968. Diploid-tetraploid-haploid cycles and the origin of variability in Dichanthium agamospecies. *Evolution* 22: 394–97.

Dempster, L. T., and F. Ehrendorfer. 1965. Evolution of the *Galium multiflorum* complex in western North America. II. Critical taxonomic revision *Brittonia* 17: 289–334.

Diels, L. 1921. Die Methoden der Phytographie und der Systematik der Pflanzen. In: Abderhaldens Handbuch der biologischen Arbeitsmethoden. Abt. 11. Berlin.

Dobzhansky, Th. 1937a. Genetic nature of species differences. *Amer. Nat.* 71: 404–20.

——. 1937b, 1941, 1951. Genetics and the Origin of Species. 1st, 2nd, and 3rd eds. Columbia University Press, New York.

——. 1950. Mendelian populations and their evolution. *Amer. Nat.* 84: 401–18.

——. 1958. Species after Darwin. In: A Century of Darwin. Ed. by S. A. Barnett. Heinemann, London.

Dobzhansky, Th., L. Ehrman, and O. Pavlovsky. 1957. *Drosophila insularis*, a new sibling species of the *willistoni* group. *Studies in the genetics of Drosophila* (Univ. of Texas), 9: 39–47.

Dobzhansky, Th., and C. Epling. 1944. Contributions to the genetics, taxonomy, and ecology of *Drosophila pseudoobscura* and its relatives. *Carnegie Inst. Washington Publ.* 554.

Dobzhansky, Th., and S. Wright. 1943. Genetics of natural populations. X. Dispersion rates in *Drosophila pseudoobscura. Genetics* 28: 304–40.

Dodds, K. S., and G. J. Paxman. 1962. The genetic system of cultivated diploid potatoes. *Evolution* 16: 154–67.

Dodson, C. H. 1956. Cytological studies in the genus Oncidium. Ph.D. thesis. Claremont Graduate School, Claremont, Calif.

Dodson, C. H., and G. P. Frymire. 1961. Preliminary studies in the genus Stanhopea (Orchidaceae). *Ann. Missouri Bot. Gard.* 48: 137–72.

Drury, W. H. 1956. The ecology of the natural origin of a species of Carex by hybridization. *Rhodora* 58: 51–72.

Duffield, J. W. 1952. Relationships and species hybridization in the genus Pinus. *Zeitschr. Forstgenetik u. Forstpflanzenzücht* 1: 93–97.

DuRietz, G. E. 1930. The fundamental units of biological taxonomy. *Svensk Bot. Tidskr.* 24: 333–428.

East, E. M. 1916. Studies on size inheritance in Nicotiana. *Genetics* 1: 164–76.

Edwardson, J. R. 1956. Cytoplasmic male-sterility. *Bot. Rev.* 22: 696–738.

Ehrendorfer, F. 1949. Zur Phylogenie der Gattung Galium L. I. Polyploidie und geographisch-ökologische Einheiten in der Gruppe des *Galium pumilum* Murray (Sekt. Leptogalium Lange sensu Rouy) im österreichi-, schen Alpenraum. *Österr. Bot. Zeitschr.* 96: 109–38.

——. 1951. Zur Phylogenie der Gattung Galium. II. Rassengliederung, Variabilitätszentren und geographische Merkmalsprogressionen als Ausdruck der raum-zeitlichen Entfaltung des Formenkreises *Galium incanum* S.S. *Österr. Bot. Zeitschr.* 98: 427–90.

——. 1953. Ökologisch-geographische Mikro-Differenzierung einer Population von *Galium pumilum* Mrr. s. str. *Österr. Bot. Zeitschr.* 100: 616–38.

——. 1958. Zur Phylogenie der Gattung Galium. VI. Ein Variabilitäszentrum als "fossiler" Hybrid-Komplex: der ost-mediterrane *Galium graecum* L.—*G. canum* Req. Formenkreis. *Österr. Bot. Zeitschr.* 105: 229–79.

——. 1959. Differentiation-hybridization cycles and polyploidy in Achillea. *Cold Spring Harbor Symp. Quant. Biol.* 24: 141–52.

——. 1961. Evolution of the *Galium multiflorum* complex in western North America. I. Diploids and polyploids in this dioecious group. *Madrono* 16: 109–40.

——. 1962. Beiträge zur Phylogenie der Gattung Knautia (Dipsacaceae). I. Cytologische Grundlagen und allgemeine Hinweise. *Österr. Bot. Zeitschr.* 109: 276–343.

——. 1964. Cytologie, Taxonomie und Evolution bei Samenpflanzen. *Vistas in Botany* 4: 99–186.

Ehrendorfer, F., F. Krendl, E. Habeler, and W. Sauer. 1968. Chromosome numbers and evolution in primitive angiosperms. *Taxon* 17: 337–53.

Ehrman, L. 1962. Hybrid sterility as an isolating mechanism in the genus Drosophila. *Quart. Rev. Biol.* 37: 279–302.

——. 1965. Direct observation of sexual isolation between allopatric and between sympatric strains of the different *Drosophila paulistorum* races. *Evolution* 19: 459–64.

Emerson, S. H. 1935. The genetic nature of DeVries's mutations in *Oenothera lamarckiana. Amer. Nat.* 69: 545–59.

Epling, C. 1947. Natural hybridization of *Salvia apiana* and *S. mellifera. Evolution* 1: 69–78.

Epling, C., and Th. Dobzhansky. 1942. Genetics of natural populations. VI. Microgeographic races in *Linanthus parryae. Genetics* 27: 317–32.

Epling, C., H. Lewis, and F. M. Ball. 1960. The breeding group and seed storage: a study in population dynamics. *Evolution* 14: 238–55.

Ernst, A. 1918. Bastardierung als Ursache der Apogamie im Pflanzenreich. Gustav Fischer, Jena.

Faegri, K., and L. van der Pijl. 1966. The Principles of Pollination Ecology. Pergamon, New York.

Fagerlind, F. 1940. Sind die *canina*-Rosen agamospermische Bastarde? *Svensk Bot. Tidskr.* 34: 334–54.

——. 1945. Die Bastarde der *canina*-Rosen, ihre Syndese- und Formbildungsverhältnisse. *Acta Horti Bergiani* (Uppsala), 14: 7–37.

Falconer, D. S. 1961. Introduction to Quantitative Genetics. Ronald Press, New York.

Fassett, N. C. 1944. *Juniperus virginiana, J. horizontalis* and *J. scopulorum.* II. Hybrid swarms of *J. virginiana* and *J. scopulorum. Bull. Torrey Bot. Club* 71: 475–83.

Fisher, R. A. 1930. The Genetical Theory of Natural Selection. Clarendon Press, Oxford.

_____. 1954. Retrospect of the criticisms of the theory of natural selection. In: Evolution as a Process. Ed. by J. Huxley. George Allen and Unwin, London.

Flake, R. H., E. von Rudloff, and B. L. Turner. 1969. Quantitative study of clinal variation in *Juniperus virginiana* using terpenoid data. *Proc. Nat. Acad. Sci.* 64: 487–94.

Focke, W. O. 1881. Die Pflanzenmischlinge. Borntraeger, Berlin.

Ford, E. B. 1964. Ecological Genetics. Methuen & Co., London.

Frankel, O., and A. Munday. 1962. The evolution of wheat. In: The Evolution of Living Organisms. Symposium, Royal Society, Victoria, Australia.

Freter, L. E., and W. V. Brown. 1955. A cytotaxonomic study of *Bouteloua curtipendula* and *B. uniflora*. *Bull. Torrey Bot. Club* 82: 121–30.

Frisch, K. 1914. Der Farbensinn und Formensinn der Biene. *Zool. Jahrb.* 35: 1–182.

_____. 1919. Über den Geruchsinn der Biene und seine blütenbiologische Bedeutung. *Zool. Jahrb.* 37: 1–238.

Fryxell, P. A. 1957. Mode of reproduction in higher plants. *Bot. Rev.* 23: 135–233.

Fürnkranz, D. 1960. Cytogenetische Untersuchungen an Taraxacum in Raume von Wien. *Österr. Bot. Zeitschr.* 107: 310–50.

Gajewski, W. 1949. Cytogenetic investigations on the genus Geum L. *Hereditas, Suppl. Vol.*, 578–79.

_____. 1953. A fertile amphipolyploid hybrid of *Geum rivale* with *G. macrophyllum*. *Acta Soc. Bot. Polon.* 22: 411–39.

_____. 1957. A cytogenetic study on the genus Geum L. *Monographiae Botanicae* (Warsaw). Vol. 4.

_____. 1959. Evolution in the genus Geum. *Evolution* 13: 378–88.

Garber, E. D. 1957. The genus Collinsia. III. The significance of chiasmata frequencies as a cytotaxonomic tool. *Madrono* 14: 172–76.

_____. 1960. The genus Collinsia. IX. Speciation and chromosome repatterning. *Cytologia* 25: 233–43.

Garber, E. D., and J. Gorsic. 1956. The genus Collinsia. II. Interspecific hybrids involving *C. heterophylla*, *C. concolor*, and *C. sparsiflora*. *Bot. Gaz.* 118: 73–77.

Gerassimova, H. 1939. Chromosome alterations as a factor of divergence of forms. I. New experimentally produced strains of *C. tectorum* which are physiologically isolated from the original forms owing to reciprocal translocation. *Compt. Rend. Acad. Sci. URSS* 25: 148–54.

Gerstel, D. U. 1953. Chromosomal translocations in interspecific hybrids of the genus Gossypium. *Evolution* 7: 234–44.

_____. 1954. A new lethal combination in interspecific cotton hybrids. *Genetics* 39: 628–39.

Gerstel, D. U., and L. L. Phillips. 1958. Segregation of synthetic amphiploids in Gossypium and Nicotiana. *Cold Spring Harbor Symp. Quant. Biol.* 23: 225–37.

Gerstel, D. U., and P. A. Sarvella. 1956. Additional observations on chromosomal translocations in cotton hybrids. *Evolution* 10: 408–14.

Gibson, J. B., and J. M. Thoday. 1964. Effects of disruptive selection. IX. Low selection intensity. *Heredity* 19: 125–30.

Gilles, A., and L. F. Randolph. 1951. Reduction in quadrivalent frequency in autotetraploid maize during a period of 10 years. *Amer. Jour. Bot.* 38: 12–16.

Gilmour, J. S. L., and J. Heslop-Harrison. 1954. The deme terminology and the units of micro-evolutionary change. *Genetica* 27: 147–61.

Goldschmidt, R. B. 1938. Physiological Genetics. McGraw-Hill, New York.

———. 1940. The Material Basis of Evolution. Yale University Press, New Haven, Conn.

———. 1955. Theoretical Genetics. University of California Press, Berkeley, Calif.

Goodspeed, T. H. 1954. The Genus Nicotiana. Chronica Botanica, Waltham, Mass.

Gottschalk, W. 1954. Die Grundzahl der Gattung Solanum und einiger Nicotiana-Arten. *Ber. Deutsch. Bot. Ges.* 67: 369–76.

Gould, F. W. 1959. Notes on apomixis in sideoats grama. *Jour. Range Management* 12: 25–28.

———. 1966. Chromosome numbers of some Mexican grasses. *Canad. Jour. Bot.* 44: 1683–96.

Gould, F. W., and Z. J. Kapadia. 1962. Biosystematic studies in the *Bouteloua curtipendula* complex. I. The aneuploid rhizomatous *B. curtipendula* of Texas. *Amer. Jour. Bot.* 49: 887–92.

——— and ———. 1964. Biosystematic studies in the *Bouteloua curtipendula* complex. II. Taxonomy. *Brittonia* 16: 182–207.

Grant, A., and V. Grant. 1956. Genetic and taxonomic studies in Gilia. VIII. The Cobwebby Gilias. *Aliso* 3: 203–87.

Grant, K. A., and V. Grant. 1964. Mechanical isolation of *Salvia apiana* and *Salvia mellifera* (Labiatae). *Evolution* 18: 196–212.

Grant, V. 1949. Pollination systems as isolating mechanisms in flowering plants. *Evolution* 3: 82–97.

———. 1950a. Genetic and taxonomic studies in Gilia. I. *Gilia capitata. Aliso* 2: 239–316.

———. 1950b. The flower constancy of bees. *Bot. Rev.* 16: 379–98.

———. 1952a. Genetic and taxonomic studies in Gilia. II. *Gilia capitata abrotanifolia. Aliso* 2: 361–73.

———. 1952b. Genetic and taxonomic studies in Gilia. III. The *Gilia tricolor* complex. *Aliso* 2: 375–88.

———. 1952c. Isolation and hybridization between *Aquilegia formosa* and *A. pubescens. Aliso* 2: 341–60.

———. 1952d. Cytogenetics of the hybrid *Gilia millefoliata* X *achilleafolia*. I. Variations in meiosis and polyploidy rate as affected by nutritional and genetic conditions. *Chromosoma* 5: 372–90.

_____. 1953. The role of hybridization in the evolution of the Leafy-stemmed Gilias. *Evolution* 7: 51–64.

_____. 1954a. Genetic and taxonomic studies in Gilia. IV. *Gilia achilleafolia. Aliso* 3: 1–18.

_____. 1954b. Genetic and taxonomic studies in Gilia. VI. Interspecific relationships in the Leafy-stemmed Gilias. *Aliso* 3: 35–49.

_____. 1954c. Genetic and taxonomic studies in Gilia. V. *Gilia clivorum. Aliso* 3: 19–34.

_____. 1956a. The genetic structure of races and species in Gilia. *Advances in Genetics* 8: 55–87.

_____. 1956b. Chromosome repatterning and adaptation. *Advances in Genetics* 8: 89–107.

_____. 1956c. The influence of breeding habit on the outcome of natural hybridization in plants. *Amer. Nat.* 90: 319–22.

_____. 1957. The plant species in theory and practice. In: The Species Problem. Ed. by E. Mayr. Amer. Assoc. Adv. Sci., Washington, D.C.

_____. 1958. The regulation of recombination in plants. *Cold Spring Harbor Symposia Quant. Biol.* 23: 337–63.

_____. 1959. Natural History of the Phlox Family. M. Nijhoff, The Hague.

_____. 1963. The Origin of Adaptations. Columbia University Press, New York.

_____. 1964a. The biological composition of a taxonomic species in Gilia. *Advances in Genetics* 12: 281–328.

_____. 1964b. The Architecture of the Germplasm. John Wiley, New York.

_____. 1964c. Genetic and taxonomic studies in Gilia. XII. Fertility relationships of the polyploid Cobwebby Gilias. *Aliso* 5: 479–507.

_____. 1965. Species hybrids and spontaneous amphiploids in the *Gilia laciniata* group. *Heredity* 20: 537–50.

_____. 1966a. Block inheritance of viability genes in plant species. *Amer. Nat.* 100: 591–601.

_____. 1966b. Linkage between viability and fertility in a species cross in Gilia. *Genetics* 54: 867–80.

_____. 1966c. Selection for vigor and fertility in the progeny of a highly sterile species hybrid in Gilia. *Genetics* 53: 757–75.

_____. 1966d. The selective origin of incompatibility barriers in the plant genus Gilia. *Amer. Nat.* 100: 99–118.

_____. 1966e. The origin of a new species of Gilia in a hybridization experiment. *Genetics* 54: 1189–99.

_____. 1967. Linkage between morphology and viability in plant species. *Amer. Nat.* 101: 125–39.

Grant, V., and A. Grant. 1954. Genetic and taxonomic species in Gilia. VII. The Woodland Gilias. *Aliso* 3: 59–91.

_____ and _____. 1960. Genetic and taxonomic studies in Gilia. XI. Fertility relationships of the diploid Cobwebby Gilias. *Aliso* 4: 435–81.

Grant, V., and K. A. Grant. 1965. Flower Pollination in the Phlox Family. Columbia University Press, New York.

Grant, W. F., M. R. Bullen, and D. Nettancourt. 1962. The cytogenetics of Lotus. I. Embryo-cultured interspecific diploid hybrids closely related to *L. corniculatus* L. *Canad. Jour. Genetics and Cytol.* 4: 105–28.

Gray, A. P. 1954. Mammalian Hybrids. Commonwealth Agricultural Bureaux. Farnham Royal, England.

———. 1958. Bird Hybrids. Commonwealth Agricultural Bureaux. Farnham Royal, England.

Greenleaf, W. H. 1941. Sterile and fertile amphidiploids: their possible relation to the origin of *Nicotiana tabacum*. *Genetics* 26: 301–24.

Grun, P. 1954. Cytogenetic studies of Poa. I. Chromosome numbers and morphology of interspecific hybrids. *Amer. Jour. Bot.* 41: 671–78.

———. 1961. Early stages in the formation of internal barriers to gene exchange between diploid species of Solanum. *Amer. Jour. Bot.* 48: 79–89.

Grun, P., M. Aubertin, and A. Radlow. 1962. Multiple differentiation of plasmons of diploid species of Solanum. *Genetics* 47: 1321–33.

Gulick, J. T. 1905. Evolution, racial and habitudinal. *Carnegie Inst. Washington Publ.* 25.

Gustafsson, Å. 1942. The origin and properties of the European blackberry flora. *Hereditas* 28: 249–77.

———. 1943. The genesis of the European blackberry flora. *Lunds Universitets Årsskrift* 39(6): 1–200.

———. 1944. The constitution of the *Rosa canina* complex. *Hereditas* 30: 405–28.

———. 1946–1947. Apomixis in higher plants. *Lunds Universitets Årsskrift* 42–43: 1–370.

———. 1948. Polyploidy, life form and vegetative reproduction. *Hereditas* 34: 1–22.

Gustafsson, Å., and A. Håkansson. 1942. Meiosis in some Rosa-hybrids. *Botaniska Notiser*, 1942, 331–43.

Haartman, J. 1751, 1764. Plantae Hybridae. Uppsala, 1751; Amoenitates Academicae, Holm, 1764.

Hagen, C. W. 1950. A contribution to the cytogenetics of the genus Oenothera with special reference to certain forms from South America. *Indiana Univ. Publ., Science Series* 16: 305–48.

Hagerup, O. 1932. Über Polyploidie in Beziehung zu Klima, Ökologie, und Phylogenie. *Hereditas* 16: 19–40.

———. 1947. The spontaneous formation of haploid, polyploid, and aneuploid embryos in some orchids. *Biologiske Meddelelser* 20: 1–22.

Hair, J. B. 1966. Biosystematics of the New Zealand flora 1945–1964. *New Zealand Jour. Bot.* 4: 559–95.

———. 1968. The chromosomes of the Cupressaceae. I. Tetraclineae and Actinostrobeae (Callitroideae). *New Zealand Jour. Bot.* 6: 277–84.

Hair, J. B., and E. J. Beuzenberg. 1961. High polyploidy in a New Zealand Poa. *Nature* 189: 160.

Håkansson, A. 1947. Contributions to a cytological analysis of the species differences of *Godetia amoena* and *G. whitneyi*. *Hereditas* 33: 235–60.

_____. 1954. Meiosis and pollen mitosis in X-rayed and untreated spikelets of *Eleocharis palustris*. *Hereditas* 40: 325–45.

_____. 1958. Holocentric chromosomes in Eleocharis. *Hereditas* 44: 531–40.

Hall, B. M. 1955. Genetic analysis of interspecific hybrids in the genus Bromus, section Ceratochloa. *Genetics* 40: 175–92.

Haller, J. R. 1962. Variation and hybridization in ponderosa and Jeffrey pines. *Univ. Calif. Publ. Bot.* 34: 123–66.

Hanelt, P. 1966. Polyploidie-Frequenz und geographische Verbreitung bei höheren Pflanzen. *Biol. Rundschau* 4: 183–96.

Hanelt, P., and D. Mettin. 1966. Cytosystematische Untersuchungen in der Artengruppe um *Vicia sativa* L. II. *Die Kulturpflanze* (Berlin) 14: 137–61.

Harberd, D. J. 1961. Observations on population structure and longevity of *Festuca rubra* L. *New Phytol.* 60: 184–206.

_____. 1962. Some observations on natural clones in *Festuca ovina*. *New Phytol.* 61: 85–100.

Harding, J., R. W. Allard, and D. G. Smeltzer. 1966. Population studies in predominantly self-pollinated species. IX. Frequency-dependent selection in *Phaseolus lunatus*. *Proc. Nat. Acad. Sci.* 56: 99–104.

Harlan, J. R. 1945. Natural breeding structure in the *Bromus carinatus* complex as determined by population analysis. *Amer. Jour. Bot.* 32: 142–48.

_____. 1949. Apomixis in side-oats grama. *Amer. Jour. Bot.* 36: 495–99.

Harper, J. L. 1957. Biological flora of the British Isles. *Ranunculus acris* L., *Ranunculus repens* L., *Ranunculus bulbosus* L. *Jour. Ecol.* 45: 289–42.

Haselton, S. C. 1951. Epiphyllum Handbook. Abbey Garden Press, Pasadena, Calif.

Hayata, B. 1921. The natural classification of plants according to the dynamic system. *Icon. Plant. Formos.* 10: 97–234.

Hayman, B. I., and K. Mather. 1953. The progress of inbreeding when homozygotes are at a disadvantage. *Heredity* 7: 165–83.

Hecht, A. 1950. Cytogenetic studies of Oenothera subgenus Raimannia. *Indiana Univ. Publ., Science Series*, 16: 255–304.

Heckard, L. R. 1960. Taxonomic studies in the *Phacelia magellanica* polyploid complex with special reference to the California members. *Univ. Calif. Publ. Bot.* 32: 1–126.

Hegi, G. 1909–1931. Illustrierte Flora von Mittel-Europa. 7 vols. Carl Hanser, Munich.

Heilborn, O. 1924. Chromosome numbers and dimensions, species formation and phylogeny in the genus Carex. *Hereditas* 5: 129–216.

_____. 1932. Aneuploidy and polyploidy in Carex. *Svensk Bot. Tidskr.* 26: 137–46.

_____. 1934. On the origin and preservation of polyploidy. *Hereditas* 19: 233–42.

_____. 1939. Chromosome studies in Cyperaceae, III–IV. *Hereditas* 25: 224–40.

Heiser, C. B. 1947. Hybridization between the sunflower species *Helianthus annuus* and *H. petiolaris*. *Evolution* 1: 249–62.

_____. 1949a. Natural hybridization with particular reference to introgression. *Bot. Rev.* 15: 645–87.

_____. 1949b. Study in the evolution of the sunflower species *Helianthus annuus* and *H. bolanderi. Univ. Calif. Publ. Bot.* 23: 157–208.

_____. 1950. A comparison of the flora as a whole and the weed flora of Indiana as to polyploidy and growth habits. *Proc. Indiana Acad. Sci.* 59: 64–70.

_____. 1951a. Hybridization in the annual sunflowers: *Helianthus annuus* X *H. debilis* var. *cucumerifolius. Evolution* 5: 42–51.

_____. 1951b. Hybridization in the annual sunflowers: *Helianthus annuus* X *H. argophyllus. Amer. Nat.* 85: 65–72.

_____. 1954. Variation and subspeciation in the common sunflower, *Helianthus annuus. Amer. Midland Nat.* 51: 287–305.

Heiser, C. B., W. C. Martin, and D. M. Smith. 1962. Species crosses in Helianthus. I. Diploid Species. *Brittonia* 14: 137–47.

Herbert, W. 1820. On the production of hybrid vegetables, with the result of many experiments made in the investigation of the subject. *Trans. Hort. Soc. London* 4: 15–50.

Hiorth, G. 1940. Eine Serie multipler Allele für Blütenzeichnungen bei *Godetia amoena. Hereditas* 26: 441–53.

_____. 1942. Zur Genetik und Systematik der *amoena*-Gruppe der Gattung Godetia. *Zeitschr. ind. Abstammungs- u. Vererbungslehre* 80: 289–349.

Hollingshead, L. 1930. A lethal factor in Crepis effective only in an inter-specific hybrid. *Genetics* 15: 114–40.

Hovanitz, W. 1949. Increased variability in populations following natural hybridization. In: Genetics, Paleontology, and Evolution. Ed. by G. L. Jepsen. Princeton University Press, Princeton, N.J.

Howard, H. W. 1961. Potato cytology and genetics, 1952–59. *Bibliographia Genetica* 19: 87–216.

Hubbs, C., and E. A. Delco. 1960. Mate preference in males of four species of gambusiine fishes. *Evolution* 14: 145–52.

Hubbs, C. L. 1955. Hybridization between fish species in nature. *Systematic Zool.* 4: 1–20.

Hubbs, C. L., and R. R. Miller. 1943. Mass hybridization between two genera of cyprinid fishes in the Mohave Desert, California. *Papers Michigan Acad. Sci. Arts and Letters* 28: 343–78.

Hultén, E. 1943. *Stellaria longipes* Goldie and its allies. *Botaniska Notiser* 1943, 251–70.

_____. 1956. The *Cerastium alpinum* complex. A case of world-wide intro-gressive hybridization. *Svensk Bot. Tidskr.* 50: 411–95.

Hurst, C. C. 1931. Embryo-sac formation in diploid and polyploid species of Roseae. *Proc. Roy. Soc., B,* 109: 126–48.

Hutchinson, J. B., and S. G. Stephens. 1947. The Evolution of Gossypium. Oxford University Press, London.

Huxley, J. S. 1942. Evolution: The Modern Synthesis. George Allen and Unwin, London.

Huziwara, Y. 1959. Chromosomal evolution in the subtribe Asterinae. *Evolution* 13: 188–93.

Iman, A. G., and R. W. Allard. 1965. Population studies in predominantly self-pollinated species. VI. Genetic variability between and within natural populations of wild oats from differing habitats in California. *Genetics* 51: 49–62.

Jackson, R. C. 1957. New low chromosome number for plants. *Science* 128: 1115–16.

——. 1962. Interspecific hybridization in Haplopappus and its bearing on chromosome evolution in the Blepharodon section. *Amer. Jour. Bot.* 49: 119–32.

Jain, S. K. 1959. Male sterility in flowering plants. *Bibliographia Genetica* 18: 101–66.

——. 1960. Cytogenetics of rye (Secale spp.) *Bibliographia Genetica* 19: 1–86.

Jain, S. K., and R. W. Allard. 1960. Population studies in predominantly self-pollinated species. I. Evidence for heterozygote advantage in a closed population of barley. *Proc. Nat. Acad. Sci.* 46: 1371–77.

—— and ——. 1965. The nature and stability of equilibria under optimizing selection. *Proc. Nat. Acad. Sci.* 54: 1436–43.

Jain, S. K., and A. D. Bradshaw. 1966. Evolutionary divergence among adjacent plant populations. I. The evidence and its theoretical analysis. *Heredity* 21: 407–41.

Jain, S. K., and K. B. L. Jain. 1962. The progress of inbreeding in a pedigree-bred population of barley. *Euphytica* 11: 229–32.

Jain, S. K., and D. R. Marshall. 1967. Population studies in predominantly self-pollinated species. X. Variation in natural populations of *Avena fatua* and *A. barbata*. *Amer. Nat.* 101: 19–33.

James, S. H. 1965. Complex hybridity of *Isotoma petraea*. I. The occurrence of interchange heterozygosity, autogamy and a balanced lethal system. *Heredity* 20: 341–53.

Jepson, W. L. 1923. The Trees of California. Associated Students Store. Berkeley, Calif.

John, B., and K. R. Lewis. 1965a. Genetic speciation in the grasshopper *Eyprepocnemis plorans*. *Chromosoma* 16: 308–44.

—— and ——. 1965b. The Meiotic System. Protoplasmatologia. Springer-Verlag, New York.

—— and ——. 1966. Chromosome variability and geographic distribution in insects. *Science* 152: 711–21.

Johnson, A. W., and J. G. Packer. 1965. Polyploidy and environment in arctic Alaska. *Science* 148: 237–39.

Johnson, A. W., J. G. Packer, and G. Reese. 1965. Polyploidy, distribution, and environment. In: The Quaternary of the United States. Ed. by H. E. Wright and D. G. Frey. Princeton University Press, Princeton, N.J.

Johnson, B. L. 1945. Cyto-taxonomic studies in Oryzopsis. *Bot. Gaz.* 107: 1–32.

Johnson, L. A. S., and B. G. Briggs. 1963. Evolution in the Proteaceae. *Australian Jour. Bot.* 11: 21–61.

Jones, D. F. 1956. Genic and cytoplasmic control of pollen abortion in maize. *Brookhaven Symp. Biol.* 9: 101–12.

Jordan, D. S. 1905. The origin of species through isolation. *Science* 22: 545–62.

Jordan, K. 1896. On mechanical selection and other problems. *Novit. Zool.* 3: 426–525.

Jowett, D. 1964. Population studies on lead-tolerant *Agrostis tenuis. Evolution* 18: 70–81.

Jukes, T. H. 1966. Molecules and Evolution. Columbia University Press, New York.

Kamemoto, H., and K. Shindo. 1964. Meiosis in interspecific and intergeneric hybrids of Vanda. *Bot. Gaz.* 125: 132–38.

Kannenberg, L. W., and R. W. Allard. 1967. Population studies in predominantly self-pollinated species. VIII. Genetic variability in the *Festuca microstachys* complex. *Evolution* 21: 227–40.

Kapadia, Z. J., and F. W. Gould. 1964. Biosystematic studies in the *Bouteloua curtipendula* complex. IV. Dynamics of variation in *B. curtipendula* var. *caespitosa. Bull. Torrey Bot. Club* 91: 465–78.

Karpechenko, G. D. 1927. Polyploid hybrids of *Raphanus sativus* L. X *Brassica oleracea* L. *Bull. Appl. Bot. Genetics Plant Breeding* (Leningrad) 17: 305–410.

Kearney, T. H., and R. H. Pebbles. 1964. Arizona Flora. 2nd ed. University of California Press, Berkeley, Calif.

Keck, D. D. 1957. Trends in systematic botany. *Survey of Biol. Progress* 3: 47–107.

Keep, E. 1962. Interspecific hybridization in Ribes. *Genetica* 33: 1–23.

Kerner, A. 1894–1895. The Natural History of Plants. 2 vols. Translation. Blackie and Son, London.

Kerster, H. W. 1964. Neighborhood size in the Rusty lizard, *Sceloporus olivaceus. Evolution* 18: 445–57.

Khoshoo, T. N. 1959. Polyploidy in gymnosperms. *Evolution* 13: 24–39.

Kihara, H. 1954. Considerations on the evolution and distribution of Aegilops species based on the analyzer-method. *Cytologia* 19: 336–57.

Kihara, H., and T. Ono. 1926. Chromosomenzahlen und systematische Gruppierung der Rumex-Arten. *Zeitschr. Zellforsch. u. mikr. Anatomie* 4: 475–81.

King, J. C. 1947. Interspecific relationships within the *guarani* group of Drosophila. *Evolution* 1: 143–53.

Knight, G. R., A. Robertson, and C. H. Waddington. 1956. Selection for sexual isolation within a species. *Evolution* 10: 14–22.

Kölreuter, J. G. 1761–1766. Vorläufige Nachricht von einigen das Geschlecht der Pflanzen betreffenden Versuchen und Beobachtungen. Reprint, 1893. In: Ostwald's *Klassiker*. Leipzig.

Koopman, K. F. 1950. Natural selection for reproductive isolation between *Drosophila pseudoobscura* and *Drosophila persimilis*. *Evolution* 4: 135–48.

Kostoff, D. 1941–1943. Cytogenetics of the Genus Nicotiana. State's Printing House. Sofia, Bulgaria.

Kruckeberg, A. R. 1951. Intraspecific variability in the response of certain native plant species to serpentine soil. *Amer. Jour. Bot.* 38: 408–19.

———. 1954. Chromosome numbers in Silene (Caryophyllaceae). I. *Madrono* 12: 238–46.

———. 1955. Interspecific hybridizations of Silene. *Amer. Jour. Bot.* 42: 373–78.

———. 1957. Variation in fertility of hybrids between isolated populations of the serpentine species, *Streptanthus glandulosus* Hook. *Evolution* 11: 185–211.

———. 1960. Chromosome numbers in Silene (Caryophyllaceae). II. *Madrono* 15: 205–15.

———. 1961. Artificial crosses of western North American Silenes. *Brittonia* 13: 305–33.

Kugler, H. 1955. Einführung in die Blütenökologie. Gustav Fischer, Stuttgart.

Kurabayashi, M. 1958. Evolution and variation in Japanese species of Trillium. *Evolution* 12: 286–310.

Lamotte, M. 1951. Recherches sur la structure genetique des populations naturelles de *Cepaea nemoralis* (L.). *Bull. biol. France et Belg., Suppl.* 35: 1–238.

Lamprecht, H. 1941. Die Artgrenze zwischen *Phaseolus vulgaris* L. und *multiflorus* Lam. *Hereditas* 27: 51–175.

———. 1944. Die genisch-plasmatische Grundlage der Artbarriere. *Agr. Hort. Genet.* 2: 75–142.

Latimer, H. 1958. A study of the breeding barrier between *Gilia australis* and *Gilia splendens*. Ph.D. thesis. Claremont Graduate School, Claremont, Calif.

Lenz, L. W. 1958. A revision of the Pacific coast irises. *Aliso* 4: 1–72.

———. 1959a. Hybridization and speciation in the Pacific coast irises. *Aliso* 4: 237–309.

———. 1959b. *Iris tenuis* S. Wats., a new transfer to the subsection Evansia. *Aliso* 4: 311–19.

Lenz, L. W., and D. E. Wimber. 1959. Hybridization and inheritance in orchids. In: The Orchids. Ed. by C. Withner. Ronald Press, New York.

Levan, A. 1937. Cytological studies in the *Allium paniculatum* group. *Hereditas* 23: 317–70.

———. 1949. Polyploidy in flax, sugar beets and timothy. *Hereditas, Suppl. Vol.* 46–47.

Levin, D. A. 1966. Chromatographic evidence of hybridization and evolution in *Phlox maculata*. *Amer. Jour. Bot.* 53: 238–45.

——. 1967. Hybridization between annual species of Phlox: population structure. *Amer. Jour. Bot.* 54: 1122–30.

——. 1968. The genome constitutions of eastern North American Phlox amphiploids. *Evolution* 22: 612–32.

Levin, D. A., and H. W. Kerster. 1967a. An analysis of interspecific pollen exchange in Phlox. *Amer. Nat.* 101: 387–400.

—— and ——. 1967b. Natural selection for reproductive isolation in Phlox. *Evolution* 21: 679–87.

—— and ——. 1968. Local gene dispersal in Phlox. *Evolution* 22: 130–39.

Lewis, D. 1943a. Physiology of incompatibility in plants. III. Autopolyploids. *Jour. Genetics* 45: 171–85.

——. 1943b. The incompatibility sieve for producing polyploids. *Jour. Genetics* 45: 261–64.

——. 1966. The genetic integration of the breeding system. In: Reproductive Biology and Taxonomy of Vascular Plants. Ed. by J. G. Hawkes. Pergamon Press, Oxford.

Lewis, H. 1953. The mechanism of evolution in the genus Clarkia. *Evolution* 7: 1–20.

——. 1962. Catastrophic selection as a factor in speciation. *Evolution* 16: 257–71.

——. 1966. Speciation in flowering plants. *Science* 152: 167–72.

Lewis, H., and C. Epling. 1959. *Delphinium gypsophilum*, a diploid species of hybrid origin. *Evolution* 13: 511–25.

Lewis, H., and M. Lewis. 1955. The genus Clarkia. *Univ. Calif. Publ. Bot.* 20: 241–392.

Lewis, H., and P. H. Raven. 1958. Rapid evolution in Clarkia. *Evolution* 12: 319–36.

Lewis, H., and M. R. Roberts. 1956. The origin of *Clarkia lingulata*. *Evolution* 10: 126–38.

Lewis, H., and J. Szweykowski. 1964. The genus Gayophytum (Onagraceae). *Brittonia* 16: 343–91.

Lewis, K. R., and B. John. 1961. Hybridisation in a wild population of *Eleocharis palustris*. *Chromosoma* 12: 433–48.

—— and ——. 1963. Chromosome Marker. Little, Brown & Co., Boston, Mass.

Lima-de-Faria, A. 1949. Genetics, origin and evolution of kinetochores. *Hereditas* 35: 422–44.

Lindsay, D. W., and R. K. Vickery. 1967. Comparative evolution in *Mimulus guttatus* of the Bonneville basin. *Evolution* 21: 439–56.

Lindstrom, E. W. 1936. Genetics of polyploidy. *Bot. Rev.* 2: 197–215.

Linnaeus, C. 1760, 1790. Disquisitio de Sexu Plantarum. St. Petersburg, 1760; Amoenitates Academicae. Erlangen. 1790.

Little, E. L., and S. S. Pauley. 1958. A natural hybrid between black and white spruce in Minnesota. *Amer. Midland Nat.* 60: 202–11.

Lorkovič, Z. 1958. Some peculiarities of spatially and sexually restricted gene exchange in the *Erebia tyndarus* group. *Cold Spring Harbor Symp. Quant. Biol.* 23: 319–25.

Lotsy, J. P. 1916. Evolution by Means of Hybridization. M. Nijhoff, The Hague.

——. 1925. Species or Linneon. *Genetica* 7: 487–506.

——. 1931. On the species of the taxonomist in its relation to evolution. *Genetica* 13: 1–16.

Löve, A. 1944. Cytogenetic studies on Rumex, subg. Acetosella. *Hereditas* 30: 1–136.

——. 1952. Preparatory studies for breeding Icelandic *Poa irrigata. Hereditas* 38: 11–32.

——. 1953. Subarctic polyploidy. *Hereditas* 39: 113–24.

Löve, A., and B. M. Kapoor. 1966. An alloploid Ophioglossum. *The Nucleus* 9: 132–38.

—— and ——. 1967. The highest plant chromosome number in Europe. *Svensk Bot. Tidskrift* 61: 29–32.

Löve, A. and D. Löve. 1949. The geobotanical significance of polyploidy. I. Polyploidy and latitude. *Portugalae Acta Biologica, Suppl. Vol.,* 273–352.

—— and ——. 1953. The geo-botanical significance of polyploidy. *Proc. 6th Internat. Grasslands Congress,* 240–46.

—— and ——. 1957. Arctic polyploidy. *Proc. Genetics Soc. Canada* 2: 23–27.

—— and ——. 1967. Polyploidy and altitude: Mt. Washington. *Biol. Zentralblatt. Suppl. Vol.,* 307–12.

Löve, A., D. Löve, and M. Raymond. 1957. Cytotaxonomy of Carex section Capillares. *Canad. Jour. Bot.* 35: 715–61.

MacKey, J. 1954. Neutron and X-ray experiments in wheat and a revision of the speltoid problem. *Hereditas* 40: 65–180.

McMinn, H. E. 1944. The importance of field hybrids in determining species in the genus Ceanothus. *Proc. Calif. Acad. Sci.* 25: 323–56.

McNaughton, I. H., and J. L. Harper. 1960a. The comparative biology of closely related species living in the same area. II. Aberrant morphology and a virus-like syndrome in hybrids between *Papaver rhoeas* L. and *P. dubium* L. *New Phytol.* 59: 27–41.

—— and ——. 1960b. The comparative biology of closely related species living in the same area. III. The nature of barriers isolating sympatric populations of *Papaver dubium* and *P. lecoquii. New Phytol.* 59: 129–37.

Magoon, M. L., D. C. Cooper, and R. W. Hougas. 1958. Cytogenetic studies of some diploid Solanums section Tuberarium. *Amer. Jour. Bot.* 45: 207–21.

Maheshwari, P. 1949. The male gametophyte of angiosperms. *Bot. Rev.* 15: 1–75.

Malecka, J. 1961. Studies in the mode of reproduction of the diploid endemic species. *Taraxacum pieninicum* Pawl. *Acta Biologica Cracoviensia* 4: 25–41.

Malheiros, N., D. De Castro, and A. Câmara. 1947. Chromosomas sem centrómero localizado. O caso da *Luzula purpurea* Link. *Agronomia Lusitana* 9: 51–71.

Malheiros-Gardé, N., and A. Gardé. 1951. Agmatoploidia no genero Luzula DC. *Genetica Iberica* 3: 155–76.

Manglesdorf, P. C. 1958. The mutagenic effect of hybridizing maize and teosinte. *Cold Spring Harbor Symp. Quant. Biol.* 23: 409–21.

Manton, I. 1950. Problems of Cytology and Evolution in the Pteridophyta. Cambridge University Press, Cambridge.

———. 1951. Cytology of Polypodium in America. *Nature* 167: 37.

———. 1958. The concept of the aggregate species. In: Systematics of Today. Ed. by O. Hedberg. Uppsala.

Marsden-Jones, E. M. 1930. The genetics of *Geum intermedium* Willd. haud Ehrh., and its back-crosses. *Jour. Genetics* 23: 377–95.

Mason, H. L. 1949. Evidence for the genetic submergence of *Pinus remorata*. In: Genetics, Paleontology, and Evolution. Ed. by G. L. Jepsen. Princeton University Press, Princeton, N.J.

Mather, K. 1947. Species crosses in Antirrhinum. I. Genetic isolation of the species *majus*, *glutinosum* and *orontium*. *Heredity* 1: 175–86.

———. 1955. Polymorphism as an outcome of disruptive selection. *Evolution* 9: 52–61.

———. 1966. Breeding systems and response to selection. In: Reproductive Biology and Taxonomy of Vascular Plants. Ed. by J. G. Hawkes. Pergamon Press, Oxford.

Mattfeld, J. 1930. Über hybridogene Sippen der Tannen. *Bibliotheca Botanica*, Vol. 25.

Mayr, E. 1942. Systematics and the Origin of Species. Columbia University Press, New York.

———. 1954. Change of genetic environment and evolution. In: Evolution as a Process. Ed. by J. Huxley. George Allen and Unwin, London.

———. 1955a. The species as a systematic and as a biological problem. In: *Biology Colloquium* (Oregon State College, Corvallis, Oregon).

———. 1955b. Karl Jordan's contribution to current concepts in systematics and evolution. *Trans. Royal Entmol. Soc.* (London), 1955: 45–66.

———. 1957. Species concepts and definitions. In: The Species Problem. Ed. by E. Mayr. Amer. Assoc. Adv. Sci., Washington.

———. 1963. Animal Species and Evolution. Harvard University Press, Cambridge, Mass.

Meglitsch, P. A. 1954. On the nature of the species. *Systematic Zool.* 3: 49–65.

Michaelis, P. 1933. Entwicklungsgeschichtlich-genetische Untersuchungen an Epilobium. II. Die Bedeutung des Plasmas für die Pollenfertilität des *Epilobium luteum-hirsutum* Bastardes. *Zeitschr. ind. Abstammungs- u. Vererbungslehre* 65: 1–71, 353–411.

Millicent, E., and J. M. Thoday. 1961. Effects of disruptive selection. IV. Gene flow and divergence. *Heredity* 16: 199–217.

Mirov, N. T. 1967. The Genus Pinus. Ronald Press, New York.

Moens, P. B. 1964. Chiasma distribution and the segregation of markers in chromosome 2 of the tetraploid tomato. *Genetics* 49: 123–33.

Mohamed, A. H., and F. W. Gould. 1966. Biosystematic studies in the *Bouteloua curtipendula* complex. V. Megasporogenesis and embryo sac development. *Amer. Jour. Bot.* 53: 166–69.

Molisch, H. 1938. The Longevity of Plants. Translation. E. H. Fulling, New York.

Mosquin, T. 1967. Evidence for autopolyploidy in *Epilobium angustifolium* (Onagraceae). *Evolution* 21: 713–19.

Muldal, S. 1952. The chromosomes of the earthworms. I. The evolution of polyploidy. *Heredity* 6: 55–76.

Muller, C. H. 1952. Ecological control of hybridization in Quercus: a factor in the mechanism of evolution. *Evolution* 6: 147–61.

Muller, H. J. 1925. Why polyploidy is rarer in animals than in plants. *Amer. Nat.* 59: 346–53.

Müntzing, A. 1930a. Outlines to a genetic monograph of the genus Galeopsis. *Hereditas* 13: 185–341.

———. 1930b. Über Chromosomenvermehrung in Galeopsis-Kreuzungen und ihre phylogenetische Bedeutung. *Hereditas* 14: 153–72.

———. 1932. Cyto-genetic investigations on synthetic *Galeopsis tetrahit*. *Hereditas* 16: 105–54.

———. 1934. Chromosome fragmentation in a Crepis hybrid. *Hereditas* 19: 284–302.

———. 1936. The evolutionary significance of autopolyploidy. *Hereditas* 21: 263–378.

———. 1938. Sterility and chromosome pairing in intraspecific Galeopsis hybrids. *Hereditas* 24: 117–88.

———. 1961. Genetics: Basic and Applied. 2nd ed. LTs Förlag, Stockholm.

Müntzing, A., and R. Prakken. 1941. Chromosomal aberrations in rye populations. *Hereditas* 27: 273–308.

Munz, P. A. 1949. The *Oenothera hookeri* group. *Aliso* 2: 1–47.

———. 1959. A California Flora. University of California Press, Berkeley, Calif.

———. 1965. Onagraceae. *North Amer. Flora*, 5: 1–278.

Naudin, C. 1863. Nouvelles recherches sur l'hybridité dans les végétaux. *Ann. Sci. Nat., Ser. Botan.* (Paris), 19: 180–203.

Navashin, M. S. 1932. The dislocation hypothesis of evolution of chromosome numbers. *Zeitschr. ind. Abstammungs- u. Vererbungslehre* 63: 224–31.

Newton, W. C. F., and C. Pellew. 1929. *Primula kewensis* and its derivatives. *Jour. Genetics* 20: 405–67.

Ninan, C. A. 1958. Studies on the cytology and phylogeny of the Pteridophytes. VI. Observations on the Ophioglossaceae. *Cytologia* 23: 291–316.

Nobs, M. A. 1963. Experimental studies on species relationships in Ceanothus. *Carnegie Inst. Wash. Publ.* 623.

Nobs, M. A., and W. M. Hiesey. 1957. Studies on differential selection in Mimulus. *Carnegie Inst. Washington Yearbook* 56: 291–92.

―――― and ――――. 1958. Performance of Mimulus races and their hybrids in contrasting environments. *Carnegie Inst. Washington Yearbook* 57: 270–72.

Nordenskiöld, H. 1949. The somatic chromosomes of some Luzula species. *Botaniska Notiser*, 1949: 81–92.

――――. 1951. Cytotaxonomical studies in the genus Luzula. I. *Hereditas* 37: 325–55.

――――. 1956. Cyto-taxonomical studies in the genus Luzula. II. Hybridization experiments in the *campestris-multiflora* complex. *Hereditas* 42: 7–73.

――――. 1961. Tetrad analysis and the course of meiosis in three hybrids of *Luzula campestris*. *Hereditas* 47: 203–38.

――――. 1962. Studies of meiosis in *Luzula purpurea*. *Hereditas* 48: 503–19.

Nygren, A. 1946. The genesis of some Scandinavian species of Calamagrostis. *Hereditas* 32: 131–262.

――――. 1951. Form and biotype formation in *Calamagrostis purpurea*. *Hereditas* 37: 519–32.

――――. 1954. Apomixis in the angiosperms, II. *Bot. Rev.* 20: 577–649.

――――. 1962a. Artificial and natural hybridization in European Calamagrostis. *Symbolae Botanicae Upsalienses* 17: 1–105.

――――. 1962b. On *Poa badensis, Poa concinna, Poa pumila, Poa xerophila* and the possible origin of *Poa alpina*. *Annales Academiae Regiae Scientiarum Upsaliensis* 1962(6): 1–29.

Oehlkers, F. 1940. Bastardierungsversuche in der Gattung Streptocarpus Lindley. III. Neue Ergebnisse über die Genetik von Wuchsgestalt und Geschlechtsbestimmung. *Ber. Deutsch. Bot. Ges.* 58: 76–91.

Ornduff, R. et al. 1959–1968. Index to plant chromosome numbers. Mimeographed or printed lists, issued annually. University of North Carolina Press, Chapel Hill, N.C. (earlier issues); and International Association for Plant Taxonomy, Utrecht (later issues).

Ownbey, M., and H. C. Aase. 1955. Cytotaxonomic studies in Allium. I. The *Allium canadense* alliance. *Research Studies State College of Washington*. Monograph Suppl. 1, 1–106.

Palmer, E. J. 1948. Hybrid oaks of North America. *Jour. Arnold Arboretum* 29: 1–48.

Pazy, B., and D. Zohary. 1965. The process of introgression between Aegilops polyploids: natural hybridization between *A. variabilis, A. ovata*, and *A. biuncialis. Evolution* 19: 385–94.

Peterson, P. A. 1958. Cytoplasmically inherited male sterility in Capsicum. *Amer. Nat.* 92: 111–19.

Pimentel, D., G. J. C. Smith, and J. Soans. 1967. A population model of sympatric speciation. *Amer. Nat.* 101: 493–504.

Powers, L. 1945. Fertilization without reduction in guayule (*Parthenium argentatum* Gray) and a hypothesis as to the evolution of apomixis and polyploidy. *Genetics* 30: 323–46.

Prazmo, W. 1960. Genetic studies on the genus Aquilegia L. I. Crosses between *Aquilegia vulgaris* L. and *Aquilegia ecalcarata* Maxim. *Acta Soc. Bot. Polon.* 29: 57–77.

———. 1961. Genetic studies on the genus Aquilegia L. II. Crosses between *Aquilegia ecalcarata* Maxim. and *Aquilegia chrysantha* Gray. *Acta Soc. Bot. Polon.* 30: 423–42.

———. 1965. Cytogenetic studies on the genus Aquilegia. IV. Fertility relationships among the Aquilegia species. *Acta Soc. Bot. Polon.* 34: 667–85.

Pryor, L. D. 1959. Species distribution and association in Eucalyptus. In: Biogeography and Ecology in Australia. *Monographiae Biologicae* 8: 461–71.

Putrament, A. 1962. Some observations on male sterility in *Geranium sylvaticum* L. var *alpestre* Schur. *Acta Soc. Bot. Polon.* 31: 723–36.

Ramsbottom, J. 1938. Linnaeus and the species concept. *Proc. Linnean Soc. London* 150: 192–219.

Randolph, L. F. 1935. Cytogenetics of tetraploid maize. *Jour. Agr. Research* 50: 591–605.

Randolph, L. F., I. S. Nelson, and R. L. Plaisted. 1967. Negative evidence of introgression affecting the stability of Louisiana Iris species. *Cornell Univ. Agr. Exp. Station Memoir* 398.

Raven, P. H. 1962. Interspecific hybridization as an evolutionary stimulus in Oenothera. *Proc. Linnean Soc. London* 173: 92–98.

———. 1969. A revision of the genus Camissonia (Onagraceae). *Contrib. U.S. Nat. Herbarium* 37: 159–396.

Raven, P. H., and D. W. Kyhos. 1965. New evidence concerning the original basic number of angiosperms. *Evolution* 19: 244–48.

Raven, P. H. and H. J. Thompson. 1964. Haploidy and angiosperm evolution. *Amer. Nat.* 98: 251–52.

Ravin, A. W. 1961. The genetics of transformation. *Advances in Genetics* 10: 61–163.

Rees, H. 1961. Genotypic control of chromosome form and behavior. *Bot. Rev.* 27: 288–318.

Remington, C. L. 1968. Suture-zones of hybrid interaction between recently joined biotas. *Evolutionary Biol.* 2: 321–428.

Renner, O. 1925. Untersuchungen über die faktorielle Konstitution einiger komplexheterozygotischer Oenotheren. *Bibliotheca Genetica* 9: 1–168.

———. 1941. Über die Entstehung homozygotischer Formen aus komplexheterozygotischen Oenotheren. *Flora* 35: 201–38.

Rensch, B. 1929. Das Prinzip geographischer Rassenkreise und das Problem der Artbildung. Borntraeger, Berlin.

Reuther, W., L. D. Batchelor, and H. J. Webber, eds. 1968. The Citrus Industry. Vol. II. Anatomy, Physiology, Genetics, and Reproduction. 2nd ed. University of California Press, Berkeley and Los Angeles, Calif.

Reuther, W., H. J. Webber, and L. D. Batchelor, eds. 1967. The Citrus Industry. Vol. I. History, World Distribution, Botany, and Varieties. 2nd ed. University of California Press, Berkeley and Los Angeles, Calif.

Rhoades, M. M., and E. Dempsey. 1966. Induction of chromosome doubling at meiosis by the elongate gene in maize. *Genetics* 54: 505–22.

Rhyne, C. L. 1958. Linkage studies in Gossypium. I. Altered recombination in allotetraploid *G. hirsutum* L. following linkage group transference from related diploid species. *Genetics* 43: 822–34.

Richardson, M. M. 1936. Structural hybridity in *Lilium martagon album* × *L. hansonii*. *Jour. Genetics* 32: 411–450.

Rick, C. M. 1963. Differential zygotic lethality in a tomato species hybrid. *Genetics* 48: 1497–1507.

Rick, C. M., and L. Butler. 1956. Cytogenetics of the tomato. *Advances in Genetics* 8: 267–382.

Rieger, R. 1963. Die Genommutationen. Veb Gustav Fischer, Jena.

Righter, F. I., and J. W. Duffield. 1951. Interspecies hybrids in pines. *Jour. Heredity* 42: 75–80.

Riley, H. P. 1938. A character analysis of colonies of *Iris fulva*, *Iris hexagona* var. *giganticaerulea* and natural hybrids. *Amer. Jour. Bot.* 25: 727–38.

Roberts, H. F. 1929, 1965. Plant Hybridization before Mendel. Reprint, 1965. Hafner, New York.

Rollins, R. C. 1944. Evidence for natural hybridity between guayule (*Parthenium argentatum*) and mariola (*Parthenium incanum*). *Amer. Jour. Bot.* 31: 93–99.

——. 1945. Interspecific hybridization in Parthenium. I. Crosses between guayule (*P. argentatum*) and mariola (*P. incanum*). *Amer. Jour. Bot.* 32: 395–404.

——. 1946. Interspecific hybridization in Parthenium. II. Crosses involving *P. argentatum*, *P. incanum*, *P. stramonium*, *P. tomentosum* and *P. hysterophorus*. *Amer. Jour. Bot.* 33: 21–30.

——. 1963. The evolution and systematics of Leavenworthia (Cruciferae). *Contrib. Gray Herbarium* (Harvard), No. 192.

Ross, H. H. 1958. Evidence suggesting a hybrid origin for certain leafhopper species. *Evolution* 12: 337–46.

Ross, J. G., and R. E. Duncan. 1949. Cytological evidences of hybridization between *Juniperus virginiana* and *J. horizontalis*. *Bull. Torrey Bot. Club* 76: 414–29.

Rousi, A. 1965. Biosystematic studies on the species aggregate *Potentilla anserina* L. *Annales Botanici Fennici* 2: 47–112.

——. 1967. Cytological observations on some species and hybrids of Vaccinium. *Der Züchter* 36: 352–59.

Russell, N. H. 1954. Three field studies of hybridization in the stemless white violets. *Amer. Jour. Bot.* 41: 679–86.

Sachs, J. 1906. History of Botany. Translation. Oxford University Press, Oxford.

Salisbury, E. J. 1942. Reproductive Capacity of Plants. G. Bell and Sons, London.

——. 1961. Weeds and Aliens. Collins, London.

Sarkar, P., and G. L. Stebbins. 1956. Morphological evidence concerning the origin of the B genome in wheat. *Amer. Jour. Bot.* 43: 297–304.

Satina, S., and A. F. Blakeslee. 1935. Cytological effects of a gene in Datura which causes dyad formation in sporogenesis. *Bot. Gaz.* 96: 521–23.

Sax, H. J. 1954. Polyploidy and apomixis in Cotoneaster. *Jour. Arnold Arboretum* 35: 334–65.

Sax, K. 1931. Chromosome ring formation in *Rhoeo discolor. Cytologia* 3: 36–53.

Schultz, R. J. 1969. Hybridization, unisexuality, and polyploidy in the teleost Poeciliopsis (Poeciliidae) and other vertebrates. *Amer. Nat.* 103: 605–19.

Schwanitz, F. 1957. Spornbildung bei einem Bastard zwischen drei Digitalis-Arten. *Biol. Zentralblatt* 76: 226–31.

Sears, E. R. 1941. Chromosome pairing and fertility in hybrids and amphiploids in the Triticinae. *Univ. Missouri Agr. Exp. Station Bull.*, No. 337.

——. 1944. Cytogenetic studies with polyploid species of wheat. II. Additional chromosomal aberrations in *Triticum vulgare. Genetics* 29: 232–46.

——. 1948. The cytology and genetics of the wheats and their relatives. *Advances in Genetics* 2: 239–70.

——. 1959. The systematics, cytology and genetics of wheat. *Handbuch der Pflanzenzüchtung* 2: 164–87.

Shindo, K., and H. Kamemoto. 1963. Chromosome numbers and genome relationships of some species in the Nigrohirsutae Section of Dendrobium. *Cytologia* 28: 68–75.

Sibley, C. G. 1950. Species formation in the red-eyed towhees of Mexico. *Univ. Calif. Publ. Zool.* 50: 109–94.

——. 1954. Hybridization in the red-eyed towhees of Mexico. *Evolution* 8: 252–90.

Simpson, G. G. 1944. Tempo and Mode in Evolution. Columbia University Press, New York.

——. 1951. The species concept. *Evolution* 5: 285–98.

——. 1961. Principles of Animal Taxonomy. Columbia University Press, New York.

Sinskaja, E. N. 1931. The study of species in their dynamics and interrelation with different types of vegetation. *Bull. Appl. Bot. Genetics Plant Breeding* (Leningrad) 25: 1–97.

Skalinska, M. 1958. Seed development after crosses of Aquilegia with Isopyrum. *Studies in Plant Physiol.* (Prague), 213–21.

——. 1967. Cytological analysis of some Hieracium species, subg. Pilosella from mountains of southern Poland. *Acta Biologica Cracoviensia, Botanica,* 10: 127–41.

Smith, D. M., and D. A. Levin. 1963. A chromatographic study of reticulate evolution in the Appalachian Asplenium complex. *Amer. Jour. Bot.* 50: 952–58.

Smith, F. H., and Q. D. Clarkson. 1956. Cytological studies of interspecific hybridization in Iris, subsection Californicae. *Amer. Jour. Bot.* 43: 582–88.

Smith, H. H. 1954. Development of morphologically distinct and genetically isolated populations by interspecific hybridization and selection. *Caryologia, Suppl. Vol.*, 867–70.

——. 1958. Genetic plant tumors in Nicotiana. *Annals New York Acad. Sci.* 71: 1163–77.

Smith, H. H., and K. Daly. 1959. Discrete populations derived by interspecific hybridization and selection in Nicotiana. *Evolution* 13: 476–87.

Smith, J. M. 1966. Sympatric speciation. *Amer. Nat.* 100: 637–50.

Smith-White, S. 1948. Polarised segregation in the pollen mother cells of a stable triploid. *Heredity* 2: 119–29.

——. 1955. The life history and genetic system of *Leucopogon juniperinus*. *Heredity* 9: 79–91.

——. 1959a. Cytological evolution in the Australian flora. *Cold Spring Harbor Symp. Quant. Biol.* 24: 273–89.

——. 1959b. Pollen development patterns in the Epacridaceae. *Proc. Linnean Soc. New South Wales* 84: 8–35.

Snyder, L. A. 1950. Morphological variability and hybrid development in *Elymus glaucus*. *Amer. Jour. Bot.* 37: 628–36.

——. 1951. Cytology of inter-strain hybrids and the probable origin of variability in *Elymus glaucus*. *Amer. Jour. Bot.* 38: 195–202.

Solbrig, O. T. 1960. Cytotaxonomic and evolutionary studies in the North America Species of Gutierrezia (Compositae). *Contrib. Gray Herbarium* (Harvard), No. 188.

Solbrig, O. T., L. C. Anderson, D. W. Kyhos, P. H. Raven, and L. Ruedenberg. 1964. Chromosome numbers in Compositae V. Astereae II. *Amer. Jour. Bot.* 51: 513–19.

Sonneborn, T. M. 1957. Breeding systems, reproductive methods, and species problems in Protozoa. In: The Species Problem. Ed. by E. Mayr. Amer. Assoc. Adv. Sci, Washington, D.C.

Sørensen, T., and G. Gudjonsson. 1946. Spontaneous chromosome-aberrants in apomictic Taraxaca. Morphological and cytogenetical investigations. *K. Danske Vidensk. Selskab. Biol.* 4: 1–48.

Soumalainen, E. 1950. Parthenogenesis in animals. *Advances in Genetics* 3: 193–253.

Sprague, E. 1962. Pollination and evolution in Pedicularis (Scrophulariaceae). *Aliso* 5: 181–209.

Squillace, A. E., and R. T. Bingham. 1958. Localized ecotypic variation in western white pine. *Forest Sci.* 4: 20–34.

Stalker, H. D. 1956. A case of polyploidy in Diptera. *Proc. Nat. Acad. Sci.* 42: 194–99.

——. 1966. The phylogenetic relationships of the species in the *Drosophila melanica* group. *Genetics* 53: 327–42.

Stanier, R. Y. 1955. Specific and infraspecific categories in microorganisms. In: *Biology Colloquium*. Oregon State College, Corvallis, Oregon.

Stebbins, G. L. 1932. Cytology of Antennaria. I. Normal species. *Bot. Gaz.* 94: 134–51.

_____. 1935. A new species of Antennaria from the Appalachian region. *Rhodora* 37: 229–37.

_____. 1938. Cytological characteristics associated with the different growth habits in the dicotyledons. *Amer. Jour. Bot.* 25: 189–98.

_____. 1942. The role of isolation in the differentiation of plant species. *Biol. Symp.* 6: 217–33.

_____. 1945. The cytological analysis of species hybrids, II. *Bot. Rev.* 11: 463–86.

_____. 1947a. Types of polyploids: their classification and significance. *Advances in Genetics* 1: 403–29.

_____. 1947b. The origin of the complex of *Bromus carinatus* and its phytogeographic implications. *Contrib. Gray Herbarium* (Harvard) 165: 42–55.

_____. 1950. Variation and Evolution in Plants. Columbia University Press, New York.

_____. 1956. Cytogenetics and evolution of the grass family. *Amer. Jour. Bot.* 43: 890–905.

_____. 1957a. Self fertilization and population variability in the higher plants. *Amer. Nat.* 91: 337–54.

_____. 1957b. The hybrid origin of microspecies in the *Elymus glaucus* complex. *Cytologia, Suppl. Vol.*, 336–40.

_____. 1958a. The inviability, weakness, and sterility of interspecific hybrids. *Advances in Genetics* 9: 147–215.

_____. 1958b. Longevity, habitat, and release of genetic variability in the higher plants. *Cold Spring Harbor Symp. Quant. Biol.* 23: 365–78.

_____. 1959. The role of hybridization in evolution. *Proc. Amer. Phil. Soc.* 103: 231–51.

Stebbins, G. L., and K. Daly. 1961. Changes in the variation pattern of a hybrid population over an eight-year period. *Evolution* 15: 60–71.

Stebbins, G. L., and L. Ferlan. 1956. Population variability, hybridization, and introgression in some species of Ophrys. *Evolution* 10: 32–46.

Stebbins, G. L., J. A. Jenkins, and M. S. Walters. 1953. Chromosomes and phylogeny in the Compositae, tribe Cichorieae. *Univ. Calif. Publ. Bot.* 26: 401–29.

Stebbins, G. L., and H. A. Tobgy. 1944. The cytogenetics of hybrids in Bromus. I. Hybrids within the section Ceratochloa. *Amer. Jour. Bot.* 31: 1–11.

Stebbins, G. L., H. A. Tobgy, and J. R. Harlan. 1944. The cytogenetics of hybrids in Bromus. II. *Bromus carinatus* and *Bromus arizonicus*. *Proc. Calif. Acad. Sci.* 25: 307–22.

Stebbins, G. L., and A. Vaarama. 1954. Artificial and natural hybrids in the Gramineae, tribe Hordeae. VII. Hybrids and allopolyploids between *Elymus glaucus* and Sitanion spp. *Genetics* 39: 378–95.

Steiner, E. 1952. Phylogenetic studies in Euoenothera. *Evolution* 6: 69–80.

Stephens, S. G. 1949. The cytogenetics of speciation in Gossypium. I. Selective elimination of the donor parent genotype in interspecific back-crosses. *Genetics* 34: 627–37.

——. 1950. The internal mechanism of speciation in Gossypium. *Bot. Rev.* 16: 115–49.

Stephens, S. G., and M. D. Finkner. 1953. Natural crossing in cotton. *Economic Bot.* 7: 257–69.

Stinson, H. T. 1953. Cytogenetics and phylogeny of *Oenothera argillicola* Mackenz. *Genetics* 38: 389–406.

Straw, R. M. 1955. Hybridization, homogamy, and sympatric speciation. *Evolution* 9: 441–44.

——. 1956. Floral isolation in Penstemon. *Amer. Nat.* 90: 47–53.

Streams, F. A., and D. Pimentel. 1961. Effects of immigration on the evolution of populations. *Amer. Nat.* 95: 201–10.

Swaminathan, M. S. 1954. Microsporogenesis in some commercial potato varieties. *Jour. Heredity* 45: 265–72.

Swaminathan, M. S., and R. W. Hougas. 1954. Cytogenetic studies in *Solanum verrucosum* variety *spectabilis. Amer. Jour. Bot.* 41: 645–51.

Swaminathan, M. S., and H. W. Howard. 1953. The cytology and genetics of the potato (*Solanum tuberosum*) and related species. *Bibliographia Genetica* 16: 1–192.

Swaminathan, M. S., and M. L. Magoon. 1961. Origin and cytogenetics of the commercial potato. *Advances in Genetics* 10: 217–56.

Swietlińska, Z. 1960. Spontaneous polyploidization in Rumex hybrids. *Acta Soc. Botan. Polon.* 29: 79–98.

Swietlińska, Z., and J. Zuk. 1965. Further observations on spontaneous polyploidization in Rumex hybrids. *Acta Soc. Botan. Polon.* 34: 439–50.

Swingle, W. T. 1967. The botany of Citrus and its wild relatives. In: The Citrus Industry. 2nd ed., Vol. 1. Ed. by W. Reuther, H. J. Webber, and L. D. Batchelor. University of California Press, Berkeley and Los Angeles, Calif.

Täckholm, G. 1922. Zytologische Studien über die Gattung Rosa. *Acta Horti Bergiani* (Uppsala), 7: 97–381.

Tahara, M. 1915 Cytological studies on Chrysanthemum. II. *Bot. Mag. Tokyo* 29: 48–50.

Tanaka, T. 1954. Species Problem in Citrus. Japanese Society for the Promotion of Science. Ueno, Tokyo, Japan.

——. 1961. Citrologia. Citrologia Supporting Foundation. Osaka, Japan. (Summarized in Swingle, 1967.)

Thoday, J. M., and T. B. Boam. 1959. Effects of disruptive selection. II. Polymorphism and divergence without isolation. *Heredity* 13: 205–18.

Thoday, J. M., and J. B. Gibson. 1962. Isolation by disruptive selection. *Nature* 193: 1164–66.

Ting, Y. C. 1967. Common inversions in maize and teosinte. *Amer. Nat.* 101: 87–89.

Tischler, G. 1935. Die Bedeutung der Polyploidie für die Verbreitung der Angiospermen. *Bot. Jahrb.* 67: 1–36.

——. 1953. Allgemeine Pflanzenkaryologie. III. Angewandte Pflanzenkaryologie. *Handbuch der Pflanzenanatomie.* Berlin.

——. 1954. Das Problem der Basis-Chromosomenzahlen bei den Angiospermen-Gattungen und -Familien. *Cytologia* 19: 1–10.

——. 1955. Der Grad der Polyploidie bei den Angiospermen in verschiedenen Grossarealen. *Cytologia* 20: 101–18.

Tobgy, H. A. 1943. A cytological study of *Crepis fuliginosa, C. neglecta,* and their F_1 hybrid, and its bearing on the mechanism of phylogenetic reduction in chromosome number. *Jour. Genetics* 45: 67–111.

Tucker, J. M. 1953. Two new oak hybrids from California. *Madrono* 12: 119–27.

Tucker, J. M., and J. Sauer. 1958. Aberrant Amaranthus populations of the Sacramento-San Joaquin delta, California. *Madrono* 14: 252–61.

Turesson, G. 1922. The genotypical response of the plant species to the habitat. *Hereditas* 3: 211–350.

——. 1925. The plant species in relation to habitat and climate. *Hereditas* 6: 147–236.

——. 1930. The selective effect of climate upon the plant species. *Hereditas* 14: 99–152.

Turner, B. L., and R. Alston. 1959. Segregation and recombination of chemical constituents in a hybrid swarm of *Baptisia laevicaulis* X *B. viridis* and their taxonomic implications. *Amer. Jour. Bot.* 46: 678–86.

Turner, B. L., and D. Horne. 1964. Taxonomy of Machaeranthera sect. Psilactis (Compositae-Asterae). *Brittonia* 16: 316–31.

U, N. 1935. Genome-analysis in Brassica with special reference to the experimental formation of *B. napus* and peculiar mode of fertilization. *Japanese Jour. Bot.* 7: 389–452.

Upcott, M. 1939. The nature of tetraploidy in *Primula kewensis. Jour. Genetics* 39: 79–100.

Vasek, F. C. 1964. Outcrossing in natural populations. I. The Breckenridge Mountain population of *Clarkia exilis. Evolution* 18: 213–18.

——. 1965. Outcrossing in natural populations. II. *Clarkia unguiculata. Evolution* 19: 152–56.

——. 1967. Outcrossing in natural populations. III. The Deer Creek population of *Clarkia exilis. Evolution* 21: 241–48.

——. 1968. The relationships of two ecologically marginal, sympatric Clarkia populations. *Amer. Nat.* 102: 25–40.

Viosca, P. 1935. The irises of southeastern Louisiana. *Bull. Amer. Iris Soc.,* No. 57.

Wagner, M. 1889. Die Entstehung der Arten durch räumliche Sonderung. Benno Schwalbe, Basel.

Wagner, W. H. 1954. Reticulate evolution in Appalachian Aspleniums. *Evolution* 8: 103–18.

Wahl, H. A. 1940. Chromosome numbers and meiosis in the genus Carex. *Amer. Jour. Bot.* 27: 458–70.

Walker, T. G. 1958. Hybridization in some species of Pteris L. *Evolution* 12: 82–92.

———. 1962. Cytology and evolution in the fern genus Pteris L. *Evolution* 16: 27–43.

Wallace, A. R. 1889. Darwinism: An Exposition of the Theory of Natural Selection. Macmillan, London.

Wallace, B. 1954. Genetic divergence of isolated populations of *Drosophila melanogaster*. *Caryologia, Suppl. Vol.*, 761–64.

———. 1959. Influence of genetic systems on geographical distribution. *Cold Spring Harbor Symp. Quant. Biol.* 24: 193–204.

———. 1966. On the dispersal of Drosophila. *Amer. Nat.* 100: 551–63.

Walters, J. L. 1942. Distribution of structural hybrids in *Paeonia californica*. *Amer. Jour. Bot.* 29: 270–75.

Walters, M. S. 1957. Studies of spontaneous chromosome breakage in interspecific hybrids of Bromus. *Univ. Calif. Publ. Bot.* 28: 335–447.

———. 1960. Rates of meiosis, spindle irregularities and microsporocyte division in *Bromus trinii* X *B. carinatus*. *Chromosoma* 11: 167–204.

Webb, D. A. 1966. Dispersal and establishment: what do we really know? In: Reproductive Biology and Taxonomy of Vascular Plants. Ed. by J. G. Hawkes. Pergamon Press, Oxford.

Webber, H. J., and L. D. Batchelor. 1943. The Citrus Industry. Vol. I. History, Botany, and Breeding. 1st ed. University of California Press, Berkeley and Los Angeles, Calif.

Weil, J., and R. W. Allard. 1964. The mating system and genetic variability in natural populations of *Collinsia heterophylla*. *Evolution* 18: 515–25.

Weismann, A. 1913. Vorträge über Deszendenztheorie. 3rd ed. Fischer, Jena.

Westergaard, M. 1940. Studies on cytology and sex determination in polyploid forms of *Melandrium album*. *Dansk Bot. Arkiv* 10: 1–131.

White, M. J. D. 1951. Evolution of cytogenetic mechanisms in animals. In: Genetics in the Twentieth Century. Ed. by L. C. Dunn. Macmillan, New York.

———. 1954. Animal Cytology and Evolution. 2nd ed. Cambridge University Press, Cambridge.

———. 1957a. Cytogenetics and systematic entomology. *Ann. Rev. Entomol.* 2: 71–90.

———. 1957b. Some general problems of chromosomal evolution and speciation in animals. *Survey of Biol. Progress* 3: 109–47.

———. 1959. Speciation in animals. *Australian Jour. Sci.* 22: 32–39.

———. 1964. Principles of karyotype evolution in animals. In: Genetics Today. Pergamon Press, Oxford.

Wiegand, K. M. 1935. A taxonomist's experience with hybrids in the wild. *Science* 81: 161–66.

Winge, Ö. 1917. The chromosomes: their number and general importance. *Compt. Rend. Trav. Lab. Carlsberg* 13: 131–275.

———. 1924. Contributions to the knowledge of chromosome numbers in plants. *La Cellule* 35: 303–24.

———. 1938. Inheritance of species characters in Tragopogon. *Compt. Rend. Trav. Lab. Carlsberg, Series Physiol.* 22: 155–94.

———. 1940. Taxonomic and evolutionary studies in Erophila based on cytogenetic investigations. *Compt. Rend. Trav. Lab. Carlsberg, Series Physiol.* 23: 41–74.

Winkler, H. 1916. Über die experimentelle Erzeugung von Pflanzen mit abweichenden Chromosomenzahlen. *Zeitschr. f. Bot.* 8: 417–531.

Wolf, C. B. 1944. The Gander oak, a new hybrid oak from San Diego County, California. *Proc. Calif. Acad. Sci.* 25: 177–88.

Wright, J. W. 1955. Species crossability in spruce in relation to distribution and taxonomy. *Forest Sci.* 1: 319–49.

Wright, J. W., and W. J. Gabriel. 1958. Species hybridization in the hard pines, series Sylvestres. *Silvae Genetica* 7: 109–15.

Wright, S. 1931. Evolution in Mendelian populations. *Genetics* 16: 97–159.

———. 1943a. Isolation by distance. *Genetics* 28: 114–38.

———. 1943b. An analysis of local variability of flower color in *Linanthus parryae. Genetics* 28: 139–56.

———. 1949. Adaptation and selection. In: Genetics, Paleontology, and Evolution. Ed. by G. L. Jepsen. Princeton University Press, Princeton, N.J.

Wulff, H. D. 1939. Die Pollenentwicklung der Juncaceen nebst einer Auswertung der embryologischen Befunde hinsichtlich der Verwandtschaft der Juncaceen und Cyperaceen. *Jahrb. Wiss. Botanik* 87: 533–56.

———. 1954. Über das spontane Auftreten einer Caninae-Meiosis bei der Mikrosporogenese der diploiden *Rosa ruga* Lindl. *Österr. Bot. Zeitschr.* 101: 539–57.

Yarnell, S. H. 1956. Cytogenetics of the vegetable crops. II. Crucifers. *Bot. Rev.* 22: 81–166.

Zobel, B. 1951. The natural hybrid between Coulter and Jeffrey pines. *Evolution* 5: 405–13.

Zohary, D., and M. Feldman. 1962. Hybridization between amphidiploids and the evolution of polyploids in the wheat (Aegilops-Triticum) group. *Evolution* 16: 44–61.

Organism Index

Abies, 173, 184
Acarina, 263, 284
Aceraceae, 317
Achillea, 109, 307; A. asplenifolia, 307; A. borealis, 109, 307; A. collina, 307; A. lanulosa, 307; A. millefolium, 109, 307; A. setacea, 307; A. virescens, 307
Acutae, 266
Adiantum, 343
Aegilops, 238, 307, 309–10; Ae. bicornis, 310; Ae. biuncialis, 310; Ae. caudata, 309–10; Ae. columnaris, 310; Ae. comosa, 310; Ae. crassa, 310; Ae. cylindrica, 310; Ae. juvenalis, 310; Ae. kotschyi, 310; Ae. ligustica, 309–10; Ae. longissima, 310; Ae. mutica, 310; Ae. ovata, 310; Ae. sharonensis, 310; Ae. speltoides, 309–10; Ae. squarrosa, 309–10; Ae. triaristata, 310; Ae. triuncialis, 309–10; Ae. umbellata, 309–10; Ae. uniaristata, 310; Ae. variabilis, 178, 310; Ae. ventricosa, 310
Aegilops-Triticum, 304, 309–11
Agrostis, 352; A. tenuis, 18
Alchemilla, 274, 281, 331, 345, 347; A. subgenus Aphanes, 330, 345; A.
arvensis, 277; A. subgenus Eualchemilla, 330, 345; A. microcarpa, 345
Allium, 58, 274, 279, 351; A. canadense, 351–52; A. carinatum, 351; A. flavum, 351; A. geyeri, 352; A. odorum, 278; A. oleraceum, 351; A. paniculatum, 351; A. paradoxum, 351; A. pulchellum, 351; A. sativum, 351
Alnus, 225, 274, 280, 282; A. rugosa, 277
Amaranthus, 195; A. caudatus, 195; A. powelii, 195; A. retroflexus, 195
Amaryllidaceae, 276
Amelanchier, 158
Amsinckia, 253
Anacardiaceae, 277
Antennaria, 58, 279–82, 331; A. alpina, 276, 279; A. carpatica, 277
Anthophora, 123
Antirrhinastrum, 89
Antirrhinum, 64, 89, 100, 124; A. glutinosum, 63; A. majus, 63; A. orontium, 89
Aphanes, Alchemilla subgenus, 330, 345
Aphidae, 263
Apogon, Iris section, 92

Aquilegia, 54, 58, 75, 87–88, 100, 124–25, 133, 155, 174, 225; A. canadensis, 88; A. chrysantha, 49, 88; A. ecalcarata, 87–88, A. formosa, 51; A. pubescens, 51; A. vulgaris, 88
Arabis holboellii, 276, 279
Araceae, 278
Arachnion, Gilia section, 95, 97, 112, 252, 311
Artemia salina, 226
Artemisia douglasiana, 240; A. ludoviciana, 240; A. suksdorfi, 240
Asarina, 89
Asclepiadaceae, 75
Asclepias, 75
Aspen, 8
Asplenium, 232, 247, 305, 343; A. monanthes, 280; A. montanum, 305; A. pinnatifidum, 305; A. platyneuron, 305; A. rhizophyllum, 305
Aster, 131
Asterales, 385
Atraphaxis frutescens, 277
Atriplex, 109; A. litorale, 108; A. praecox, 108; A. sarcophyllum, 108
Aurantioideae, 337, 340
Avena barbata, 107; A. fatua, 11–12

Bagworm moth, 254
Balanophora, 280; B. globosa, 276; B. japonica, 276
Balanophoraceae, 276
Baptisia, 52, 168; B. leucophaea, 165; B. sphaerocarpa, 165
Berberidaceae, 225
Berinia, 103
Betula, 54, 225
Betulaceae, 134, 225, 277, 317
Biscutella laevigata, 242, 244
Blackberry, 196, 286, 289, 325–29
Blepharodon, Haplopappus section, 318
Bombus, 243
Bouteloua, 335, 337, 343; B. curtipendula, 305, 331–37; B. distans, 334; B. disticha, 333–34; B. juncea, 334; B. media, 334; B. pedicillata, 334; B. purpurea, 334–35; B. reflexa, 334; B. triaena, 334; B. uniflora, 333–34, 336; B. vaneedenii, 334; B. warnockii, 334–35

Brassica, 69, 97, 101, 190, 221, 224; B. campestris, 109, 224, 240; B. napus, 224, 240; B. nigra, 109; B. oleracea, 192, 224, 240; B. rapa, 16
Brine shrimp, 254
Bromopsis, 313, 315
Bromus, 77, 101, 314; B. arizonicus, 313, 315; B. carinatus, 313; B. catharticus, 313; B. section Ceratochloa, 97, 313, 315; B. marginatus, 313
Buddleia, 232, 316; B. colvilei, 226, 232
Bulbostylis, 268
Burmannia, 281; B. coelestis, 276
Burmanniaceae, 276
Buxaceae, 278

Cactaceae, 90, 278, 350
Cactus, 196
Caesalpinaceae, 225
Calamagrostis, 279–82, 284, 286, 331; C. arundinacea, 286; C. canadensis, 276; C. epigeios, 286; C. purpurea, 286
Calochortus, 147
Caltha, 109; C. palustris, 109
Camissonia micrantha, 100–101
Campanula rotundifolia, 109
Caninae, Rosa section, 372, 274, 376
Capillares, Carex section, 266
Capparidaceae, 278
Capparis frondosa, 278
Capsicum, 76
Cardamine bulbifera, 352
Carex, 155, 264–69; C. section Capillares, 266; C. caryophyllea, 266; C. ericetorum, 267; C. membranacea, 266; C. paludivagans, 196, 266; C. physocarpa, 266; C. pilulifera, 267; C. rostrata, 195, 266; C. rotundata, 195, 266; C. siderosticta, 266; C. stenantha, 266
Caryophyllaceae, 93, 350
Catalpa bignonioides, 49; C. ovata, 49
Cattleya, 89
Ceanothus, 58, 82–84, 100–101, 124–25, 133, 154, 156, 251, 340; C. cuneatus, 82–83; C. foliosus, 85; C. jepsonii, 83; C. masonii, 85; C. pumilus, 83
Celastraceae, 278
Celastrus scandens, 278

Author Index

Subject Index

Adventitious embryony, 9, 272 ff., 284, 340

Agamic complex, 297–98, 321 ff.

Agamospermy, 6, 9 ff., 186–87, 271 ff., 296

Agmatoploidy, 262 ff.

Allopolyploidy, *see* Amphiploidy

Alternate disjunction, 360

Amphiploidy, 30 ff., 186–87, 190 ff., 198, 213, 221, 234 ff., 246 ff., 296, 300 ff.

Aneuploidy, 117–18, 131, 224, 233, 262 ff., 288–90, 307, 316–19, 331 ff., 352–53

Apogamety, 9, 275 ff., 343

Apomixis, 6, 40, 186–87, 271, 290, 346; *see also* Agamospermy, Vegetative propagation

Apospory, 274 ff., 283

Autogamy, 6, 10 ff., 97 ff., 178, 184, 203, 215, 252–53, 298–99, 357, 370–71

Autoploidy, 234 ff., 247

Autosegregation, 289

Balanced lethals, 361

Bees, 15, 75, 78, 88–90, 123, 144, 194–95, 243

Block inheritance, 69 ff.

Breeding system, 6, 11, 178, 187, 203, 252–54, 280–81, 298–99

Bud sports, 289, 342

Chromosome numbers, 64, 219 ff., 226 ff., 315 ff., 330 ff., 351–53; trends in, 131, 315 ff.

Chromosome rearrangements, 67 ff., 77, 101–104, 116, 126 ff., 178 ff., 190 ff., 199 ff., 251

Chromosome size, 257–58

Clonal complex, 297–98, 349 ff.

Clonal senescence, 354

Clone, 6, 40

Complementary genes, 63

Convergence, 165, 183, 385–86

Deletion, 67 ff.

Differentiation-hybridization cycles, 383–84

Diffuse centromere, 263, 268

Diplospory, 272 ff., 283

Dispersal, 13 ff., 353

Ecological niche, 27, 38–39, 164

Environmental isolation, 73 ff., 158

Epistatic interactions, 63